The Brewer's Handbook

The Complete Book to Brewing Beer

The Brewer's Handbook

The Complete Book to Brewing Beer

Ted Goldammer

Apex®

The Brewer's Handbook
The Complete Book to Brewing Beer
By Ted Goldammer

Apex Publishers
6731 Rock Brook Drive
Clifton, Virginia 20124
http://www.beer-brewing.com

Copyright © 2008 Apex Publishers

Printing History

July 1999:   First Edition
September 2008:   Second Edition

Notice of Rights

All rights reserved. No part of this book may be reproduced or transmitted in any form by any means, electronic, mechanical, photocopying, recording, or otherwise, without the prior written permission of the publisher. For information on getting permission for reprints and excerpts, go to http://www.beer-brewing.com.

Notice of Liability

The information in this book is distributed on an "As Is" basis, without warranty. While every precaution has been taken in the preparation of the book, neither the author nor Apex Publishers shall have any liability to any person or entity with respect to any loss or damage caused or alleged to be caused directly or independently by the information contained in this book.

ISBN-10: 0-9675212-3-8
ISBN-13: 978-0-9675212-3-7

Printed and bound in the United States of America

RepKover. This book uses RepKover™, a durable and flexible lay-flat binding.

Table of Contents

List of Figures .... *xi*

List of Tables .... *xiii*

Introduction .... *xv*

Chapter 1 — U.S. Beer Industry *1*
   History 1
   Industry Concentration 2
   Classification of Brewers 4
   Beer Sales Domestic and Imported 7
   Beer Segments 12
   Advertising 15
   Beer Distribution 15
   Demographics 18
   References 19

Chapter 2 — Barley Malts *21*
   Barley 21
   Malting 23
   Malt Modification 24
   Malt Constituents 25
   Malt Analysis 26
   Types of Barley Malts 31
   Other Malted Grains 36
   Malt Extracts 37

Calculations 37
References 38

Chapter 3 — Hops *41*

Hop Constituents 41
Hop Varieties 43
Whole Hops 44
Hop Products 46
Hop Stability 53
Bitterness Levels in Beer 54
Hop Utilization 55
Dry Hopping 57
References 58

Chapter 4 — Yeast *61*

Ale Yeast 61
Lager Yeast 62
Yeast Life Cycle 62
Nutritional Requirements 64
Yeast Byproducts 67
Yeast Strain Selection 71
Pure Culture Maintenance 75
Yeast Propagation and Scale-up 77
Culture Contamination 81
Yeast Washing 82
Yeast Viability and Replacement 84
Yeast Storage 86
References 87

Chapter 5 — Brewing Water *93*

Evaluation of Brewing Water 93
pH, Alkalinity, and Water Hardness 94
Minerals in Brewing Water 96
Water Treatment 100
References 106

Chapter 6 — Brewing Adjuncts *109*

Uses of Adjuncts 109
Types of Adjuncts 110

Classification of Cereal Adjuncts 111
Syrup and Sugar Adjuncts 116
References 118

Chapter 7 — Brewery Cleaning and Sanitation *121*

Cleaning Detergents 121
Additives 125
Sanitizing Agents 128
Cleaning Methods 133
Cleaning and Sanitation Manual 137
Material Corrosion Resistance 137
References 139

Chapter 8 — Malt Milling *143*

Malt Handling and Storage 143
Types of Malt Milling 145
Sieve Analysis of Crushed Malt 149
Safety 150
References 151

Chapter 9 — Mashing *153*

Chemical Changes at Mashing 153
Factors Affecting Mashing Conditions 157
Mashing Systems 160
Mashing Equipment 170
Calculations 174
References 176

Chapter 10 — Wort Separation *181*

Mash Tun 181
Lauter Tun 183
Mash Filters 189
Strainmaster 190
Calculations 191
References 191

Chapter 11 — Wort Boiling *195*

Biochemical Changes 195
Formation of Hot Break 199

Kettle Additives 201
Wort Boiling Systems 206
Hot Wort Clarification 209
Calculations 216
References 216

Chapter 12 — Wort Cooling and Aeration *221*

Wort Cooling Systems 221
Formation of Cold Break 222
Removal of Cold Break 224
Aeration of Chilled Wort 227
References 230

Chapter 13 — Fermentation *233*

Pitching Yeast 233
Lager Fermentations 236
Ale Fermentations 239
Diacetyl Rest 240
High Gravity Fermentations 242
Yeast Collection 242
Fermentation Systems 244
Calculations 250
References 251

Chapter 14 — Conditioning *255*

Maturation 255
Clarification 260
Stabilization 263
Conditioning Tanks 268
References 270

Chapter 15 — Beer Filtration *275*

Filtration Methods... 275
Powder Filters 276
Powder Filter Aids 280
Colloidal Stabilization 283
Sheet (Pad) Filters 284
Cartridge Membrane Filters 286
Deep-Bed Filtration 289

Sterile Filtration	289
References	290

Chapter 16 — Beer Carbonation ... 295

Principles of Beer Carbonation	295
Methods of Carbonation	297
Safety Procedures	301
References	301

Chapter 17 — Bottling ... 303

Sterilization of Beer	303
Bottle Feeding	305
Bottle Rinsing	307
Bottle Filling	308
Tunnel Pasteurization	315
Bottle Labeling	316
Case Packing	318
References	319

Chapter 18 — Kegging ... 321

Keg Styles	321
Keg Racking Machines	322
Sterilization of Beer	324
Kegging Operations	325
References	330

Chapter 19 — Beer Spoilage Organisms ... 333

Microorganisms — Brewing Stages	333
Bacteria	333
Wild Yeast	340
Molds	342
Microbiological Quality Assurance	342
References	343

Chapter 20 — Wastewater and Solid Waste Management ... 345

Brewery Wastewater	345
Solid Wastes	347
Brewery Wastewater Treatment	348
Aerobic Wastewater Treatment	353

    Anaerobic Wastewater Treatment 356
    References 358

Chapter 21 — Beer Styles *361*

    American Beer Styles 362
    Belgium Beer Styles 368
    British Beer Styles 376
    Czech Republic Beer Styles 386
    French Beer Styles 387
    German Beer Styles 388
    Irish Beer Styles 401
    Scottish Beer Styles 402
    References 405

Chapter 22 — Government Regulations *411*

    Federal Regulations 411
    State Regulations 416
    Local Regulations 418
    References 419

Appendixes

    Appendix A — Hop Varieties 423
    Appendix B — Carbon Dioxide Volume Table 429
    Appendix C — State Beer Excise Tax Rates 431
    Appendix D — Conversion Factors 433

Glossary *441*

Index *465*

List of Figures

Figure 4.1 — Laminar Flow Hood *78*
Figure 4.2 — Carlsberg Flask *78*
Figure 4.3 — Yeast Propagation Plant *80*
Figure 7.1 — Stationary CIP System *134*
Figure 7.2 — Static and Dynamic Spray Balls *136*
Figure 8.1 — Two-roll Mill *146*
Figure 8.2 — Hammer Mill *148*
Figure 9.1 — Mash Tuns *161*
Figure 9.2 — Mash Tun Paddles *173*
Figure 9.3 — Mash/Lauter Vessel and Kettle *174*
Figure 10.1 — Lauter Tun *184*
Figure 10.2 — Lauter Tun Rakes *185*
Figure 11.1 — Kettles *207*
Figure 11.2 — Centrifuge *213*
Figure 11.3 — Whirlpools *214*
Figure 12.1 — Wort Cooler *222*
Figure 13.1 — Cylindroconical Fermenters *245*
Figure 13.2 — Yorkshire Squares *248*
Figure 13.3 — Open Square Fermenters *250*
Figure 14.1 — Bright Beer Tanks *269*

Figure 15.1 — Plate and Frame Filter *277*
Figure 15.2 — Horizontal Leaf Filter *279*
Figure 15.3 — Sheet Filter *284*
Figure 15.4 — Cartridge Membrane Filter *287*
Figure 15.5 — Deep Bed Filter *289*
Figure 16.1 — In-Line Carbonator *298*
Figure 16.2 — Ceramic Stone Carbonator *299*
Figure 16.3 — Carbonation Tank *300*
Figure 17.1 — Case Unpacker/Packer *306*
Figure 17.2 — Rinser/Filler Bloc *308*
Figure 17.3 — In-Line Bottle Filler *310*
Figure 17.4 — Rotary Bottle Filler *311*
Figure 17.5 — Rotary Cold Glue Labeller *317*
Figure 17.6 — Pick and Place Case Packer *318*
Figure 18.1 — Sankey Valve (SVK) Keg *322*
Figure 18.2 — In-Line Kegging System *323*
Figure 18.3 — Flash Pasteurizer *324*
Figure 18.4 — Complete Kegging Line *326*

List of Tables

Table 1.1 — U.S. Beer Sales 9
Table 1.2 — Top Ten U.S. Beer Brands 10
Table 1.3 — Per Capita Beverage Consumption 11
Table 1.4 — U.S. Beer Sales by Segments 12
Table 1.5 — U.S. Beer Brands by Segment 13
Table 5.1 — Mineral Content of Classic Brewing Waters 94
Table 6.1 — Gelatinization Temperatures of Various Cereals 111
Table 7.1 — CIP Cleaning and Sanitation Program 135
Table 8.1 — Standard Sieve Sizes 150
Table 19.1 — Beer Spoilage Microorganisms – Brewing Stages 334
Table 19.2 — Beer Spoilage Bacteria 335

Introduction

In the past two decades we have witnessed unprecedented changes in the U.S. beer industry. The emergence of craft brewers and consumers' newfound appreciation for quality beer have resulted in what is often called the "Craft Beer Renaissance." Beer has taken on a new excitement and relevancy to the average person. It is no longer thought of only in the context of large commercial brewers with their mass-marketed beers. The marked increase in the number of brew pubs and microbreweries and the burgeoning growth in the number of home brewers indicate how deeply brewing beer has captured the popular imagination.

The Brewer's Handbook is intended to provide an introduction to brewing beer, and to give a balanced, reasonably detailed account of every major aspect of the brewing process. This book not only discusses brewing beer on a large-scale commercial basis, it has made every effort to address brewing practices typically used by craft brewers. Thus its applicability extends to home brewers and to individuals working in the brewing industry and related fields.

It is written in a language that can be easily understood by anyone not having a background in brewing beer. However, the material is not so elementary that it insults your intelligence, nor is it so difficult that its makes you lose interest in the subject. Clarity is the touchstone that has been employed throughout this book.

Chapter One

U.S. Beer Industry

Beer the largest alcohol segment nationwide, accounting for roughly 85% of all alcohol volume sold in the United States and annually generating over $91.6 billion in retail sales. The industry has continued to grow and increase its profitability despite economic and flat consumption trends. The beer market is the most concentrated of the three alcohol sectors with three brewers—Anheuser-Busch, South African Breweries' Miller (SAB-Miller), and Molson Coors Brewing Company—accounting for about 79% all beer sales. Expanded market share, price increases and improved production efficiencies are the keys to improving operating margins for large national brewers in the U.S. beer market. Although there are still some traditional regional brewers that continue to operate, they continue to decline in numbers, often closing or being sold to a larger national brewer. After going through several decades of brewery consolidation, a number of pioneering brewers in the 1980s started producing traditional more full-flavored "craft" beers using only traditional ingredients and brewing practices.

History

The U.S. beer industry got its start in the 1840s and 1850s with the introduction of lager style beers, brought by German immigrants. Before that point, beers were heavily oriented toward ale, porter, and stout and were mostly brewed at home. At about the same time, several technological advances occurred that led to the development of the U.S. beer industry as we know it today. Mechanical refrigeration greatly aided in the production as well as the storage of beer. Pasteurization was also adopted during this period, which opened the way for wide-scale bottling and off-premise consumption of beer. By 1850 there were about 430 breweries in the United States, producing about 750,000 barrels of beer annually. Commercial brewers began to grow in size and number, and by the late nineteenth century there were almost 1,300 breweries.

Early in the twentieth century, the Eighteenth Amendment was ratified, which enacted national prohibition of alcoholic beverages. This led to the closure of most breweries, and by 1933 there were fewer than 35 in operation. Only a few breweries, mainly the largest, were able to stay in business by manufacturing near beer, malt syrup, dairy products or other non-alcohol grain products, in addition to soft drinks such as colas and root beers. The malt syrup was used to make "home brew."

When Prohibition was repealed in 1933, federal and state governments tightened controls, and brewers, distillers, and vintners adopted policies of self-regulation. The Federal Alcohol Administration (FAA) Act was put into place soon that provided for regulation of those engaged in the alcohol beverage industry, and for protection of consumers.

After the repeal of the Eighteenth Amendment, breweries began reopening, and within a year almost 800 were back in operation throughout the country. However, the demand for beer was unexpectedly low after repeal and many breweries went out-of-business. In 1947, about 400 breweries were operating in the United States, with beer sales at about 87 million barrels, up from 32 million barrels in 1934. Beer demand did not grow significantly until the 1960s and 1970s. The number of barrels of beer sold increased from about 88 million in 1960 to 123 million in 1970, and then to 169 million in 1980. By the mid-1980s, the market demand for beer had stabilized, with sales of 175 million barrels in 1985. The increased number of young people, the lowering of drinking age requirements, and the enhanced acceptability of beer among women all contributed to the increase in beer sales in the 1960s and 1970s. The shrinking in the number of "dry" areas in the United States was a factor too.

Then, in the 1980s, several entrepreneurs began "crafting" beer on a small scale. Because of their size, the new enterprises were called "boutique" or "micro" breweries. These brewers returned to the brewing practices of the past by brewing specialty beers such as ales, porters, stouts, and dark lagers. Typically, craft beers feature traditional ingredients, such as malted barley, not rice or other adjuncts. Owing to their limited production capacity, these new enterprises have not made significant contributions to industry shipments; however, they have had a significant impact on the variety of beer styles available to the American public leading to a change in consumer preferences.

Industry Concentration

During the 1950s and through to the 1980s, the U.S. beer industry experienced a period of consolidation as the large national brewers grew larger, and numerous traditional regional breweries went out of business. The decline in the number of regional breweries can be attributed to many reasons. Consumer preferences were shifting away from dark, strongly flavored beers to lighter, drier beers. But it was the regional, smaller brewers who generally specialized in the darker, more flavorful beers, while national and some regional brewers generally produced lighter beers.

Other reasons attributed to the decline in number of breweries include technological improvements, economies of scale, the rise in advertising expenditures by national breweries, mergers, and price wars.

Technological Improvements

Beer prior to World War II was distributed to wholesalers and then to retailers within an extremely limited geographic region, by today's standards. Beer was a relatively expensive product to transport considering its value, so any brewers wishing to expand their sales area had to consider freight costs in shipping to other markets. The chief means of selling beer at that time was in draft form and in refillable bottles. Both of these packages were expensive to ship and return and, as a result, this limited brewers' access to other markets. However, after the World War II the use of aluminum cans, along with the use of one-way glass bottles, enabled brewers to ship beer more cheaply and expand their markets accordingly. Many regional breweries found it very difficult to compete, thus shipping breweries gained a competitive advantage not formerly available to them.

Economies of Scale

For national brewers, internal expansion led to increasing the size of plants, improvements in packaging, introduction of automated breweries, the development of special fermentation processes, and the building of additional plants. The new breweries were more modern and efficient than any plant that could be obtained through acquisition. In addition, they were built wherever population and consumption patterns dictated, thus minimizing shipping costs. As a result, regional brewers usually found it difficult to compete with the lower-priced beer produced by the national brewers.

Advertising Expenditures

The rise in concentration can also be attributed to the skyrocketing growth in advertising expenditures—during the 1950s and early 1960s as well as during the 1970s and early 1980s—undertaken by the national brewers. Regional brewers were vulnerable, since they could not undertake national advertising campaigns.

Mergers

Mergers played a role also, but they accounted for only a small share of the increase in concentration due to very strict enforcement of antitrust laws by the Justice Department. Indeed, the early enforcement of the anti-merger laws was partly responsible for the emphasis on internal growth and development of new brands by the leading brewers. While some regional breweries were able to merge with others, the antitrust laws prevented many from achieving economies of scale in production, marketing, and distribution.

Price Wars

Another behavior employed by the big brewers was price wars, especially in the 1970s. Many small regional brewers responded with price cuts to attract customers, which put further strain on the weaker firms. Regional brewers were thrust into a war of attrition in which many failed or were purchased by other brewers.

Classification of Brewers

Although much has changed in recent years brewing companies are still divided it what can be considered national, regional, and specialty craft brewers. The first segment includes national brewers who brew beer on a mass scale and who sell in all (or almost all) states. This group includes Anheuser-Busch, SAB-Miller, and Molson Coors Brewing Company. The next segment is regional brewers who are distinguished by their geographic scope of operation within the United States serving a multi-state area—no more than two or three states. Among the regional brewers still in existence are Pittsburgh Brewing Company, Yuengling, Latrobe Brewing Company, and High Falls Brewing Company. The third segment is specialty craft brewers, which consists of brewpubs and microbreweries that serve a single metropolitan area or perhaps a single state or region.

National Brewers

The U.S. brewing industry is dominated by three firms—Anheuser-Busch, SAB-Miller, and Molson Coors Brewing Company—who together account for about 79% of beer shipments in 2007—Anheuser-Busch (49%), South African Breweries' Miller (SAB-Miller) (18%), and Molson Coors Brewing Company (11%). The large national brewers generally offer a homogeneous selection of beers designed for mass appeal. These beers, principally light-bodied lagers, are brewed using low cost mass-production techniques and low cost adjuncts, such as rice and corn, and relatively fewer hops. National brewers are those with annual shipments more than 15 million barrels (31gal/barrel).

Several of these brewers are either vertically integrated themselves or make up divisions of large-diversified corporations that are vertically integrated. For instance, Anheuser-Busch also owns Busch Agricultural Resources, Inc., which owns a number of barley elevators, hops farms, malt plants, rice mills, seed facilities, etc. Molson Coors owns barley farms and malting plants. Most if not all are publicly owned.

The large brewing companies have become highly globalized. Difficulties associated with the international transport and distribution of beer was the motivating forces behind their globalization. Beer, on account of its volume, weight, and limited shelf life, is inherently costly to ship long distances. As a result, the large brewing companies often brew and market their brands abroad through licensed-brewing agreements (LBAs), joint ventures (JVs), and brewery acquisitions. The large brewing companies also establish export distribution agreements (EDAs) and trading companies abroad in order to facilitate the marketing, distribution, and sale of their exported brands. Recently, large brewing companies have also begun purchasing equity shares in foreign breweries in order to obtain market presence and influence in markets otherwise difficult to penetrate.

Anheuser-Busch

Anheuser-Busch (A-B) is currently the largest brewer in the United States. A-B's growth in production capacity has come from internal expansion rather than from acquisitions. Today, A-B produces some of the leading brands in the industry and also competes in the specialty beer market. There are many reasons for A-B's continued success. A-B was one of the first brewers to nationally advertise and invest in marketing research. A-B was one of the first to offer a diversified line of brands. Finally, it has lower production costs, since it was large enough to attain economies of scale, not to mention it produces beer in a number of locations thus cutting transportation costs.

In 2008, Belgian beer giant InBev purchased Anheuser-Busch—the two will form the world's largest brewer by volume, passing current leader SAB-Miller. The combined entity will henceforth be known as Anheuser-Busch InBev, with St. Louis, Missouri being the North American headquarters and home of the flagship Budweiser brand. The deal will open the way for InBev to sell Budweiser in emerging markets such as China and Brazil.

South African Breweries' Miller (SAB-Miller)

South African Breweries (SAB-Miller), formerly known as Miller Brewing Company, was initially purchased by Philip Morris in the 1970s. Philip Morris in turn expanded production capacity, increased advertising spending, and initiated a brand-proliferation strategy. This event is said to have sparked the great "beer wars" during that time. Of all the brands introduced by Miller, Miller Lite was a major success even to the point that a light-beer segment was established. This strategy increased their market share making Miller the second-largest brewer in the United States, surpassing Coors and other brewers. A-B and Coors eventually followed suit introducing their own lite beer brands. Philip Morris eventually sold Miller Brewing Company to South African Breweries (SAB), a London-based firm, and is now known as SAB-Miller.

Molson Coors Brewing Company

Molson Coors Brewing Company, formerly known as Adolf Coors Company, unlike other brewers emphasized engineering inventiveness and product quality over marketing and brand development up through the late 1970s. Initially, Coors brewed just one brand of beer, available in a limited number of states, with production in just one single plant. Coors was the first major company to ship unpasteurized beer in cans due to their development of a cold-filtration system. However, the cans required refrigeration and distributors were required to pull cans off the shelf just after 30 days since the beer was unpasteurized. In the mid-1980s, Coors started investing heavily in advertising and introduced several brands. Coors also built other plants and expanded into other markets.

In 2004, Adolph Coors Co. and Molson, Inc., the biggest brewer in Canada, created a new company, Molson Coors Brewing Co. In 2007, SAB-Miller and Molson Coors merged their businesses in a new joint venture called MillerCoors.

Regional Brewers

Regional brewers are sometimes distinguished by their geographic scope of operation. Regional brewers are those with annual shipments fewer than 15 million barrels, but greater than 2,000,000 barrels. Most regional brewers are privately held by single plant brewing companies. Among the regional brewers still in existence are the Pittsburgh Brewing Company, Yuengling, the Latrobe Brewing Company, and the High Falls Brewing Company. Price competition combined with increasing vertical integration, and the inherent production economies of the market leaders has made it very difficult for regional brewers to compete with the national brewers.

Specialty Craft Brewers

Specialty craft brewers consist of brewpubs, microbrewers, and regional craft brewers that are small, independent brewers whose predominant product is brewed with only traditional brewing processes and ingredients. Craft beers are full-flavored beers brewed with quality hops, malted barley, yeast and water without adjuncts such as rice, corn or stabilizers or water dilution used to lighten beer for mass-production and consumption. Among the well-known specialty craft brewers are the Anchor Steam Brewing Company, the

Boston Beer Company, the Sierra Nevada Brewing Company, and the New Belgium Brewing Company.

A brewpub as defined by the Brewers Association is a restaurant-brewery that sells the majority of its beer on-premise, a common practice in Europe. Annual production for brewpubs rarely exceeds 5,000 barrels. A microbrewery is a brewery that produces fewer than 15,000 barrels (17,600 hectoliters) of beer per year. A regional craft brewery is a designation given to a brewery with an annual beer production of between 15,000 and 2,000,000 barrels.

In 2007, the number of craft breweries was estimated at 1,420 with brewpubs dominating the craft brewery segment at 975 followed by microbreweries at 392, and regional craft breweries at 53. Although craft brewing is largely a West Cost phenomenon, craft breweries can now be found in every state of Union. They are primarily concentrated in California, Colorado, Washington, Oregon, Michigan, Wisconsin, Pennsylvania, Florida, and New York in decreasing order.

Factors Contributing to Growth in Craft Breweries

The growth in craft breweries is generally attributed to several factors. First, changes in government policy have benefited the craft brewing industry. In 1977, the government cut excise taxes for smaller brewers. Before the cut, all brewers paid a federal tax of $9 per barrel. Under the new law, brewers selling fewer than 2 million barrels annually paid an excise tax rate of $7 per barrel on the first 60,000 barrels and $9 per barrel on additional sales. The tax advantage for small brewers was enlarged in 1991. The federal excise tax rate doubled from $9 to $18 per barrel, but brewers with annual sales fewer than 2 million barrels continued to pay $7 per barrel on the first 60,000 barrels sold annually. Second, the increased preference for craft beers is part of an overall consumer trend toward gourmet, handcrafted, locally-produced items. Third, craft brewers convincingly promoted the notion that beer "made in small batches" from "all natural ingredients" is better than mass-produced products. Fourth, the craft breweries successfully appealed to regional loyalties and a general nostalgia for the days when small beer companies tended to be the rule rather than the exception. Lastly, consumers' interest in beer making and brewing history has flourished, as demonstrated by the growth of related industries such as home brewing, beer festivals, and consumer publications devoted to beer as well as the increasing popularity of beer celebrities.

Beer Sales Domestic and Imported

After rising more or less steadily since the repeal of prohibition, U.S. beer sales have been relatively flat since the mid-1980s. The period of booming sales of the 1970s is now but a distant memory for the beer industry. In 2007, total U.S. beer sales were 212.9 million 31.5 gallon barrels, slightly up from the previous year's level of 209.8 million barrels as shown in Table 1.1. Sales of domestic beer production were about 183.0 million barrels in 2007, slightly up from the previous year's level of 180.5 million barrels, but down from the 1990 record level of 184.4 million barrels. Imported beer volume increased by 1.4% to nearly 30 million barrels, making 2007 the first-year imports under performed in the U.S. beer market since 1991 (which also happens to be the last year they declined). In 2006, imported beer volume advanced by almost 15% and domestic volume barely grew at all. However, looked at over a longer time frame, imports have done quite well approaching a 14% share of volume in 2007 compared to a market share of 0.7%, or only 900,000 barrels in 1970. Moreover, in the five-year period from 2002 to 2007, imports achieved a compound annual growth rate (CAGR) greater than 5%, while domestic beer's CAGR was essentially flat.

Beer Sales by National Brewers

Sales for the country's three biggest brewers—Anheuser-Busch, Inc., SAB-Miller, and Molson Coors Brewing Company—were about 169.6 million barrels or about 79% of total domestic beer sales in 2007. Anheuser-Busch outperformed the industry with beer sales at about 105.2 million barrels, followed by SAB-Miller at an estimated 40.2 million barrels, and Coors at approximately 24.3 million barrels.

Anheuser-Busch gained the most in market share, increasing from 28% in 1980 to an estimated 49% in 2007. SAB-Miller's market share declined from 21% to about 18% from 1980 to 2007 while Molson Coors increased from 8% to about 11% for the same time period.

Because it is difficult for any one firm to continue sales growth in this mature market unless it takes sales away from a competitor, tremendous competition for market share has become the primary objective of national brewers.

Top Beer Brands

The top ten beer brands in 2007 had combined shipments estimated at 142.5 million barrels, up from last years level at 141.5 million barrels (see Table 1.2). Some of the notables include Bud Light, Budweiser, Miller Lite, and Coors Light. Inovative strategies in pricing, merchandising, and supply chain continue to be the key elements driving success of these brands.

Off- and On-Premise Sales

Approximately three-quarters of the volume of beer marketed in the United States are for off-premise consumption, whereas only a quarter is consumed on-premise. The beer sold for on-premise consumption, however, contributes nearly half of all retail dollar sales. Most on-premise sales take place in bars, restaurants, ball parks, stadiums, and airplanes; while most off-premise sales

Table 1.1
U.S. Beer Sales
Domestic and Imported Volume, Share, and Change
2002 – 2007

Year	Millions of Barrels Domestic	Millions of Barrels Imported	Millions of Barrels Total	Share of Market Domestic	Share of Market Imported	% Share of Volume Domestic	% Share of Volume Imported	% Share of Volume Total
2002	180.4	23.1	203.5	88.6%	11.4%	--	--	--
2003	179.4	23.5	202.9	88.4%	11.6%	-0.6%	1.7%	-0.3%
2004	181.3	23.8	205.1	88.4%	11.6%	1.1%	1.3%	1.1%
2005	179.6	25.6	205.2	87.5%	12.5%	-0.9%	7.1%	0.0%
2006	180.5	29.3	209.8	86.0%	14.0%	0.5%	14.8%	2.2%
2007	183.0	29.9	212.9	86.0%	14.0%	1.4%	1.4%	1.4%
Five-year compound growth rates:						0.3%	5.3%	0.9%

CY 2007 represents preliminary numbers.

Source: Beer Institute

Table 1.2

Top Ten U.S. Beer Brands
Volume, Share, and Change
2006 – 2007

Brand	2007 Rank	Millions of Barrels 2006	Millions of Barrels 2007	Share of Volume 2006	Share of Volume 2007	% Change 2006/07
Bud Light	1	41.2	42.0	19.6%	19.6%	1.9%
Budweiser	2	25.8	24.6	12.2%	11.5%	–4.7%
Miller Lite	3	18.0	18.4	8.5%	8.6%	2.2%
Coors Lite	4	16.7	17.3	7.9%	8.1%	3.6%
Natural Lite	5	9.0	9.1	4.3%	4.3%	1.1%
Corona Extra	6	8.5	8.3	4.0%	3.9%	–2.4%
Busch Light	7	6.1	6.3	2.9%	2.9%	3.3%
Busch	8	6.2	6.2	2.9%	2.9%	0.0%
Heineken	9	5.1	5.2	2.4%	2.4%	2.0%
Miller High Life	10	4.9	5.0	2.3%	2.3%	2.0%
Subtotal		141.5	142.5	67.2%	66.6%	0.7%
All Others		69.4	71.4	32.8%	33.4%	2.9%
Total		210.9	213.9	100%	100%	1.4%

Source: Beer Marketer's Insights

occur in liquor stores, large beverage retail chains, grocery stores, convenience/drug stores, and gas stations.

Factors Affecting Beer Sales

The U.S. beer industry has encountered a number of challenging economic as well as powerful demographic and cultural trends that have hurt sales and will likely continue for years to come.

The most important single factor affecting the current beer market is the aging of America, particularly the Baby Boomer generation, a variable that is significant because of this group's tremendous size. The effect of the aging population on the beer industry is readily apparent given the fact that as the average age of the population increases, per capita consumption figures for beer decrease. The industry has not seen much growth since the baby boomers were in their prime drinking years during the 1970s.

Beer marketers are increasingly finding themselves in competition not only with one another but also with the purveyors of other beverage products. Besides spirits and wine—the other traditional alcoholic beverages—"New

Age" products such as ciders, bottled waters, bottled ice teas, fancy coffees, and fruit-based soft drinks have entered the market.

On the social front, stagnant beer sales can be partly attributed to the furtherance of public policies regarding alcoholic beverages. Because of the increased public concern with drunk driving, the federal government has mandated a legal drinking age of twenty-one throughout the United States. This raising of the legal drinking age has (at least in theory) decreased the size of the domestic beer market. There are other public measures that have also been implemented to deter alcoholic consumption in the past decade. Some examples include lowering the blood alcohol content (BAC) levels for drunk driving, the increase in federal and state excise taxes, stiffer penalties imposed by states for driving while under the influence, and higher insurance premiums or cancellation of policy if convicted of DUI charges.

Finally, there has been growing national focus on health and fitness that has also pushed alcohol consumption to a more moderate level. Health-conscious baby-boomers, fretting about waistlines and heart murmurs, are eschewing high-carb beer for cardiac-friendly merlot.

Per Capita Consumption

Among commercial beverages in 2006, beer ranks fourth in per capita consumption behind carbonated soft drinks, bottled water, and coffee followed by milk and fruit drinks/cocktails (see Table 1.3). Per capita beer consumption rose rapidly during World War II, declined during the 1950s and the early 1960s, increased before peaking in the early 1980s, and has generally leveled-off thereafter. In 2006, U.S. per-capita beer consumption for the total population was about 21.6 gallons, down from 23.2 gallons in 1985. In the United States, demand varies from state to state with North Dakota having the highest per capita consumption of 32.4 gallons and Utah

Table 1.3
Per Capita Beverage Consumption

Beverage	Rank	Gallons	% Change 2005/06
Carbonated Soft Drinks	1	50.6	−1.9%
Bottled Water	2	27.6	8.2%
Coffee	3	24.4	0.8%
Beer	4	21.6	0.5%
Milk	5	21.0	0%
Fruit Drinks/Cocktails	6	13.5	−2.9%
Source: USDA/Economic Research Service				

the lowest of 13.4 gallons. By world standards, the United States is a major beer-drinking nation. In 2007, the United States ranked fourteenth among developed countries in beer consumption.

Beer Segments

The market for beer in the U.S. consists of essentially the following ten segments (see Table 1.4): imports, domestic specialties, super-premium, premium regular, light, ice, maltalternatives, malt liquor, popular regular, and others. Table 1.5 lists popular brands by each segment.

Although the beer category has a whole has seen consistently modest growth for several years certain beer segments are still performing extremely well. Among the beer segments, the greatest growth has been in light, imports, domestic specialties, and super-premium. Growth in the other beer segments have been flat to down.

Table 1.4

U.S. Beer Sales by Segments
Volume, Share, and Change
2006 – 2007

Segment	2007 Rank	Millions of Barrels 2006	Millions of Barrels 2007	Share of Volume 2006	Share of Volume 2007	% Change 2006/07
Light	1	103.7	105.6	49.2%	49.4%	1.8%
Premium Regular	2	31.6	30.4	15.0%	14.2%	−3.8%
Imports	3	29.3	29.6	13.8%	13.6%	1.0%
Popular Regular	4	16.8	16.5	7.9%	7.7%	− 0.8%
Domestic Specialties	5	8.1	9.0	3.8%	4.2%	11.1%
Ice	6	7.3	7.6	3.5%	3.6%	4.1%
Malt Liquor	7	5.8	6.0	2.7%	2.8%	3.4%
Maltalternatives	8	4.4	4.3	2.1%	2.0%	−2.3%
Super-premium	9	3.0	3.8	1.4%	1.8%	26.7%
Subtotal		209.7	212.4	99.4%	99.3%	1.3%
Others		1.2	1.5	0.6%	0.7%	25.0%
Total		210.9	213.9	100%	100%	1.4%

Source: Beer Marketer's Insights

Table 1.5
U.S. Beer Brands by Segment

Beer Segments	Brands
Ice	Bud Ice, Natural Ice, Molson Ice
Malt Liquor	Colt 45, Olde English 800, Mickey's, King Cobra
Domestic Specialties	Sierra Nevada Pale Ale, Harpoon IPA, Acnchor Steam California Common
Super-premium	Michelob, Lowenbrau
Premium Regular	Budweiser, Miller Genuine Draft, Coors Original
Popular Regular	Miller High Life, Milwaukee's Best, Pabst
Imports	Heineken, Corona, Guiness Extra Stout
Maltalternatives	Smirnoff Ice, Bacardi Silver, Mike's Hard Lemonade
Light	Bud Light, Miller Lite

Light

As far as domestic segments go, the most significant and unstoppable trend in the beer business is the continued momentum of light beer. Since the creation and marketing of the first light brands in the mid-1970s consumers have increasingly been attracted to these beers for their smooth, mild taste and lower calories. Today the light beer segment accounts for about 49% of all beer sales in the Unites States led by the seemingly indomitable Bud Light, which continues to increase sales and remains the best-selling beer in the United States by far. Although the light beer segment has grown in popularity, the gains have come largely at the expense of the premium-regular and popular-regular segments as consumers trade up and look for products with a healthier image and lighter taste profile.

Imports

Imports have made significant incursions into the U.S. beer market, mostly at the expense of domestic brands. Indeed, imports made a major impact on the U.S. beer market in 2007 and did so despite uncharacteristically slow volume growth that actually lagged that of the market overall. In 2007, the market share of imports was 14.0% with a volume of 29.9 million barrels.

The primary countries that ship beer to the United States are Mexico, Canada, Germany, Ireland, Netherlands, and the United Kingdom with Mexico leading the way with Corona Light accounting for 8.5 million barrels or 29% of all beer imports.

Clearly, U.S. demographic trends play a part of this growth. Hispanic-Americans now comprise about 14% of the U.S. population, representing about $700 billion in buying power, and all projections are that the group will continue expanding. These consumers have pre-established beverage preferences and, especially for younger adult Hispanics, beer (particularly Mexican beer) is part of their lifestyle. Indeed, across the United States, consumers have been connecting to things Mexican, particularly food.

Import brand growth, in many cases, also has been self-perpetuating. The more brands have grown, the more money marketers have been able to plow back into them, building distribution and awareness among consumers.

Domestic Specialties

Although growth in domestic beer segment has been relatively flat except for light beer, the domestic specialty segment has shown strong double-digit growth that should continue in the foreseeable future. This segment includes brewpubs, microbreweries, and regional special breweries. Domestic specialties like that of imports has not only benefited from American consumers trading up for more expensive "fuller flavor" beer, but also from consumers who favor small local producers over national or multinational corporations. In spite of rapid growth and notoriety, this segment has a market share of only about 4.2%, comparable to that of the ice beer segment.

Big Brewers Get Crafty

In response to the significant growth in the craft-beer industry, the three major brewers—Anheuser-Busch, SAB-Miller, and Molson Coors Brewing Co.—have all entered this fast-growing market, either through developing their own specialty craft beers or by acquiring or forming partnerships with existing craft brewers.

The major brewers generally avoid using the parent company's name on the labels for their craft beers. Some beer aficionados and industry executives deride the brews made by big companies as "faux" crafts, regardless of their quality or whether they are using traditional ingredients.

Super-Premium

Like imports and domestic specialties, the super-premium segment has experienced strong growth over the past several years. This too can be attributed to the fact that Americans are becoming more yuppified and "trading up" to better products across the board in food and beverage. Millions of consumers have in recent years become connoisseurs (that is, insufferable snobs) when it comes to coffee, cheese, chocolate, wine, and for that matter beer. Also contributing to the rise in beer sales in this segment, just like imports and domestic specialties, is the willingness of consumers to pay more. The surge in the numbers of specialty craft beers and imports in the past ten years has conditioned consumers to pay higher prices for what they perceive to be high-quality products.

Advertising

Beer brewers spent about $975 million in 2007 on advertising promoting their brands in major media channels. Not surprisingly, the Big Three accounted for most of the expenditures, with Anheuser-Busch leading the pack at $378 million, followed by SAB-Miller at $175 million and Molson Coors Brewing Co. at $151 million. The Big Three spend most of their advertising dollars on premium and light beers.

Beer is advertised on television, radio, and billboards; in the written media; through promotional merchandise; at point-of-sale; and through brewing company sponsorship of sports teams and of numerous sports, recreational, cultural, and community events. As might be expected in today's electronic age, the majority of beer ad spending was allocated to television, primarily network broadcasts. The large brewing companies often contract with several advertising agencies at once, each being responsible for a particular brand.

Advertising among craft brewers, on the other hand, varies considerably with some brewers investing heavily in advertising while others disdain advertising altogether.

Although most of the industry's advertising focuses on brand introduction and promotion, recent advertising campaigns designed to improve or change consumers' image of beer and its intended use are also important. In an effort to address the negative publicity regarding alcohol abuse, the industry has invested substantial advertising resources in the promotion of responsible beer serving and drinking.

Beer Distribution

The industry is organized into a so-called "three-tier" distribution system: 1) brewers and importers, 2) wholesalers, and 3) retailers. Under this system, brewers and importers sell their products to wholesalers who in turn sell to the retailers. (Supplier, wholesaler, and retailer are the three tiers.) As explained above, when states took over regulation of these three tiers after Prohibition, they chose to either license or directly control sales and distribution. These laws (known as "tied house" laws), adopted at the state and federal levels, regulate how alcoholic beverages are marketed and how the various tiers of the industry interact.

The three-tier system dates back to the repeal of prohibition in 1933. After the 21st Amendment was repealed, the federal government established rules for the marketing of beer in the Federal Alcohol Administration (FAA) Act. The Act granted U.S. state governments virtual plenary powers to regulate the production, importation, distribution, sale, and consumption of alcoholic beverages, including beer, within their borders. As a result, the states enacted numerous provisions to regulate the distribution and sale of beer, including various types of ownership, franchise, exclusive territory, primary source, "at rest," and price posting provisions.

Brewers and Importers

This tier includes firms that brew beer and other malt beverages, and firms that import malt beverages for consumption in the United States. The brewing tier also includes company-owned packaging and wholesaling operations. Brewers include major multistate multi-operational brewing companies, regional brewers, microbrewers, and brewpubs.

Beer imports enter the United States through import companies which serve as main distributors selling the brands to wholesalers in local markets. The largest import brands are distributed locally by major beer wholesalers (i.e., those handling leading national or dominant regional beers) or, in some markets, by major wine and liquor wholesalers. Smaller import brands, unable to attract major distribution, are sold through smaller wholesalers.

Wholesalers

Beer wholesalers serve as the middle tier in the three-tier system. They purchase beer from suppliers and importers and, in turn, sell beer to retailers. In general, a brewer sets the wholesale price at the brewery (freight on board, f.o.b.) but may vary price by region of the country in response to different demand and levels of competition. Although most beer wholesalers handle beer exclusively, a growing number also distribute food products and other beverages—primarily soft drinks, iced teas, and bottled waters—in conjunction with beer.

Most of the leading beer wholesalers in the country fall into one of two general categories: exclusive wholesalers or independent multi-brand wholesalers located in one or more regions. In most states, each distributor is awarded an exclusive sales area by the brewer and is primarily responsible for building relations with the retail and other consumer outlets in order to build the sales of the product. Unlike wine and spirits wholesalers, which are generally multistate operators, exclusive beer wholesalers tend to operate within the boundaries of a single state since they are regulated by the state in which they do business. In the last few years, however, more wholesalers have been venturing outside their home states. These moves are being accelerated by the support of leading brewers and "super retailers." In states with exclusive territories, retail customers, including grocery chains, bars, and restaurants, have only one source of supply and are therefore at the mercy of distributor pricing. In some states, such as Indiana, exclusive territories are not allowed, and competition is fierce among wholesalers.

The primary function of wholesalers is to provide a local warehouse facility for quick delivery to retailers, to purchase and finance inventory, and to sell the product to retailers. Not only are wholesalers responsible for distribution, they may also be responsible for setting up point-of-sale promotions. These activities include creating, setting up, and maintaining signs and in-store retail displays. Wholesalers are also involved in retrieving and recycling empty bottles and cans; tracking data and retail sales, inventories, and other market information; and fulfilling state, federal and, sometimes, local tax reporting requirements.

Wholesaler Consolidation

Since 1995, the number of distributors has dropped from 5,500 to about 2,000 as smaller multi-brand wholesalers have been bought out. Rising costs as well as the need for scale and for broad, well-balanced portfolios has driven consolidation among U.S. beer wholesalers. Today, consolidation has been driven, in part, by increased operational costs—particularly health care and insurance costs—which have put pressure on smaller distributors, who do not have the same ability to negotiate with insurance companies that larger distributors have, and cannot allocate their costs across as wide a range of operations as larger distributors. Additionally, smaller companies also face challenges in acquiring a broad portfolio of attractive brands. Given these and other challenges, smaller distributors have opted to sell their franchises to larger, more efficient distributors that are looking to diversify their portfolios.

> **Note | Anheuser-Busch-InBev Merger Impact on Wholesalers**
>
> The merger of Anheuser-Busch and InBev will likely affect wholesalers more than retailers since the combination of MillerCoors and AB InBev represents 80 percent of the category volume. The merger could make it difficult for some imports and independent craft brands to maintain shelf space, as both MillerCoors and AB InBev bring cross-segment portfolios to the retailer.

Retailers

Finally, the third tier of the industry directly sells products to the consumer either by on-premise or off-premise sales. On-premise retailers were those retailers that sold beverages to be consumed at that location. These include restaurants, hotels, bars, etc. Off-premise retailers were those that sold beverage for consumption in locations other than those in which they are sold. These include traditional liquor stores, grocery stores, and convenience stores.

Retail Consolidation

Over the last decade or so, the off-premise retail segment for beer has changed dramatically. While traditional liquor stores and mom-and-pop operations continue to account for a significant share of off-premise beer sales but in ever decreasing numbers, national retail chains—supermarkets, drug stores, membership clubs, or convenience stores—are increasingly accounting for an increasing share of total market volume. Part of this trend is due to the American consumer's search for convenience and the competitive pricing and brand selection these stores offer. Similarly, while independently owned taverns, restaurants, and night-clubs account for a sizeable share of on-premise beer sales, national restaurant/hotel chains are growing in importance.

This increasing share of total market volume by fewer larger retail chains changes the playing field ... and impacts the relative leverage among the three tiers. Suppliers (i.e., brewers and importers) are faced with an increasingly powerful retail sector—a component of which has become skeptical of the three-tier system as a concept. Retailers have responded by demanding more from suppliers and wholesalers in terms of various operational and category management services. Suppliers have responded in turn by demanding more from their wholesaler networks, and by reengineering their own sales systems to become more specialized by account type. Suppliers are increasingly involved in headquarter sales calls on national and regional chains—and even in local accounts.

Demographics

Beer consumption is overwhelmingly male-dominated, with men accounting for more than 80% of the volume consumed. A large number of these beer drinkers are white and favor domestic light beer, followed by domestic draft beer. Of all the beer types, light beer has the strongest following among women consumers. Women beer drinkers are more attracted to specialty microbrewed beers than they are to the big brands, possibly because of their greater variety. The appeal of craft-brewed beer is stronger among white beer drinkers than among African-Americans.

Over two-thirds of all beer is consumed away from home, with clubs and bars being the most preferred away-from-home drinking locations. Unlike consumers of distilled spirits and wine, beer drinkers are relatively young. About half of all beer drinkers are from the ages of 25 to 44. Beer drinkers have diverse income levels, unlike drinkers of distilled spirits and wine, whose incomes are likely to be higher. Among different beer types, regular beer is very popular with consumers with low incomes. On the other hand, the largest audience for light beer is those with higher incomes. Geographically, consumers in the Northeast spend the most on beer, although those in the South and West spend close to the national average. Consumers in rural areas spend less annually on beer than do urban consumers.

References

1. Backus, Richard. 1997. "The Robert Weinberg Interview." *The New Brewer* 14 (2).
2. Barsby, Steve L. and Associates. 1999. *Beer Wholesalers: Their Role and Economic Performance*. Alexandria, Virginia: National Beer Wholesalers Association.
3. Charlier, Marj. 1994. "Specialty Beers' Success Prompts Big Brewers To Try Out the Niche." *The Wall Street Journal* January 5.
4. Demetrakakes, Pan, ed. 1997. "Getting Crafty." *Food Processing* 58 (2).
5. Edgar, David. 1997. "1996 in Review." *The New Brewer* 14 (3).
6. Elzinga, Kenneth. 2005. "Beer." *The Structure of American Industry*, edited by Walter Adams and James Brock. Upper Saddle River, New Jersey: Pearson Hall.
7. Finnegan, T., ed. 1997. *Modern Brewery Age Blue Book*. Modern Brewery Age Publishing: Stamford, Connecticut.
8. Fried, Eunice. 1996. "Has Beer Gone Upscale." *Black Enterprise* 26 (11).
9. McGowan, Richard A. 1997. *Government Regulations of the Alcohol Industry: The Search for Revenue and Common Good*. Westport, Connecticut: Greenwood Publishing Group, Inc.
10. Harney, Amy K. 1995. *Malt Beverages*. Washington, D.C.: Office of Industries, United States International Trade Commission, USITC Publication 2865.
11. Jones, Lester. 2008. Personal Communication. Washington, D.C., Beer Institute.
12. Katz, P.C. 1991. "Brewing Industry in the United States." *Brewers Almanac*. Washington, D.C., The Beer Institute.
13. Khermouch, Gerry. 1995. "This Year's Model." *Brandweek* 36 (29).
14. Khermouch, Gerry. 1996. "Micros Scramble Amid A-B Purge." *Brandweek* 37 (19).
15. Khermouch, Gerry. 1996. "Even as A-B Backs Off, Small Brews Still Face Distribution Bottleneck." *Brandweek* 37 (38).
16. Khermouch, Gerry. 1997. "Shakeout Looming, Craft Brewers Grapple with Marketing." *Brandweek* 38 (13).

17. Lee, Byunglak, and Victor J. Tremblay. 1992. "Advertising and the U.S. Market Demand for Beer." *Applied Economics* 24 (January).

18. Levitt, Craig. 1997. "Another Round." *Discount Merchandiser* 37 (3).

19. MBA Statistical Study. 1997. "Imported Beer Focus." *Modern Brewery Age* 48 (28).

20. Maxim, Lenderman. 1996. "Mind Games." *Beverage World* 115 (1623).

21. McDowell, Bill. 1996. "New A-B Specialty Beer Targets Northern Calif." *Advertising Age* 67 (46).

22. Melcher, Richard A. 1995. "From the Microbrewers Who Bought You Bud, Coors..." *Business Week* (3421).

23. Palmer, Jay. 1996. "Brewing Storm." *Barron's* 76 (44).

24. Prince, Greg W. 1994. "Little Giants." *Beverage World* 113 (1581).

25. Prince, Greg W. 1996. "This Seat is Taken." *Beverage World* 115 (1610).

26. Reid, Peter V.K., ed. 1997. "Year in Review: 1996." *Modern Brewery Age* 48 (11).

27. Reid, Peter V.K. 1997. "The Great and Powerful Bob." *Modern Brewery Age* 48 (11).

28. Sass, Tim R. and David S. Saurman. 1998. *Brand Equity Created by Beer Wholesales through Local Promotional Activities*. Alexandria, Virginia: National Beer Wholesalers Association.

29. Shepard, Eric. 2008. Personal Communication. West Nyack, New York., Beer Marketer's Insights.

30. Sherer, Michael. 1996. "High-Flying Imports." B*everage and Food Dynamics* 107 (2).

31. Sherer, Michael. 1996. "Lights Still Shine." *Beverage and Food Dynamics* 107 (6).

32. Staff Report of the Federal Trade Commission Bureau of Economics. 1978. *The Brewing Industry*. (December).

33. Tremblay, Victor J. and Carol Horton Tremblay. 2005. *The U.S. Beer Industry: Data and Economic Analysis*. Cambridge, Massachusetts: The MIT Press.

34. Weinberg, Robert S. 1999. "Watching the Market." *Modern Brewery Age* 48 (11).

Chapter Two

Barley Malts

Barley malt is to beer as grapes are to wine. It is ideally suited to brewing for many reasons. Malted barley has a high complement of enzymes for converting its starch supply into simple sugars and contains protein, which is needed for yeast nutrition. Other grains, such as wheat and rye can be malted and used to brew beer too, but they are not widely used.

Barley

Barley, a cereal grain that has been cultivated for millennia grows in two-row, four-row, or six-row form, as distinguished by the number of seeds on the stalk of the plant. Four-row barley is unsuitable for brewing. European brewers traditionally use the two-row type because it has a better starch/husk ratio and because of its malty flavor. Americans often preferred the six-row type because of the higher levels of diastatic enzymes and protein, which makes it better suited for mashing adjuncts, such as corn or rice.

Six-Row Barley

Generally, six-row barley has a higher enzyme content for converting starch into fermentable sugars, more protein, less starch, and a thicker husk than two-row barley. The higher level of diastatic enzymes makes six-row barley desirable for conversion of adjunct starches (those that lack enzymes) during mashing. On the down-side, the higher protein content can result in greater amount of break material (protein-polyphenol complexes) during wort boiling and cooling, as well as possibly increased problems with haze in the finished beer. The husk of the malt is high in polyphenols (tannins) that contribute not only to haze, but also imparts an astringent taste.

Six-row barley is grown mainly in North Dakota, South Dakota, and Minnesota, and in Canada. The main six-row barley varieties are Robust and

Stander and, to a lesser extent, Foster and Excel. The Robust variety remains the industry standard in terms of malting quality. Six-row malt is grown in Europe, with its use limited to increasing diastatic activity, adding nitrogen, and for specialty malts (e.g., chocolate and black).

Two-Row Barley

Generally, two-row barley has a lower enzyme content, less protein, more starch, and a thinner husk than six-row barley. American two-row barley has greater enzyme potential than most European two-row barley. The protein content of U.S. two-row barley is comparable to that of continental Europe, while barley grown in the U.K. is generally lower in protein.

In comparison to six-row barley, two-row has a higher starch content—the principal contributor to extract. Extract is a major economic concern for many large-scale brewers because the amount of brewhouse extract obtained determines the amount of beer that can be produced from a given amount of malt. Small-scale brewers, however, are generally less concerned about extract yield and may not consider this as important a criterion in their malt choice.

> **Note | Extract**
>
> Extract is the total amount of dissolved materials in the wort after mashing and wort separation. Typically it consists mostly of carbohydrates (fermentable sugars and some dextrins) and nitrogenous matter which determines the starting gravity of the wort.

The thinner husk associated with two-row barley makes for mellower (less astringent) beers due to lower levels of polyphenols. However, the lower percentage of husk can make filtering more difficult during sparging and lautering. Extra care is often required by brewers to ensure adequate filtering of the husks. These malts are very popular with craft brewers.

Two-row barley is grown mainly in Europe and the United Kingdom. The two-row Alexis barley variety is very popular among German and continental brewers, and Optic is the number one variety in Britain. In the United States, two-row malting barley is grown primarily in Idaho and Montana, but it is not as extensively planted as six-row barley. The ratio is changing gradually to more two-row because of six-row *Fusarium* disease. Popular two-row varieties grown in the United States include Harrington, Galena, and Crystal. Harrington is the most widely produced two-row malting barley cultivar in North America. Harrington is high in diastatic power, is well modified, and makes excellent base malt. Though it is best for American lagers, it lends itself well to all beer styles. The European counterpart can be regarded as equivalent.

Identification

The number of rows of kernels makes for easy identification of two- and six-row varieties. In six-row varieties, two-thirds of the kernels are twisted in appearance because of insufficient space for symmetrical development. Since they must overlap, they twist as they grow. In two-row barley there are no lateral kernels—all kernels being straight and symmetrical. The kernels of two-row barley are broader than the central kernels of six-row barley and do not taper as sharply.

Malting

Malting serves the purpose of converting insoluble starch to soluble starch, reducing complex proteins, generating nutrients for yeast development, and the development of enzymes. The three main steps of the malting process are steeping, germination, and kilning.

Steeping

The purpose of steeping is to evenly hydrate the endosperm mass and to allow uniform growth during germination. Steeping begins by mixing the barley kernels with water to raise the moisture level and activate the metabolic processes of the dormant kernel. The water is drained, and the moist grains are turned several times during steeping to increase oxygen uptake by the respiring barley. A wide variety of steeping regimes are used, depending on the steep vessel configuration, barley variety, barley quality, kernel size, brewery specifications, and maltster preferences (21). Generally, the steeping times range from 25 to 60 hours or more with water applications ranging from a single immersion followed by water sprays to a steeping schedule of several separate immersions (21). Draining is done to remove dissolved carbon dioxide and to reintroduce oxygen-rich water. Steeping is complete when the white tips of the rootlets emerge, which is known as chitting.

Germination

In the next step, the wet barley is germinated by maintaining it at a suitable temperature and humidity level until adequate modification has been achieved. The techniques most commonly used for germinating barley are floor malting and pneumatic malting. Floor malting is an old process in which the chitted malt is spread on the floor to a height of 10 to 20 cm. Pneumatic malting uses forced air for the germination process. There are numerous mechanical designs for pneumatic malting such as drum malting, compartment malting, continuous malting, and tower and circular malting, and flexi malting.

The process of germination develops a small amount of sugars, soluble starch, and starch-to-sugar-converting enzymes. The most significant enzymes are alpha-amylase, beta-amylase, and limit dextrinase. Alpha-amylase is absent in raw barley but is formed during malting. Alpha-amylase and beta-amylase together are referred to as diastatic power. Degradation of complex proteins to soluble peptides and free amino acids occurs, too. The peptides and amino acids will nourish the yeast during fermentation as well as add a "mouthfeel" and texture to finished beer. Another metabolic change that occurs during germination is synthesis of S-methyl methionine (SMM), the main precursor of dimethyl sulfide (DMS) in brewing. The SMM content is influenced by the barley variety and the kilning procedure (25).

Kilning

The final step is to dry the green malt in the kiln, which is done at different temperatures. The temperature regime in the kiln determines the color of the malt and the number of enzymes which survive for use in the mashing process. For example, low temperature kilning is more appropriate for malts when it is essential to preserve enzymatic (diastatic) power. These malts are high in extract but low in coloring and flavoring compounds. Pilsner and pale ale malts are examples of malts kilned at low temperatures. Malts kilned at intermediate temperatures, such as Munich and Vienna malts, are lower in enzymes but higher in coloring and flavoring compounds. Malts kilned at high temperatures, such as crystal and chocolate malts, have little if any enzymes thus are lower in extract. On the other hand, these malts are high in coloring and flavoring compounds. Color and flavor compounds, known as melanoidins, are the result of Maillard reactions generated at kilning temperatures. After kilning, the malts are cooled and undergo a rest for conditioning before use.

Malt Modification

In general, modification refers to the extent to which the endosperm breaks down. During the malting process, enzymes break down the cell structure of the endosperm, releasing nutrients necessary for yeast growth and making the starch available for enzyme degradation during mashing. Modification of the endosperm correlates with growth of the acrospire. As the acrospire grows, chemical changes are triggered that result in the production of numerous enzymes, which are organic catalysts. Their function is to break down the complex starches and proteins of the grain.

One traditional measure of modification is the length of the acrospire. Generally, the growth of the seedling is allowed to progress until the acrospire is just over three-fourths of the length of the grain. At this point, the acrospire is considered fully modified. Shorter or longer acrospire lengths are referred to as "under" and "over" modified malts, respectively. Depending on the needs of the brewer, the maltster determines the level of modification. For example, the acrospire should only reach a length equal to about two-thirds to three-fourths of the seedling length for Pilsner malt but should be greater than three-fourths the seedling length for dark malt (22).

Full modification will result in lower yield per weight of barley because a lot of food energy (weight of kernel, potential yield) goes into the growth of the acrospire. Full modification also results in the conversion of very long chains of proteins to usable yeast nutrients (amino acids). The brewer who uses well-modified malt has less concern about haze problems created by the raw proteins that are more prevalent in under-modified malt.

> **Note | Modification**
>
> If the malt is too under-modified it can lead to a host of problems from poor extract recovery, slow wort separation, cloudy worts, poor hot break formation, filtration problems, to low fermentability of the wort. On the other hand, if the malts are over-modified it can result in large quantities of hot and cold breaks, filtration problems, and poor head retention.

Under modification will result in more yield per weight of barley, but the proteins are more complex, with fewer yeast nutrients (amino acids). Low levels of amino acids can result in slow or stuck fermentations, and protein haze problems in the finished beer may occur.

Generally, American malts are more modified than continental malts but not as modified as those produced in England or Ireland. American six-row malted barley has as much higher nitrogen content than traditional lager malt, and its enzyme strength is correspondingly greater. Therefore, high kiln temperatures can be held longer for American malts than with continental malts without risking a serious depletion of the malt's enzyme strength. The lesser degree of conversion is responsible for the lower enzyme strength and greater nitrogen complexity of continental malt. Traditionally, continental malts were not as modified and thus required decoction mashing. Today, continental malts are more modified, and decoction mashing is mostly replaced with other mashing regimes.

Malt Constituents

Malt is largely made up of carbohydrates, which are composed of starch, sugars, insoluble cellulose, and soluble hemicellulose. Starch, the most important constituent, accounts for about 60 to 65% of the malt's weight, is composed of amylose, which is reduced to maltose and maltotriose and amylopectins that decompose into glucose. Glucose, a monosaccharide, accounts for about 1 to 2% of the total starch found in a barley kernel. Cellulose constituents do not contribute to fermentable extract or desirable flavors in the malt. Hemicellulose is a constituent of the endosperm cell walls, which consist largely of beta-glucan.

In addition to carbohydrates, malt is made up of proteins and contains vitamins, principally those of the vitamin B complex. Other significant malt constituents include polyphenols, phosphates, lipids, and fatty acids, as well as silica and trace minerals. Polyphenols are acidic precursors of tannins and give beer a bitter and astringent taste. Phosphatases or phytases are major factors in the acidification of the mash. Fatty acids are essential in the yeast cell as reserves, but in excess they reduce the foam's stability and result in stale flavors in beer—"cardboardy," "goaty," and "soapy."

Malt Analysis

Malt analysis provides guidance on the effectiveness of the malting process and the suitability of the malt for brewing. The brewer judges malt quality by referring to the malt analysis provided by the maltster. Maltsters generally publish "typical" malt analyses, which are provided before purchase, and a lot analysis of the malt, which is provided at the time of purchase. Malt is analyzed in accordance with standard industry tests such as the IoB (Institute of Brewing), EBC (European Brewing Convention), and ASBC (American Society of Brewing Chemists) methods of analysis. A malt analysis provides very useful information, listing a number of parameters. Some of the more important parameters are listed below.

Moisture Content

Moisture content (MC) is shown as a percentage of the total malt weight; the higher the MC content, the less extract yield per kilogram. The moisture content can be expressed as "extract dry" or "extract as is"—which includes the moisture content. The brewer needs to take into account the moisture content of each lot analysis in calculating the weight of malt required since the moisture content influences the amount of extract yield. Generally, the higher the moisture content, the lower the extract yield for the malt. Malts that are more highly kilned (thus darker in color) tend to have a lower, moisture content. For example, ale malt tends to have a lower percentage of moisture than lager malt. If the moisture content is too low it may lead to breakage and husk damage, potentially affecting brew house performance.

Typical MC values for standard malts are (31);

    Standard ale malt: 2–3% moisture
    Standard lager malt: 4–6% moisture

Color

During kilning, chemical reactions take place, producing color compounds. The higher the kilning temperature, the more color compounds produced. American maltsters quote color in Standard Reference Method (SRM) units or degrees Lovibond. Degree Lovibond is an older method of visual measurement and is used because of old habits. SRM is a more sophisticated method, involving the use of light meter analyzers to assign a number to light intensity. This measurement system has largely replaced the older Lovibond color rating system in the brewing industry. Degrees SRM and degrees Lovibond are approximately equivalent. Degrees SRM for light-colored barley malts are usually very low, ranging from 1 to 5° SRM, whereas degrees SRM for most crystal or caramel malts range from 10 to 100. For chocolate malts, degrees SRM average about 300, and for very black malts (dark roasted) they range from 500 to 700. European brewers express color in European Brewing Convention (EBC) units.

To convert between these two units, the formula is:

Degrees EBC = (2.65° SRM) − 1.2

Fine-Grind Extract

Fine-grind, dry basis extract (FGDB) values are a measure of the maximum potential yield of extractable material under ideal mashing conditions. The higher the FGDB extract, the better the malt. Values of 78 or higher are considered acceptable (26).

Coarse-Grind Extract

Coarse-grind dry basis extract (CGDB) values are a measure of the yield that malt is capable of giving under a coarser milling regime not unlike that used in a brew house. However, since CGDB values are on a dry basis, brew house yields will be lower. In addition, the laboratory crush and mash are always more efficient than those under brew house conditions. Usually, brew house efficiency (BHE) is 80 to 95% of what the coarse grind predicts (26).

Fine Grind-Coarse Grind Difference

Fine grind-coarse grind difference (FGCG) is another measure of malt modification, and can be used to predict extract yield and rate of wort separation. FGCG is the difference between the extract obtained from a finely ground malt and that from the same malt ground more coarsely. Low FCCGs are associated with malts that are highly modified. Steely under-modified malt has a greater fine grind-coarse grind difference than well-modified mealy malts. As a rule, malts with a FGCG difference of 1 to 1.5% are suitable for infusion mashing, and malts with a FGCG difference of 1.5 to 2.2%

would be more suitable for decoction or step mashing. As an estimate of modification, the FGCG difference is not entirely accurate since it represents a small difference between two large numbers.

Hot-Water Extract (HWE)

The hot-water extract (HWE) value is the single most important measurement in judging malt quality. Malts that are more finely ground and have a smaller fine grind-coarse grind difference (associated with highly modified malts) will produce greater extract yields. British maltsters quote hot-water extract in liter-degrees per kilogram (l°/kg), expressing how many liters of wort each kilogram of malt will yield. The hot water extract can be converted to fine-grind extract (FGDB) by dividing HWE by 386.

Cold-Water Extract (CWE)

Cold-water extract measures the amount of sugars broken down during the malting process and thus the level of modification. The higher the CWE value, the higher the malt modification. CWE values of 18–20% indicate well-modified malts. British maltsters quote cold-water extract (CWE) instead of FGCG values.

Diastatic Power

The enzymatic power of the malt enzymes that break-down complex carbohydrates into sugars is measured as DP (diastatic power in degrees Lintner). The principal enzymes in breaking down starch are alpha-amylase and beta-amylase. The higher the number of degrees Lintner, the greater the diastatic power of the malt. High enzyme activity is often associated with well-modified barley malts. Malts with high diastatic power will convert faster in the mash and are capable of converting a greater quantity of adjuncts too. Insufficient diastatic power in malt can be compensated for by prolonging the mashing time at temperatures that will allow the action of amylases to take place (8). The "maltose equivalent" is the diastatic power multiplied by four.

The DP is typically 35 to 40 for standard ale malts, but can be as high as 100 to 125 for lager malts (30). North American two-row lager malts typically have diastatic powers between 125 and 135°L (10). Six-row lager malts have diastatic powers between 140 and 160°L. For European malts, the diastatic power generally ranges from 90 to 100°L.

With EBC the diastatic power is expressed in degrees WK (Windisch-Kolbach units). Degrees WK may be as low as 100 for ale malts or as high as 600 for green malt (26). Degrees WK can be converted to degrees Lintner by using the following formula:

$$°Lintner = (°WK + 16)/3.5$$

Alpha-amylase (DU)

Alpha amylase hydrolyses starch to produce shorter chains and reduces the viscosity. The activity of the alpha-amylase enzyme is measured by the length of time required to break down a standard starch solution to a specific color standard using an iodine indicator, which is expressed as dextrinizing units (DU). Low DU values are indicative of under-modified malts, whereas well-modified malts have greater DU values.

Total Protein

The total protein content of barley is probably the single most important guide for predicting general malt quality and is expressed as a percentage of the total weight of the malt. Low protein malts are almost synonymous with high extract. Acceptable protein levels in American malting barley are 11 to 13% for six-row and 10 to 12% for two-row barley. For British two-row malts, 8 to 10% is typical. The amount of protein in the malt can vary from season to season. Protein contains approximately 16% by weight of nitrogen. To determine total nitrogen, divide percent total protein by 6.25.

Total Soluble Nitrogen

Total soluble nitrogen (TSN), consists of free amino acids and peptides and is expressed as a percentage of the total nitrogen (TN). TSN values should be sufficiently high so that the body and mouth-feel of the beer is adequate, not to mention sufficient enough for stable "head" (beer foam) (6).

The TSN/TN ratio is referred to as the soluble nitrogen ratio (SNR), or more commonly as the ratio of soluble protein to total protein (S/T) in the malt. S/T is an important indicator of malt modification, with higher values indicating greater modification—greater breakdown of proteinaceous material during malting. Malts with S/T values from 37 to 44 are suitable for infusion mashing, and those with S/T values from 30 to 40 are suitable for decoction mashing. Total soluble protein has a big influence on fermentation, beer foam, body, color, and ultimately beer flavor. The Kolbach index is a widely used measure of soluble protein in Europe.

Free Amino Nitrogen

Free amino nitrogen (FAN) is a measure of the portion of the soluble protein that has been further broken down into free amino acids and small peptides. FAN values must be sufficiently high to ensure that the lack of nitrogenous yeast nutrients does not limit fermentation (6).

Mealiness

Mealiness is a classification system used to indicate the level of malting. Often mealiness is expressed as "mealy," "half-glassy," or "glassy" (steely of vitreous). Mealy kernels are those in which the endosperm is not more than 25% glassy. Half-glassy kernels are between 25 and 75% glassy, while glassy kernels are more than 75% glassy. Where mealiness is expressed as mealy/half-glassy/glassy, values should be of 95/4/1 percent or better for infusion mashing and 92/7/1 percent or better for decoction or step mashing (26).

Friability

Friability, expressed as a percent, refers to the softness of the kernel and is determined by the amount of malt that passes through a rotating screen. Friability is greatly influenced by overall malt modification, malt protein, and malt moisture.

Sieving Tests

Sieving tests are used to determine the distribution of kernel size within a sample of malt. Malt size is most often expressed as screen separations or as plump/thin, and some analyses list homogeneity instead of sizes. Brewers generally demand that malt homogeneity be of 95% or greater for a uniform crush. This can be determined by using screens. In the United States, the standard screen sizes are 7/64, 6/64, and 5/64 inches, with the first two used for testing uniformity.

Dimethyl Sulfide

Dimethyl sulfide (DMS) usually provides a good indication of the intensity of a beer's malty/sulfury taste. The kilning stage of malting determines the extent to which DMS is formed; higher kilned malts generally have lower DMS levels. For example, there is very little DMS in British style ales because the pale ale malt is kilned at a very high temperature, which in effect converts the DMS precursors—S-methylmethionine—to DMS, which is volatilized. DMS is a normal flavor component of many lagers, particularly continental lagers, but is not present in ales in sufficient quantity to affect flavor. Lager beer tends to be insipid without such flavor components.

Beta-Glucan

Beta-glucan content is important because it can increase viscosity of the wort, which leads to slower runoffs during lautering and results in beer filtration problems. In addition, it can cause problems and may contribute to chill haze. High levels of beta-glucan are often associated with lesser degrees of malt modification. In general, six-row barley usually has higher beta-glucan content than two-row barley (21). The beta-glucan content of most barley cultivars is between 4 and 7% (35). Oats and rye are especially high in beta-glucans.

Viscosity (cP)

Viscosity is a measure of the "thickness" of the wort solution. It is expressed in centipoise units (cP), a measure of the breakdown of beta-glucans. The higher the values, indicates lesser degrees of modification, hence the slower the sparging, lautering, and filtering. Malt that has a high viscosity may need to be step mashed or even decoction mashed.

Types of Barley Malts

Malt influences the flavor of beer more than any other ingredient. The malt types selected for brewing will determine the final color, flavor, mouth feel, body, and aroma. Depending on the style of beer desired and the type of malt, it takes from 15 to 17 kg of malt to produce a hectoliter of beer.

There is no universal system used in classifying malts since maltsters categorize and market their products differently. However, most often malts are classified either as base malts or specialty malts. Base malts usually account for a large percentage of the total grain bill, with specialty malts accounting for a much smaller proportion of the total grist bill. The only exception is wheat malt, which can make up to 100% of the total grain bill in brewing wheat beers.

Base Malts

Base malts provide most of the enzymatic (diastatic) power to convert starches into fermentable sugars. The base malts provide the highest extract potential.

Pilsner Malt

Pilsner malt or lager malt is the standard malt type used for most lager production. These malts are produced to retain maximum enzyme activity and to preserve certain sulfur-based flavor precursors characteristic of light-colored lagers. They produce less color and flavor than standard base malts and are very useful in producing beers in which other flavors and aromas are to be achieved. Traditionally, they are less modified than pale ale malt, as indicated by the harder kernels.

There are both two-row and six-row varieties. The two-row varieties have a higher extract, slightly lower enzyme content, and less protein content than six-row varieties. Two-row malt beers generally require longer aging to develop the mellow flavor of lager beer. Six-row malt can produce a slight phenolic, husky taste, especially in malt beers.

European brewers and U.S. craft brewers predominantly use two-row malts, while large brewers in the United States and Canada predominantly use six-row malts. Because of their high enzyme content, American six-row malts are suitable for high-adjunct pale American pilsners. The German two-row pilsner malt produces smooth, grainy flavors and is used in the finest German lagers. The Belgian two-row pilsner malt is not only excellent base malt for full-flavored lagers, but also for Belgian ales and European wheat beers. It is most preferred by Belgian brewers for producing Trippels and light-colored beers such as white beers and various Specials.

Pale Ale Malt

This is the standard malt type used for most ale production. Pale ale malt is traditionally more fully modified, with a lower protein content and more easily accessible starch than Pilsner malt. Pale ale malt is usually kilned at high temperatures, which gives it a darker color and more malt character than Pilsner malt. Pale ale malt is perfectly suited to infusion mashing.

British pale ale malt is made from two-row barley, whereas most American pale malts come from two- and six-row varieties. British pale ale malts are lower in diastatic power and in protein. Many brewers of full-flavored ales prefer the British malt. Belgian pale ale malts impart a definite "maltiness" to the finished beer. They are low in protein (approximately 10%), are well-modified, and have low SMM levels. Because of the lower enzyme content of British and Belgian malts, brewers generally limit adjuncts to about 10 to 15% of total grist.

Mild Ale Malt

Mild ale malt is kilned at a higher temperature than pale ale malt but still has enough diastatic power to be used as a base malt or as a substitute for a portion of the base malt. Mild ale malt imparts a deeper color (golden to amber) to the finished beer. In addition, it provides a fuller malt flavor and aroma with rich nutty, toffee and caramel flavors of true dark beers, such as mild and brown ales. English ales, German Märzen, and other festival beers make use of this malt.

Vienna Malt

Vienna malt is very close to pale ale malt but is kilned at a higher temperature to emphasize the production of melanoidins that are responsible for differing flavors and aromas. It gives beer a richer, maltier flavor, some degree of fullness, and a golden color. It retains sufficient enzyme power for use as base malt ranging from 60 to 100% of total mash grist. The German Vienna malt (Weiner) is high in diastatic power and can be used to make up 100% of the total grist for a fuller, deeper malt flavor and aroma. Vienna malt is typically used to flavor Vienna-style, Märzen or Oktoberfest beers.

Munich Malt

Munich malt, like Vienna malt, is kilned at high temperatures to emphasize the production of melanoidins. It gives an amber color to the beer; however, its most important contribution is a nutty, rich malty aroma and flavor. Although Munich malt has about half the normal enzyme complement of lower-kilned malts, it can still be used as a base malt. While it is a malt of choice for Vienna and Munich beer styles, it is well-suited for adding richness to almost any kind of beer—Pilsners, German bocks, dark beers, and strong beers. Munich malt is an excellent choice for dark and amber lagers when blended with Pilsner malt at the rate of 10 to 60% of the total grist (34). It can also be used at the rate of 20 to 75% of the total malt content in lagers for its full, malty flavor and aroma. Darker grades of Munich malt are available from continental maltsters. Munich malt has a much lower content of high-molecular-weight proteins and is much less inclined to form protein hazes in dark beers.

Specialty Malts

The specialty malts are designed to contribute a unique characteristic to beer, such as color, flavor, midsized proteins for foam improvement, body, or other accentuating characteristics. Unlike base malts, specialty malts provide little or no enzymatic (diastatic) power but do contain some extractable material. Specialty malts are used in relatively small quantities compared to base malts. Depending on the style of beer brewed, the brewer may use only one or two types of malts, or as many as seven or eight different types of specialty malts.

Specialty malts are classified into four groups with reference to the way they are produced.

1. Light-colored malts
2. Caramel (crystal) malts
3. Dry-roasted malts
4. Unmalted barley

Light-Colored Malts

Light-colored specialty malts are kilned at higher temperatures than base malts and impart a deeper color and a fuller malt flavor and aroma to the finished beer. Enzyme levels are lower than for base malts. Dextrin malts and honey malts are examples of specialty light-colored malts.

Dextrin Malt – Dextrin malt is the lightest in color, are high in unfermentables, and contribute to the beer's body without affecting color. Dextrin malts are similar to caramel malts but have been dried at a lower temperature to prevent the formation of color and flavor compounds. Caramelization of the sugars and darkening of the husk are avoided, thus the malts possess very little caramel flavor (residual sweetness). Dextrin malt does not impart the characteristic flavor and reddish hue of caramel malts, but adds smoothness, fullness of body, and initial head retention without affecting the color of pale or light lagers. Dextrin malt tends to impart a light flavor described as "sweet biscuit" and "full," but with little caramelized-flavor.

During manufacture, the malting conditions are manipulated so that sugars and dextrins are changed into non-fermentable isomers that cannot be attacked by malt enzymes—alpha- and beta-amylase. Thus, little or no breakdown occurs in the dextrin malt in the mash tun. In addition, virtually all the carbohydrates extracted from the dextrin malt are unfermentable. These characteristics make dextrin malt an almost ideal "body-builder," and enhance mouthfeel and head retention.

Beers have been formulated with up to 30% dextrin malt with no residual sweetness, though some brewers believe that using high proportions (>10%) of malt can impart a starchy flavor note (cloying sweetness) to the beer. Kunze reports that 8 to 12% is sufficient to improve foam and palate fullness (22). Dextrin malt is most useful in lager beers because lager yeast uses more of the triple sugars in wort. The results are a lower terminal gravity and a lighter body compared to ale yeast in wort of similar starting gravity.

Honey Malt – Honey malt has a sweet, "honey-like" flavor; however, unlike caramel malts, it lacks the roasty and astringent flavors. Honey malt has been patterned after German Bruhmalt and is well suited for use in brown ale and strong ales. Generally, makes up 15 to 30% of the total grist bill (15).

Caramel (Crystal) Malts

Continental lager brewers traditionally use caramel malts, whereas British ale brewers favored crystal malts. Today, most maltsters no longer make a distinction between caramel and crystals malts and, more often than not, use "caramel malt" when referring to these malts. Other names that can be used when referring to caramel malts include CaraMunich, CaraVienne, Special B, Carastan, Cara, and Extra Special.

Caramel malt is made from green malt that is produced by roasting the wet germinated barley (no kilning) at controlled temperatures, causing the starches to convert to sugars and caramelize. The major variable in the process is the roasting temperature, which determines the depth of the color and the degree of caramel flavor.

Caramelized malts come in a wide range of colors, from light to very dark amber, and with flavors ranging from a mild sweet candy-like caramel to caramel/burnt sugary flavor. Light caramel malts accentuate the soft malt flavor, while darker caramel malts promote a caramel, slightly sweet taste, European in flavor. It is primarily known for its color control but because the malt contains non-fermentable sugars it gives a fullness, or mouth feel, to the beer. It also contains a considerable amount of midsized proteins, which will provide longer-lasting foam to the beer and pyrazines (15). Pyrazines act as oxygen scavengers thus extending the shelf life of the bottled beer.

Caramel malt is used in export pale ale, India pale ale, and barley wines. Generally, caramel malt makes up as little as 2–3% of the total grist for British ales and as much as 10% for dark beers. Astringent notes can be produced if the level of crystal malt is too high. Reportedly, ales can have up to 15% caramel malt and even more when adjuncts are used. These malts have little or no enzyme activity, due to the roasting or kilning process.

Dry-Roasted Malts

Dry-roasted malts are produced by kilning at very high temperatures followed by roasting. The heat and duration of the roasting determine the color and flavor of malt. Dry-roasted malts include amber malt, biscuit malt, brown malt, black malt, chocolate malt, and dark chocolate malt.

Amber Malt – Amber malt is a British variety, well described by its name that is produced by gently roasting mild ale malt. It is used in a few amber and dark ales and is generally used to produce premium bottled ales. It is reported to impart a warm, pleasant biscuit flavor with toffee undertones and gives beer a ruby-red hue. Amber malt making up 10% of the total grist is used to give a nutty flavor to nut brown ales (15). This malt is rare, even in Great Britain.

Biscuit Malt – Biscuit malt has a flavor described as warm bread crust, biscuit, and earthy. These malts are used in Kölshs, Scottish ales, brown ales, and alts at a rate of 2 to 5% of the total grist (15).

Brown Malt – Brown malt is sometimes kilned over a hardwood fire, which imparts a smoky flavor to the beer. Traditionally, it was used in dark ales, but it is very hard to find because it is often sold under different names, and is rarely made commercially. Its uses are generally restricted to mild ales, brown ales, and sweet stouts.

Chocolate Malt – Chocolate malt is not roasted quite as long as black malt; consequently, it is lighter in color—more dark brown—and retains some of the aromatics and flavor of malt's sweetness. It imparts a nutty, roasted flavor to the beer but does not make it as bitter as black malt. There are no enzymes in chocolate malt. Chocolate is an essential ingredient in porters and stouts and can be used in mild ales, brown ales, and old ales, and can even be incorporated into the grist of dark lagers. British chocolate malt is a little darker than American chocolate malt, yet it has a slightly smoother character in the roast flavor and aroma profiles. Typical addition rates are 1.5 to 3.0% of the grist. It is also added in small quantities to cask-conditioned beers to give a deeper color and to produce a fuller flavor with a "bite."

Black Malt – Making black malt involves roasting the malted barley at temperatures so high that they drive off all of the aromatics (malt flavor).

There are no enzymes in black malt. In excess, black malt will contribute a dry, burnt flavor to the beer that may be perceived as a bitterness different from that derived from hops. It is quite appropriate for dark ales but too sharp for dark lagers. The darkest of all malts, it is used sparingly to add deep color and roast-charcoal flavor. Black malts are used in porters and stouts, ranging from 7 to 15%, in Alts at a rate from 1 to 3%, and in Oktoberfest beers at a rate of 0.1 to 0.25% (15).

Unmalted Barley

Two other specialty products made from unmalted barley which are roasted barley and black barley. There are no enzymes in either of these products.

Roasted Barley – Roasted barley is unmalted barley roasted at high temperatures. Roasted barley is not black in appearance; it is rather a rich, dark brown. It has an assertive, roasted flavor, similar to roasted coffee beans, with a sharp, acrid after-palate, and is especially used in the making of dry stouts and porters. It contributes significantly to the color of the beer, enhances head production and stabilization, and whitens the head on the beer. Roasted barley produces a stronger, drier, a more bitter taste than roasted malt and is less aromatic and drier, with a more intense burnt flavor than black malt. In contrast to roasted malts roasted barley gives no hint of sweetness (6).

Use 10 to 12% to impart a distinct, roasted flavor to dry and sweet stouts. Other dark beers also benefit from smaller quantities (2–6%). Small amounts are acceptable in porters, provided they don't overpower the chocolate/caramel notes. Roasted barley is rarely used in any Belgian ales or German lagers.

Roasted barley, because it is an unmalted product, can contribute to beta-glucans. This factor, combined with its tendency to shatter during milling, can make lautering difficult (24).

Black Barley – Black barley is unmalted barely roasted like black malt used primarily in dry stouts for its coffee note.

Other Malted Grains

Wheat Malt

Wheat malt, for obvious reasons, is essential in making wheat beers making up to 100% of the grists, including the German *Weissbier* (white beer) and *Weizenbeer* (wheat beer). Wheat is also used in malt-based beers (3–10%) because its protein gives the beer a fuller mouthfeel and enhanced beer head stability. Other benefits claimed are improved beer clarity and palate fullness (6). On the down side, wheat malt contains considerably more protein than barley malt, often 13 to 18%, and consists primarily of glutens that can result in beer haze. Compared to barley malt it has a slightly higher extract, especially if the malt is milled somewhat finer than barely malt. European wheat malts are usually lower in enzymes than American malts, probably because of the malting techniques or the varieties of wheat used.

Most wheat malts are produced from relatively low-protein soft white spring and hard red spring wheat. In general, white wheat is lower in protein than red, and soft wheat varieties contain less protein than hard wheat. Consequently, soft white wheat varieties have lower protein levels than hard red wheat.

Wheat malt has a thinner husk than barley malt. The thinner husk is low in polyphenols, which enables the brewer to sparge and re-circulate longer during wort separation at higher than normal temperatures without having to worry about leaching out astringent polyphenols from the husks.

Rye Malt

Rye malt, like wheat malt, is huskless. It yields less extract than the other malts previously discussed and is slightly darker than barley or wheat malt. Rye malt has a very pronounced flavor and can be overpowering if too mush is used in brewing the beer. Additions of rye malt as low as 5% impart a nutty finish to the beer. Rye malt may give a red tinge to beer.

Malt Extracts

Malt extracts are made by griding the malt, mashing it, with or without adjuncts and supplementary enzymes, and then separating the wort. Many different types of malt extracts can be produced depending on the malt, the mashing regime, and the evaporation conditions used. Typically hopped or unhopped malt extracts are used by home brewers and small-brewers because they lack the facilities for milling the malt, mashing, and wort separation. Malt extracts can be used as a sole source of fermentable sugar, or they can be combined with barley malt.

The malt extract comes in the form of syrup or dried powder. If the final product is a dried powder, the malt extract has undergone a complete evaporation process by means of "spray-drying," thus removing almost all of the water. For simplicity, use an 85% conversion factor when substituting dried malt for syrup. Syrups are more popular than dried malt extract, possibly because they are less trouble to store. A common problem noticed in malt extract beers is the thin, dry palate, which correlates with a low terminal gravity. Another common problem is the lack of a true "dark malt" flavor in dark beers.

Calculations

Brewhouse Yield

The brew house yield (BHY) can be calculated from the CGDB by accounting for the moisture content and brew house efficiency. The formula is:

BHY = [(CGDB/100)/(1 + moisture content) − 0.02](%BHE/100)

For example, if CGDB is 82% and the moisture content is 4%, with 90% brew house efficiency (BHE), the result would be:

BHY = [(82/100)/(1 + 0.04) − 0.02](90/100)

BHY = 0.692, or 69.2%

Dry basis should be used when comparing malts that have different moisture levels. To convert to percent extract dry basis, use the following formula:

Extract Dry Basis = % Extract As Is/(100 − % Moisture)

References

1. Adamic, E.B. 1977. "Barley and Malting." *The Practical Brewer*, edited by H. M. Broderick. Madison, Wisconsin: Master Brewers Association of the Americas.
2. Axcell, B., D. Jankovsky and P. Morrall. 1984. "Malt Specifications." *MBAA Technical Quarterly* 21 (3).
3. Bamforth, C.W. and A.H.P. Barclay. 1993. "Malting Technology and the Use of Malt." *Barley: Chemistry and Technology*, edited by A.W. MacGregor and R.S. Bhatty. St. Paul, Minnesota: American Association of Cereal Chemists.
4. Basler, Jim. Personal Communication. 1999. Briess Malting Company, Inc.
5. Bemment, D. W. 1985. "Specialty Malts." *The Brewer* 71 (854).
6. Briggs, Dennis E., Boulton, Chris A., Brookes, Peter A., and Stevens, Roger. 2004. *Brewing: Science and Practice*. Cambridge, England: Woodhead Publishing Limited.
7. Burger, W. C. and D. E. Laberge. 1985. "Malting and Brewing Quality." *Barley*, edited by D. C. Rasmusson. Madison, Wisconsin: American Society of Agronomy.
8. De Clerck, Jean. 1957. *A Textbook of Brewing*. Volume 1. Translated by K. Barton-Wright. London: Chapman-Hall, Ltd.
9. Fix, George J. 1989. *Principles of Brewing Science*. Boulder, Colorado: Brewers Publications.
10. Fix, George J. 1993. "Belgian Malts: Some Practical Observations." *Brewing Techniques* 1 (1).
11. Fix, George and Laurie Fix. 1997. *An Analysis of Brewing Techniques*. Boulder, Colorado: Brewers Publications.
12. Freeman, Peter L. 1999. "Barley and Malting." *The Practical Brewer*, edited by John T. McCabe. Wauwatosa Wisconsin: Master Brewers Association of the Americas.

13. Gruber, Mary Anne. 1994. "Inspecting Your Malt." *The New Brewer* 11 (2).

14. Gruber, Mary Anne and Jerrold Hilton. 1995. "Malt and Hops Primer." *The New Brewer* 12 (6).

15. Gruber, Mary Anne. 2006. "Specialty Malts." *Raw Materials and Brewhouse Operations*. Volume 1, edited by Karl Ockert. St. Paul, Minnesota: Master Brewers Association of the Americas.

16. Hobson, Derek. 1980. "Influence of Brewhouse and Raw Materials." *Brewers' Guardian* 109 (10).

17. Hough, J.S. et al. 1982. *Malting and Brewing Science*. Volume 2. London, England: Chapman-Hall, Ltd.

18. Jupp, D.H. 1994. "Coloured Malts." *The Brewer* 80 (961).

19. Kendall, N.T. 1995. "Barley and Malt." *Handbook of Brewing*, edited by William A. Hardwick. New York, New York: Marcel Dekker, Inc.

20. Klimovitz, Raymond J. 1992. "Is This the Malt You Ordered?" *Quality Brewing Share the Experience*, Brewery Operations, Volume 9, edited by Virginia Thomas. Transcripts from the 1992 National Microbrewers Conference. Boulder, Colorado: Brewers Publications.

21. Kramer, Paul. 2006. "Barley, Malt, and Malting." *Raw Materials and Brewhouse Operations*. Volume 1, edited by Karl Ockert. St. Paul, Minnesota: Master Brewers Association of the Americas.

22. Kunze, Wolfgang. 1996. *Technology Brewing and Malting*, Translated by Dr. Trevor Wainwright. Berlin, Germany: VLB Berlin

23. Lewis, Michael J. and Tom W. Young. 1995. *Brewing*. London, England: Chapman & Hall.

24. Mallet, John. 1994. "Small Brewers and Malt." *The New Brewer* 11 (2).

25. Moll, Manfred, ed. 1995. *Beers and Coolers*. New York, New York: Marcel Dekker, Inc.

26. Noonan, Gregory J. 1996. *New Brewing Lager Beer*. Boulder, Colorado: Brewers Publications.

27. Noonan, Gregory J. 1997. "Understanding Malt Analysis Sheets – How to Become Fluent in Malt Analysis Interpretation." *The 1997 Brewer's Market Guide*. Eugene, Oregon: Brewing Techniques Magazine.

28. O'Rourke, Timothy. 1994. "Making the Most of Your Malt." *The New Brewer* 11 (2).

29. O'Rourke, Timothy. 1999. "Barley and Malting (part 1)." *Brewers' Guardian* 128 (1).

30. O'Rourke, Timothy. 1999. "Barley and Malting (part 2)." *Brewers' Guardian* 128 (2).

31. O'Rourke, Timothy. 2002. "Malt Specifications and Brewing Performance." *The Brewer International* 2 (10).

32. Palmer, Geoffry H. 2006. "Barley and Malt." *Handbook of Brewing*, edited by Fergus G. Priest and Graham G. Stewart. Boca Raton, Florida: CRC Press, Taylor & Francis Group.

33. Pfisterer, Egbert and Robert Gentry. 1999. "Improving Brewing Material Yield and Efficiency." *The New Brewer* 16 (1).

34. Rhodes, Christine P. and Pamela B. Lappies, eds. 1995. *The Encyclopedia of Beer*. New York, New York: Henry Holt and Company.

35. Schwarz, Paul and Richard Horsley. 1996. "A Comparison of North American Two-Row and Six-Row Malting Barley." *Brewing Techniques* 4 (6).

36. Sherwood, Sonja, ed. 1997. "World of Malts – A Survey of the Lingua Franca of Brewing." *The 1997 Brewer's Market Guide*. Eugene, Oregon: New Wine Press.

37. Steele, Todd. 1997. "Shredding the Wheat Myths – Beer Made from 100% Wheat." *Brewing Techniques* 5 (6).

38. Stewart, G.G. and I. Russell. 1985. *"Modern Brewing Biotechnology."* Food and Beverage Products, edited by Murray Moo-Young. Oxford, England: Pergamon.

39. Thomas, David and Geoffrey Palmer. 1994. "Malt." *The New Brewer* 11 (2).

40. Wilson, W.W. 1985. "Production and Marketing in the United States and Canada." *Barley*, edited by D.C. Rasmusson. Madison, Wisconsin: American Society of Agronomy.

Chapter Three

Hops

Hops, a minor ingredient in beer, are used for their bittering, flavoring, and aroma-enhancing powers. Hops also have pronounced bacteriostatic activity that inhibits the growth of Gram-positive bacteria in the finished beer. Hops, when in high enough concentrations, aids in the precipitation of the more water-insoluble proteins in the kettle.

Hop Constituents

Hops contain hundred of components, but most of the brewing value of the hops is found in the resins and hop oils.

Resins

Hop resins are subdivided into hard- and soft-resins based on their solubility. Hard resins are of little significance, as they contribute nothing to the brewing value, while soft resins contribute to the flavoring and preservative properties of beer. Alpha and beta acids are two compounds present as soft resins and are responsible for bitterness. Alpha acids are responsible for about 90% of the bitterness in beer. Magnesium, carbonate, and chloride ions can accentuate hop bitterness too.

Alpha Acids

Alpha acids are the precursors of beer bitterness since they are converted into iso-alpha acids in the brew kettle. The three major components of alpha acids are humulone, cohumulone, and adhumulone. Humulone contributes the major proportion of the bittering power and the antiseptic properties of the hop. There is a widely held belief that hops with high levels of cohumulone have a harsher bitter character than hops with low cohumulone levels, while humulone and adhumulone tend to be more rounded and mellower.

The cohumulone content is a good indicator of varietal type. High levels of cohumulone are seen in most high-alpha hop varieties, such as Brewers Gold, Bullion, Cluster, Nugget, Eroica, and Galena. However, one exception is the aroma variety Cascade, with high cohumulone. Aroma hops, such as Hallertauer, Tettnanger, Hersbrucker, Saaz, and Fuggle, contain significantly less cohumulone than the high-alpha varieties. Generally, cohumulone makes up from 15 to 50% of the alpha-acid fraction, whereas humulone makes up from 45 to 70%. Adhumulone occurs in minor amounts compared to the other two components of alpha acids. The amount of adhumulone appears to be constant at around 15% of the total.

Hops that are bred to be high in alpha acids are referred to as high-alpha-acid hops or bittering hops. Alpha acid is listed as a percentage weight of the total hop weight (w/w) based on a range of values at harvest. A rating below 5% alpha acid content indicates low bittering values; 5 to 8% alpha acid indicates an assertive medium range of bitterness; and 8 to 14% alpha acid indicates a very powerful bittering value. It has become conventional practice to use the alpha acid content of hops to calculate the amount of required hops for the brew. Large breweries usually choose hops that have high alpha acid contents because their use lowers production costs. For craft breweries, this is not always a factor since the cost of hops is typically negligible.

Isomerization – When hops are added to the boiling wort in the kettle, their alpha acids go through a chemical change known as isomerization. Isomerized alpha acids, commonly referred to as "iso-alpha acids," are quite soluble in wort, so the beer stays bitter. A higher pH results in better isomerization, but the bitterness at a lower pH is always considered to be more balanced and finer (18). Losses of iso-alpha acids occur at many stages during the brewing process.

Beta Acids

Hops also contain a second group of acids known as the beta acids. The beta acids (lupulone, colupulone, and adlupulone) are only marginally bitter. Their bitterness is generally perceived as milder and is more pronounced in aroma-type hops. Unlike alpha acids, beta acids don't isomerize in normal wort boiling. The majority of the beta acids are lost in the brewing process because of precipitation or adsorption onto solids. If oxidized, they can taint the beer with an unpleasant, spoiled-vegetable taste. Beta acid is listed as a percentage weight of the total hop weight (w/w) based on a range of values at harvest.

Oils

Although hops with high alpha acid content are preferred for their bittering and flavoring properties, hops are also selected for the character of their oils. Oils are largely responsible for the characteristic aroma of hops and, either directly or indirectly, for the overall perception of hop flavors. Hops selected for character of their oil content are often referred to as aroma or "noble" type hops. Oils also tend to make a beer's bitterness a little more pronounced as well as enhance the body or mouth feel of the beer.

The total oil content is expressed as a percentage of total hop weight (w/w) based on a range of values at harvest. Hops may contain from 0.5% to 3.0% oils, which are made up of hydrocarbons (23). Myrcene, humulene, and caryophyllene are three of the major hydrocarbon components. Aroma hops usually are low in myrcene but high in humulene. Those hops containing higher proportions of myrcene contribute a harsher and more unpleasant aroma and flavor (18). Bullion and Cluster are good examples.

These oils are volatile and will be almost entirely vaporized from the kettle if added at the start of wort boiling. Consequently, many brewers who want to preserve a "hoppy" character in beer add selected aroma varieties into the kettle before the end of the boil.

Hop Varieties

Although there is only one hop species (*Humulus lupulus*) that is used for brewing beer, there are a number of varieties (technically known as "cultivars") in that species, each with its own spectrum of characteristics. Varieties of hops are chosen for the properties of bitterness, flavor, or bouquet that they will lend to the beer. Hop varieties can be roughly divided into two classes, bittering hops and aroma hops, although there are hops that can be considered dual-purpose.

Bittering Hops

As their name implies, bittering hop varieties are those that impart bitter flavor to beer and have high alpha acid levels. Bittering hops usually have a high alpha acid content. The bitterness qualities that hops add are from the alpha acids in the lupulins. Generally, bittering hops are added at the beginning of the boil, a process referred to as "kettle hopping" or "bitter hopping," while aroma hops are added in the final stages of the boil for their aromatic and flavoring properties, a process referred to as "late hopping." Bittering hop varieties are grown predominantly in the United States.

Aroma Hops

Aroma hops, with low- to medium alpha levels, mainly impart characteristic hop aromas to beer. In Europe, aroma hops are mostly grown, with a smaller but growing emphasis on bittering hop varieties. The aroma varieties that are grown in Germany often are referred to as "noble" hops. The three recognized noble hops are Hallertauer Mittelfrüh, Tettnang Tettnanger, and Spalt Spalter.

Czech Saaz is the primary aroma variety grown in the Czech Republic. Noble hops commonly have an alpha/beta ratio of about 1:1 and are relatively low in alpha and beta acids (2–5%). They also have low cohumulone content, and the myrcene content in the hop oil is low (typically below 50%). The humulene content in the oil is usually high; the ratio of humulene to caryophyllene is usually above three. Noble hops have relatively poor storage characteristics, being more prone to oxidation than other varieties.

Dual-Purpose Hops

Dual-purpose hops tend to have intermediate levels of alpha acids together with desirable aroma properties. Challenger and Northdown are two such varieties that have both bittering and aroma properties.

Growing Regions

Traditionally, hops have been named after their place of origin. For example, Hallertau hops are named after the Hallertau district north of Munich, Germany. Saaz hops come from the Zatec area in western Bohemia, near Pilsen. However, today it is common for hops with the same name to be grown in different regions. The most famous growing regions in the world are the Hallertau, Tettnang, Spalt, and Hersbruck regions in Germany; the Pacific Northwest in the United States; the Kent and Hereford regions in Great Britain; and the Saaz region in the Czech Republic.

Cultivars

Different cultivars are important for the production of specific styles of beers. For instance, Bullion (a strong odor, high myrcene, and high alpha acid hop) is suitable for strongly bittered ales, but would not be suitable for final hopping of light lager. Goldings is suitable for pale ales.

Refer to Appendix A for a list of hop cultivars along with their technical parameters and recommended uses in brewing.

Whole Hops

When hops are used in their raw or unprocessed form, directly from the bale, they are designated as whole, raw, or leaf hops. Although whole hops are not as popular today as in the past, some of the world's major brewers, as well as some craft brewers, use them in the belief that whole hops provide the best aroma. Brewers who use only whole hops believe hop products cause unacceptable flavor change in the beer. Processing the hops is believed to change the chemical composition and, therefore, the nature of the product.

Whole hops require a longer boiling time to extract the oils and resins. The relatively slow-release of oils from whole glands of the hops allows time for oxidation of the major hydrocarbons, which is thought to be responsible for good hop aroma in beer.

In spite of some of the purported advantages in using whole hops, it is considered to be inefficient in brewing beer. Whole hops tend to give the poorest alpha acid utilization during boiling (estimated at from 15 to 24%) since they have higher quantity of vegetative material and the alpha acid is less readily available (31). Also, if one considers the removal of wort in spent hops—even after pressing—it can be said that leaf hops is probably the most inefficient brewing material of all. To this must be added the fact the storage of whole hops requires a considerable amount of space, it is unstable in storage, subject to a range of contaminants and bulky and expensive to ship, handle and dispose of.

> **Note | Quantity of Hops Used In Brewing**
>
> As might be expected the quantity of hops used in brewing to make a barrel of beer can depend on a number of factors including the beer style, the type of hops used, and the brewers preference. If you're brewing a traditional lager, the quantity of whole hops required could be as low as 90 grams per barrel and as high as 0.45 kg per barrel for some of your craft brews.

Physical Characteristics in Evaluating Whole Hops

In using whole hops, it is very important for the brewer to use fresh and well-cared for hops. The examination of a sample of hops by hand can reveal much about the sample that is not readily accessible by chemical means. Whole hops should be relatively intact. Pressing down the palm of the hand on the sample should show the elasticity or spring of the hops. Old hops, and in particular those poorly stored, become brittle and lose their "bounce."

The aroma can be evaluated by smelling a small amount of crushed hops. Fresh hops should have a crisp, "piney" smell when they are rubbed in the palm of the hand, but this scent can vary a good deal depending on the variety. Poor hops, regardless of variety, have very little aroma, are not sticky, and/or possess a strong "cheesy" smell, or other odors not characteristic of hops.

Hop color is an indication of maturity, quality, and variety. Maturity can be indicated accurately by the color change. Young, immature hops are grayish-green; mature hops are yellowish-green with a silky sheen; and aged hops are reddish-brown in color. Generally, the color range for low-quality hops is from pale to dark green, for high quality hops from yellowish-green to greenish-yellow, and for deteriorated hops from yellow to brownish-green.

Oxidation will turn the lupulin glands a dry brown color. Of course, the color of the hops will depend largely on the particular variety. Goldings have primrose-yellow tips on the petals when ripe, whereas the higher alpha content types tend to be much darker all over. Some imported British hops have been sulfured in order to help preserve freshness. They may have a gold appearance.

For whole hops, the brewer should also evaluate the bale for stem and leaf material. The presence of a few stems, seeds, and leaf parts is normal. Hop seeds contain fatty oils, which can negatively affect beer foam and act as precursors to staling compound in beer (30). Hops from Europe tend to have a higher stem and seed content than those grown in the United States. Hop varieties grown in Oregon reportedly have higher seed content than those grown in the Yakima Valley in Washington State.

Hop Products

Besides whole hops (hop cones), other hop products are used in brewing beer, which fall into three general categories:

1. Non-isomerized hop products
2. Isomerized hop products
3. Hop oil products

Non-isomerized Hop Products

The alpha acids are unchanged during processing non-isomerized products. As a result, the products can only be added to the kettle only during wort boiling. Products included in this group are listed below.

Hop Plugs

Hop plugs are used in the United Kingdom as a convenient means of dry hopping cask-conditioned ales and by brewers with traditional hop backs who cannot use powder pellets. The hop plug is technically a Type 100 pellet, and is made by compressing whole hops (unground) into a plug that measures between 10 and 30 mm in thickness, depending on the weight. Weights specified by the brewer for this type of pellet are usually 7, 14, and 28 grams. As with other pellets, resins hold the plug together. Type 100 pellets preserve the quality of oil better than whole hops do since they are packed under vacuum—a factor that contributes to producing a more consistent flavor in the beer. Although hop plugs reduce the volume of leaf hops and can be vacuum-packaged, utilization of hop materials is not significantly improved. Also, it can be said the disadvantage in using plugs, as with whole hops, is the long contact times required for developing flavor in beer.

Hop Pellets

Hop pellets are nothing more than whole hops mechanically processed by removing foreign material, milling in a hammer mill, blending batches of hops together for product consistency, pelleting through a standardized pellet die, cooling, and packing in aluminum based foil packs. The pellets are held together naturally by their own oils, resins, and moisture content. The alpha acid rating (and, if available, the oil content rating) is determined after the pelletizing process. Craft brewers commonly use hop pellets to the exclusion of hop extracts.

> ### Note | Hop Pellet Packaging
>
> Hop pellets are usually packaged in laminated foil-plastic bags for better storage life. The bags may be fully evacuated and sealed (hard pack); or back filled with a nitrogen gas (soft pack), which results in much improved stability of brewing constituents. One advantage of hard packs is their low volume; another is that the integrity of the packaging is immediately noticeable. The disadvantage is that the vacuum packing may lead to agglomeration of pellets into solid blocks, particularly if the resin content is too high. Back flushing may be necessary in order for the pellets to remain in pourable form. European pellets are usually soft-packed using a carbon dioxide-nitrogen atmosphere. In either case they should be stored at a minimum of $-3°C$.
>
> The most common package is a foil and aluminum bag sealed under vacuum or inert gas contained within cardboard boxes. Net content 5 kg (11 lb.), 20 kg (44 lb.) or 150 kg (330 lb.).

There are many advantages in using pelletized hops. The pellets readily disintegrate to a powder in the boiling wort. The process of pelletizing bursts most of the lupulin glands, so pellets will give a better extraction rate of alpha acids than whole hops for shorter boil or contact times. Thus the hopping rate can be reduced by as much as 10 to 20% because of improved dispersion of hop acids in the kettle (10). For longer times (more than 30 minutes); however, the difference is negligible. Brewers are able to stock a fair large supply of different varieties without taking up much storage space resulting in reduced storage costs. The freshness of the hops is more easily controlled since oxygen can only reach the surface layer of resins and oil. In addition, pellets are easier to use than whole hop cones, in that they do not need to be strained out after the boil. The powder is later removed along with the trub by sedimentation in a coolship, by a settling tank, a centrifuge or more commonly by a whirlpool or a combined kettle/whirlpool. Hop pellets are much more practical for dry hopping and, finally, they allow blends of hops to be processed into one pellet. Hops of the same variety but with differing alpha contents are often blended to give a standard product with a constant alpha acid appropriate for each variety and growing season.

While it can be said the ruptured glands result in a higher level of isomerization and utilization of the alpha acids, it is less than favorable for essential oil utilization. The ruptured glands in hop pellets may lose the vast majority of these hydrocarbons by volatilization before the oxidation products have a chance to form. It is possible to overcome this loss by late additions of pellets but this is wasteful of the alpha acids.

Pellets should be mostly intact, with very little powder in the bag. Hop powder should be used first because of the likelihood of deterioration. Pellets will have some aroma when broken apart and rubbed between the fingers, but it will usually not be very intense, especially if the hops are cold.

Regular or Standard Hop Pellets (Type 90) – One of the most common hop pellet on the market is the regular or standard (Type 90) pellet, which are whole hops ground and pelletized. The Type 90 pellet refers to the weight of pellets produced expressed as a percentage of the weight of hops processed, i.e., 90% recovery on a weight basis. Type 90 pellets can be considered as being virtually the same as leaf hops. Nothing is added and nothing is taken away except moisture and any debris (stems, leaves, stones, soil, etc.). Hop utilization for regular or standard hop pellets is estimated at from 30 to 35% (31).

Concentrated Hop Pellets (Type 45) – Another type of pellet on the market is the concentrated (Type 45) hop pellet, also called "enriched" pellets. This pellet is derived from a process in which the coarser "waste" fraction is mechanically separated from resins and oils at subfreezing temperatures (– 35°C). By metering a proportion of the "waste" fraction back into the line, standardization of the alpha acids level is assured. Concentrated hop pellets contain less vegetable material in relation to brewing value, a trait that some brewers consider desirable from a flavor standpoint since it involves a reduction in polyphenols.

Typically, Type 45 pellets have twice the alpha acid level and half the original weight of Type 90 pellets for a given weight of whole hops. The weight advantage may be useful to the brewer where there is a limitation in the capacity of the whirlpool in the brewhouse. In addition, for a given weight, Type 45 pellets contain only half the polyphenols, nitrates, heavy metals, and pesticide residues of Type 90 pellets. The lower level of undesirable substances is an important advantage for the brewer. Today, some Type 90 pellets are higher in alpha acid than Type 45 pellets. Type 45 pellets often are sticky, and they clump together if the alpha acid content is greater than 20% (15). Hop utilization for concentrated hop pellets is estimated at from 30 to 35% (31).

Stabilized Hop Pellets – In stabilized pellets, magnesium oxide is mixed in with the hop powder and the mixture is then pelletized. The addition of magnesium oxide gives the pellets greater storage stability than Type 90 pellets. This is because the alpha acids are converted to their magnesium salts, which are less prone to degradation. These salts also show a greater propensity to isomerize in wort boiling than the alpha acids themselves thus improving utilization. Stabilized pellets are isomerized faster during kettle boiling and so can be added later in the boil. Essential oils are not stabilized by the process so if it is desired to preserve the aroma the pellets should be cold stored. On storage of stabilized pellets there is little of no deterioration of the alpha acid but some isomerization to alpha acids may occur. Hop utilization for stabilized hop pellets is estimated at from 35 to 45% (31).

Hop Extracts

Hop extracts are made by passing a solvent through a packed column of milled hops to collect the resin components followed by the removal of the solvent to yield a whole or "pure" resin extract. The solvents typically used in making hop extracts are ethanol alcohol (predominantly used in Europe) and "organic" solvents (hexane). These solvent extracts are increasingly falling out of favor worldwide due to perceived problems with the residues.

Alternatively, carbon dioxide is gaining favor as it is seen as a "natural" solvent in producing carbon dioxide extracts—supercritical and liquid. Carbon dioxide extracts are also growing in use not only because they are considered safer but more environmentally friendly and a better preservative of hop aroma than other extracts. This combination has been found to produce beers with flavors very close to those made from whole hops.

Supercritical carbon dioxide is more selective than the organic solvents extracting less of the tannins and waxes but less selective than liquid carbon dioxide extraction. Liquid carbon dioxide is the most selective solvent producing the most pure whole resin and oil extract. It extracts none of the hard resins or tannins, much lower levels of plant waxes, no plant pigments, and less water, and water soluble materials. The resin's active ingredients remain intact over extended periods, assuring the brewer consistent brewing characteristics. When stored in sealed containers at cool temperatures (0–5°C), these extracts will maintain their integrity for years.

It is possible to produce fractionated extracts, which are either rich in oil components for aroma or rich in alpha acids for bitterness, by using both liquid and supercritical carbon dioxide. Hop resin extracts are available as undiluted "pure resin extract" (P.R.E.) with an alpha acid content anywhere from 30 to 65% (depending on the variety of the hops). The P.R.E. is then diluted to the required standard before final packaging by the addition of either glucose or corn syrup. Diluted extracts also are available with tannins or without tannins. Extract products are commonly labeled by the type of solvent used in extraction.

Some of the major advantages of extracts are reduced bulkiness, improved stability in storage, standardization, consistency, improved hop utilization, and reduced wort losses. Many brewers believe the only real way to ensure the "clean" addition of hop essences is with hop extracts, as they are not only free of contaminating bacteria, but also nitrate and pesticide residues. The improvement in hop utilization varies from brewery to brewery; but in most cases utilization is in the range of 35 to 45%. The major disadvantage of extracts is the slightly higher cost per bitterness unit compared to whole hops or pellets but considering the higher utilization rate this in not an issue. The cost for organic solvent extracts is generally lower than that for carbon dioxide forms.

Extracts are available in a wide range of package sizes and types, including cans (0.5–4.0 kg), polyethylene pails, and 200 kg steel drums lined with high quality, food-grade lining. Shelf-life is approximately four years or longer if the extract is kept at or below 0°C.

Isomerized Hop Products

With isomerized hop products, the alpha acids have been isomerized to iso-alpha acids to improve alpha-acid utilization and minimize boil times, both of which reduces costs. As a result, they can be added to the kettle not only during wort boiling, but also—unlike non-isomerized products—during the conditioning phase. Products included in this group widely used by brewers are as follows:

Isomerized Hop Pellets

Isomerized hop pellets are basically conventional Type 90 pellets in which the alpha acids have been converted to iso-alpha acids. Isomerized pellets can be added at any stage of boiling while maintaining high utilization. Consequently, late additions of aroma hops optimize utilization of the alpha acids not usually converted with typical hop aroma late kettle additions. Isomerized pellets can be added to the kettle 15 minutes prior to knockout if the goal is to add hop flavor and aroma to the beer (30).

The major benefit to the brewer of using isomerized hop pellets is a significant increase in utilization of the bittering component. Hop utilization for isomerized hop pellets is estimated to be in the range of from 45 to 60% (3). The use of isomerized hop pellets also reduces boil times due to the rapid extraction of the iso-alpha acids into the hot wort. The stability of iso-alpha acids in isomerized hop pellets is very good compared to non-isomerized hop pellets, such as Type 90 pellets.

A notable disadvantage of this product perceived by some brewers is the heating of the hops and its effect on the essential oils. Also, since this product is "chemically processed" it falls outside the terms of the "Reinheitsgebot" or German Beer Purity Law.

Isomerized Extracts

Isomerized extracts represent the most recent technology in the development of hop products. The processes for producing isomerized extracts are sophisticated chemical processes that convert the alpha acids to iso-alpha acid compounds and formulate them in concentrated form. The main advantage of using isomerized hop extracts is the significant increase in utilization that can be achieved. Utilization in the range of from 80 to 90% can be expected, and although these products are relatively expensive to purchase, significant cost savings can be achieved. Another big advantage is that these products can be added at various stages of the brewing process. Other advantages associated with use of isomerized extracts include improvements in consistency and quality of bitterness, use of a purer product, reduced wort losses, and increased plant flexibility. The two types of isomerized extracts available on the market are isomerized hop extracts and isomerized kettle extracts.

Isomerized Hop Extracts – Isomerized hop extracts can be used as a partial replacement of kettle hops, as a means of improving utilization, or as a final adjustment of beer bitterness prior to filtration. The extract may be added to beer in the fermenter after the bulk of the yeast has been removed, or during transfer from the fermenter to the maturation vessel. Isomerized hop extracts contain only iso-alpha acids and do not contain any essential oils or other hop constituents (15). They are available as potassium and magnesium salts of iso-alpha acids and as free iso-alpha acids in aqueous propylene glycol. Extracts have the highest possible utilization of hop alpha acids (80–90%) and often are used as a supplement in the cellar. The iso-alpha solution usually is metered into a flowing beer stream during transfer before final filtration. Uniform addition is desirable to avoid locally high concentrations.

Isomerized hop extracts can be used for partial or total replacement of bittering and for final adjustment of bitterness. If added to the wort, they will be subject to the same fermentation losses as iso-alpha acids from the hops. If added after fermentation, almost all of the added iso-alpha acids remain in the beer. Flavor matching may be difficult with isomerized hop extracts, resulting in thinner hop character and less fullness than normally hopped beer (15). Some brewers use beta fraction (also called "base extract") to obtain a fuller hop character because isomerized hop extracts lack hop constituents, which contain beta acids, uncharacterized soft resins, and essential oils (15).

In addition, there are reduced isomerized hop extracts that have been chemically altered to minimize the light sensitivity of beer. Light sensitivity is particularly a problem when a beer is packaged in clear or green glass that is transparent to the damaging wavelengths of light. This exposure can result in a "sun-struck" or "skunky" flavor. Many major brewers use these chemically altered extracts to bitter their normal beers so they can be packaged in such bottles. As in the case of isomerized hop extracts, flavor matching may be a challenge. Like isomerized hop extracts, reduced isomerized extracts can be used at any stage in the brewing process; however, it is preferable to add them post fermentation to achieve better utilization.

The term "reduced isomerized hop extract" is a collective term that includes three different products: rho-iso-alpha acids (can be referred to as dihydroiso-alpha acids), those containing tetrahydroiso-alpha acids, and those containing hexahydroiso-alpha acids. Tetrahydro-iso extract may produce a softer bitterness; hexahydro-iso extract provides much greater foam improvement (18).

Isomerized Kettle Extracts – Isomerized kettle extracts contain iso-alpha acids, other soft resins (including beta acids), and essential oils. If the oils are removed prior to processing, the isomerized extract is used only for kettle bittering unless it is reconstituted (oil added back after processing), in which case it can be used for both bittering and aroma (15). The major benefit to the brewer in using isomerized kettle extracts is a significant increase in utilization of the bittering constituents. Utilization of iso-alpha acids should be in the range of from 45 to 60% (3). A disadvantage with standard extracts is the unknown effect of processing on non-alpha acid components (15). The reconstituted extract can be used to impart substantial late hop aroma without the unknown effects of submitting the oils to processing.

Hop Oil Products

Pure Hop Oils

Pure hop oils usually have a very pale green/yellow appearance. They have been on the market for many years, and give beer hop aroma without imparting any bitterness. Traditionally, hop oils are added to the kettle late in the boil, which is known as "late hopping," or they are used for dry hopping. It is generally agreed that the addition of hop oil to filtered beer does not produce a satisfactory flavor since some contact time with yeast is necessary to remove "raw hop," "tobacco," and "grassy" notes.

Hop oils can be derived from a single hop variety, or produced from a blend of varieties. Hop oils are blended so that a consistent product can be produced from year to year and batch to batch. This is especially important to commercial breweries, which want to make a very consistent beer. The disadvantage of blends is that they do not resemble a specific hop aroma profile. Most of the oil products currently available are designed for dry hopping and include various whole hop oils, oil-rich liquid carbon dioxide extracts, and oil fractions (15).

All hop oils are insoluble in wort and in beer. If hop oils are not pre-dissolved, the wort and beer will have an oil "slick." Consequently, various methods are available for adding these oils to beer, including ethyl alcohol solutions, dosing as emulsions, and injection of hop oil in liquid carbon dioxide into a flowing beer stream. Emulsions are water-based formulations in which the oil remains in suspension.

Oil-Rich Extracts

Oil-rich extracts are usually yellow/brown in color and less viscous than pure resin extracts. These products originally were developed in the United Kingdom as a replacement for dry hopping. A problem associated with dosing oil-rich extract is the solubility of hop oil in beer. This can be overcome by use of a carrier that facilitates the dispersion and solubilization of hop oil in the beer.

Hop Stability

Hop components (such as alpha acids and essential oils) undergo changes as soon as they are harvested. The rate of loss is dependent in part on storage conditions, and on the variety, and condition of the hops when baled. For example, alpha-acid stability is expressed as the percentage of alpha-acids remaining in baled leaf hops after six months of storage at ambient temperature.

Although hops deteriorate with storage, the aging process can be slowed with refrigeration, anaerobic packaging, and minimizing contact with light. Of these three, refrigeration is the most important, followed by anaerobic packaging. For optimum preservation, hops should be stored at temperatures between 0 and 2°C. The hops should be packaged in barrier bags that are either vacuum packed or flushed with an inert gas such as nitrogen or carbon dioxide. Badly stored hops can cause the alpha acids to oxidize, and some of the oxidation components are responsible for off tastes in the beer, such as "rancid" or "cheesy" odors or "oily" or "sour" flavors.

Under the same conditions, some varieties are more susceptible to deterioration than others. Fuggle is considered very stable. Clusters are also very stable and undergo very little deterioration in alpha acids. Bullion, on the other hand, is quite unstable, and has significant losses in alpha acids in the first few weeks of ambient storage. Other hops with poor storage characteristics include Brewer's Gold, Eroica, Cascade, Hallertauer, and Saaz.

Bitterness Levels in Beer

International Bitterness Units

Beer bitterness is expressed as International Bitterness Units (IBUs), which represent a measurement of the intensity of the bitterness of the beer. It is not a measurement of iso-alpha acids in beer, as is thought by many brewers. However, 1 IBU is usually assumed to be equivalent to 1 mg of iso-alpha acid in 1 liter of water or beer. The beers brewed today show a wide range in their levels of bitterness. Lagers brewed in the United States range in bitterness from 12 to 15 IBUs for the large brewers, while UK ales range in bitterness from about 16 to 50 IBUs.

The amount of IBUs in a beer does not always equal the same amount of perceived bitterness. For example, rich, full-bodied stout with 20 IBUs will be perceived by the tongue as having far less bitterness than 20 IBUs in a light American-style lager beer. Consequently, light beers take fewer hops to be bitter, whereas high-gravity beers require more hops for balance.

There are a number of formulas used in estimating IBUs, all providing only rough estimates. One formula used to calculate IBUs in metric units is as follows:

IBU = [(weight grams × %utilization × %alpha-acid)/batch size in liters]1,000

For example:

IBU = [(4,500g × 0.25 x 0.047)/1,200 liters] = 0.0444g/l
IBU = (0.0444g/l)1,000
IBU = 44.1 mg/l (or ppm)

Utilization is determined after a number of test brews by assessing the IBUs in the finished beer through laboratory analysis. Any recipe that includes more than one hop addition will require a separate calculation of IBUs for each addition of hops. The equation for estimating the weight of hops given a batch size, an alpha acid content, and a utilization factor derived by the boil time gives the following:

Weight (g) = [(IBU × batch size in liters)/(%alpha acid × 1,000 × %utilization factor)]

HomeBrew Bitterness Units (HBUs)

Homebrewers and some craft brewers lacking sophisticated equipment for estimating bitterness levels have devised a system for estimating the bitterness for a given volume of beer that is expressed as homebrew bitterness units (HBUs). HBUs are the same as Alpha Acid Units (AAUs). The major drawback of the HBU system is that it is a measurement of the amount of hops being added to the beer, not the amount of bitterness in the final product. HBUs can be calculated using the following equation:

HBUs = (%alpha acid of hops × ounces of hops)/volume

For example, 2 ounces of 5% alpha acid Hallertauer hops used in a 10 gallon batch would result in 10 HBU.

Controlling Alpha Acid Levels

Commercial brewers measure the alpha acid content in a laboratory; then, after a few test batches in the pilot brewery, they adjust the hopping rate to the International Bitterness Units (IBUs) desired in the beer. To ensure consistency, large brewers purchase huge lots of hops and blend for uniform consistency throughout the year. Furthermore, virtually all large breweries blend beer to even out the fluctuations in bitterness from batch to batch, and add hop extracts to further adjust the bitterness.

Small craft brewers and homebrewers, on the other hand, don't have these luxuries. Therefore, they cannot expect to achieve the level of accuracy or reproducibility seen in large brewers' beers. Small brewers, however, can determine their utilization rates by having representative beer analyzed for IBUs and calculating back from added alpha acids.

Hop Utilization

Hop utilization is the percentage of alpha acids that is isomerized and remains in the finished beer. The utilization of the bitter substances rarely exceeds 40% in commercial breweries and is often as low as 25% (31). Hop utilization can be calculated using the following formula:

% Utilization = (iso-alpha acids present in beer/alpha acids added to brew kettle)100

For example, the hop utilization rate if 50 kg of hops having an alpha acid content of 3.7 kg is added to a 500 hl brew yielding 475 hl of beer at 17 mg/l iso-alpha acid is calculated as follows:

Amount of iso-alpha acids = 47,500 l × 17 mg/l
Amount of iso-alpha acids = 807,500 mg or 0.8075 kg

Utilization = (0.8075/3.7)100 = 21.8%

Factors Affecting Utilization

Not all of the bitterness potential from the alpha acid in the hop is utilized, which can be attributed to a number of reasons:

- Form of hops
- Mash conditions

- Boil conditions
- Hopping rate
- Fermentation conditions
- Maturation and filtration conditions

Form of Hops

The isomerization rate is initially affected by the form of hops. Isomerization is slower and at a much lower rate with whole hops or plugs, slightly faster with standard pellets, and greatest with extracts. When whole hops or plugs are added to boiling wort, the isomerization rate of alpha acids is slower since the lupulin glands are intact. These glands contain the alpha acids and the essential oils. Isomerization starts sooner with standard hop pellets or with stabilized pellets, since their alpha acids are partially converted to calcium and magnesium salts. Eventually, the isomerization rate of whole hops or plugs will catch up to the rate of the pellets.

Boil Conditions

Boil conditions can affect isomerization in a number of ways. For instance, the longer the boil continues, the more isomerization takes place, though eventually the reaction reverses itself, degrading the iso-alpha acids. No one knows exactly how long it takes for this to occur, but some researchers claim that this point is reached in about two hours, while some say it takes longer (8). Since most brewers only boil their wort between 60 and 90 minutes, it is usually not a main concern.

The temperature of the boil has an effect on the isomerization reaction, and therefore on utilization. At higher elevations the utilization rate is lower due to the lower boil temperature, thus the hop rate needs to be adjusted upward.

Other factors affecting utilization during the boil is the pH of the wort. The main reason for the low utilization has to do with low pH of the wort (i.e., pH of 5.0–5.5), with more efficient utilization occurring at a higher pH (i.e., pH of 10.0–11.0).

Wort gravity affects hop utilization resulting in lower utilization rates with high gravity worts. This phenomenon can be compensated for by using high hop rates than would be needed for a normal strength beer of similar bitterness.

Hopping Rate

Isomerization is also affected by hopping rate; as the hopping rate increases, the rate of isomerization decreases. This effect can be partially offset by adding bittering hops in stages.

Fermentation Conditions

Fermentation conditions can affect the amount of iso-alpha acids that remain in the beer in a number of ways. Loss of iso-alpha acids also occurs during fermentation as they are adsorbed onto the yeast cell walls. The amount of loss is dependent on the total growth of the yeast crop and, more importantly, on the amount of time the yeast stays in suspension. The faster the yeast flocculates, the less time there is for the iso-alpha acids to attach themselves to the cell walls, resulting in a higher concentration and a more bitter beer. However, if the yeast stays suspended for a long period of time, more iso-alpha acids will be adsorbed; thus, assuming the yeast drops out, the beer will be less bitter.

The surface area of the fermenter also plays a role in hop utilization since some of the iso-alpha acids adhere to the fermenter walls and top covering surfaces. This explains why small batches sometimes do not scale up well to larger batches, because the larger containers usually have a smaller surface area-to-volume ratio than smaller vessels.

Maturation and Filtration Conditions

After fermentation, maturation and filtration conditions affect the extent to which not only bitterness, but also other hops components survive in the finished beer. For example, the addition of finings reduces the amount of iso-alpha acids as well as filtration.

Dry Hopping

Dry hopping is the process of adding hops to the primary fermenter, the maturation tank, or the casked beer to increase the aroma and hop character of the finished beer. Some brewers believe dry hopping should not be done during primary fermentation because of the risk of contaminating the beer with microorganisms. Dry hopping adds no bitterness to the beer, and any lingering bitterness will dissipate in a few weeks. This is because alpha acids are only slightly soluble in cold beer. It should also be mentioned that a beer that has been dry hopped is also usually late hopped in the kettle. British brewers use this method to give a special hop character to cask-conditioned ales.

Varieties most commonly used for dry hopping are Goldings and Fuggles, Challenger, Target, Northdown, and Whitbread. Styrian Goldings and Super Styrians also are prized for their dry hopping characters.

In the United Kingdom, dry hopping usually is done when the beer is racked from the primary fermenter into the cask. The cask then is left to "condition" for 7 to 14 days at around 13°C. The oil is extracted slowly from the hops and disseminated throughout the beer. Dry hopping is not favored by continental

European brewers, most likely due to the prevalence of bottled beer and the fact that kegs are pressurized under carbon dioxide.

Pellets are popular for dry hopping because the hop oils are readily available and will get into the beer faster than with whole hops, and probably with more efficiency. Using whole hops is reportedly microbiologically unsound and gives variable results, in particular because the aroma quality of hops deteriorates during storage and because of variations between individual batches of hops (22). In addition, longer contact times are required when using whole hops, and the utilization rates are very low. Hop oils and extracts are used for dry hopping also. Extracts can be used as a substitute but first must be diluted with ethanol (four parts ethanol to one-part extract) (30).

References

1. Ashurst, P.R. 1971. "Hops and Their Use in Brewing." *Modern Brewing Technology*, edited by W.P.K. Findlay. Cleveland, Ohio: The Macmillian Press.

2. Clarke, B.J. 1986. "Hop Products." *Journal of the Institute of Brewing* 92 (2).

3. European Brewery Convention (EBC). 1997. *Hops and Hop Products, Manual of Good Practice*. Nürenberg, Germany: Getränke-Fachverlag Hans Carl.

4. Fix, George. 1989. *Principles of Brewing Science*. Boulder, Colorado: Brewers Publications.

5. Fix, George and Laurie Fix. 1997. *An Analysis of Brewing Techniques*. Boulder, Colorado: Brewers Publications.

6. Gardner, David. 1992. "Hop Products – Their Use and Function in the Brewing Process." *Brewers' Guardian* 121 (7).

7. Gardner, David. 1997. "Advances in Brewing Technology – Hops." *The Brewer* 83 (989).

8. Garetz, Mark. 1994. *Using Hops – The Complete Guide to Hops for the Craft Brewer*. Danville, California: Hop Tech.

9. Grant, Herbert L. 1977. "Hops." *The Practical Brewer*, edited by H. M. Broderick. Madison, Wisconsin: Master Brewers Association of the Americas.

10. Grant, Herbert L. 1995. "Hops." *Handbook of Brewing*, edited by William A. Hardwick. New York, New York: Marcel Dekker, Inc.

11. Gruber, Mary Anne and Jerrold Hilton. 1995. "Malt and Hops Primer." *The New Brewer* 12 (6).

12. Hilton, Jerrold F. and G. H. Salazar. 1993. "Hops – The Essence of Beer." *Brewers Digest* 68 (10).

13. Hudston, H. Randulph. 1997. "Wort Boiling." *The Practical Brewer*, edited by H. M. Broderick. Madison, Wisconsin: Master Brewers Association of the Americas.

14. Hop Growers of America, Inc. *U. S. Hops Resource Guide*. Yakima, Washington.

15. Hysert, David W. 1994. *Processing and Processed Hop Products: Hop Storage*. Hop Growers of America, Inc. European Seminar Series.

16. Hysert, David W. 1999. Personal Communication. John I. Haas, Inc.

17. Kollnberger, Peter. 1986. "Techniques and Problems of Hopdosing in Modern Brewing Plants." *MBAA Technical Quarterly* 23 (4).

18. Kunze, Wolfgang. 1996. *Technology Brewing and Malting*, translated by Dr. Trevor Wainwright. Berlin, Germany: VLB Berlin.

19. Lemmens, Gerard. 1996. "Hops in America: A 20-Year Overview." *Brewing Techniques* 4 (6).

20. Lemmens, Gerard. 1999. Personal Communication. Yakima Chief, Inc.

21. Lewis, Michael J. and Tom W. Young. 1995. *Brewing*. London, England: Chapman & Hall.

22. MacDonald, J., et al. 1984. "Current Approaches to Brewery Fermentations." *Progress in Industrial Microbiology*, edited by M. E. Bushell. Guildford, United Kingdom: Elsevier.

23. Moir, M. "Hop Aromatic Compounds." 1994. EBC Monograph XXII. *Symposium on Hops, Zoeterwoude*. Nürenberg, Germany: Verlag Hans Carl.

24. Noonan, Gregory J. 1996. *New Brewing Lager Beer*. Boulder, Colorado: Brewers Publications.

25. O'Rourke, Timothy. 1999. "Hops and Hops Products." *Brewers' Guardian* 128 (4).

26. Peacock, Val. 1998. "Techniques and Problems of Hopdosing in Modern Brewing Plants." *MBAA Technical Quarterly* 35 (1).

27. Reed, David Allan. 1991. *An Analysis of the U.S. Hop Industry's Statistical Market Information*. M. A. thesis, Department of Agricultural Economics, Washington State University, Pullman, Washington.

28. Roberts, Trevor R. and Richard J.H. Wilson. 2006. "Hops." *Handbook of Brewing*, edited by Fergus G. Priest and Graham G. Stewart. Boca Raton, Florida: CRC Press, Taylor & Francis Group.

29. Rosendal, I. 1985. "Hops and Hop Products Terminology." *Journal American Society of Brewing Chemists* 43 (1).

30. Sidor, Larry. 2006. "Hops and Preparation of Hops." *Raw Materials and Brewhouse Operations*. Volume 1, edited by Karl Ockert. St. Paul, Minnesota: Master Brewers Association of the Americas.

31. S.S. Steiner, Inc. *Hopsteiner Hop Pellets*. Yakima, Washington.

32. Strauss, Karl. 1986. "Hop Extracts – Past, Present, and Future." *MBAA Technical Quarterly* 23 (4).

33. Stewart, G.G. and I. Russell. 1985. "Modern Brewing Biotechnology." *Food and Beverage Products*, edited by Murray Moo-Young. Oxford, England: Pergamon.

34. Tinseth, Glenn. 1994. "The Essential Oil of Hops: Aroma and Flavor in Hops and Beer." *Brewing Techniques* 2 (1).

35. Tomlinson, Thom. 1994. "India Pale Ale, Part II: The Sun Never Sets." *Brewing Techniques* 2 (3).

36. Westwood, Keith. 1994. "Hop Products and Their Effect on Bittering Quality." *Brewers' Guardian* 123 (12).

Chapter Four

Yeast

Yeast is one of the most important ingredients in brewing beer responsible for metabolic processes that produce ethanol, carbon dioxide, and a whole range of other metabolic byproducts that contribute to the flavor and finish of beer. There are literally hundreds of varieties and strains of yeast. In the past, there were two types of beer yeast: ale yeast (the "top-fermenting" type, *Saccharomyces cerevisiae*) and lager yeast (the "bottom-fermenting" type, *Saccharomyces pastorianus*, formerly referred to as *Saccharomyces carlsbergensis* or *Saccharomyces uvarum*). Top fermenting yeasts produce beers that are more estery, fruity, and sometimes malty, whereas bottom-fermenting yeasts give beers a characteristic sulphurous aroma. Some other notable differences also include fermentation temperatures and flocculation characteristics. Top-fermenting yeasts are used for brewing ales, porters, stouts, Altbier, Kölsch while bottom-fermenting yeasts are used for brewing lagers such as Pilsners, Dortmunders, Märzen, and Bocks are fermented with bottom-fermenting yeasts.

Nowadays, the differentiation in ale and lager yeasts is not as distinct given the fact that beers are widely fermented in similar vessels (cylindroconical tanks) and because of other advances in brewing science.

Ale Yeast

Ale yeast strains are best used at temperatures ranging from 10 to 25°C, though some strains will not actively ferment below 12°C (33). Ale yeasts are generally regarded as top-fermenting yeasts since they rise to the surface during fermentation, creating a very thick, rich yeast head. That is why the term "top-fermenting" is associated with ale yeasts. Fermentation by ale yeasts at these relatively warmer temperatures produces a beer high in esters and higher alcohols, which many regard as a distinctive character of ale beers. Some brewers use a mixture of two or more strains of ale yeast to ensure both rapid fermentation and high attenuation. It has been reported that mixed ale yeast cultures—which ferment quickly and at relatively high temperatures—are less liable to infection and may tend to remain more uniform in composition from generation to generation. Mixed yeast cultures can also result in mutual flocculation and better head formation than any single strain of yeast.

Lager Yeast

Lager yeast strains are best used at temperatures ranging from 7 to 15°C. At these temperatures, lager yeasts grow less rapidly than ale yeasts, and with less surface foam they tend to settle out to the bottom of the fermenter as fermentation nears completion. This is why they are often referred to as "bottom" yeasts. The final flavor of the beer will depend a great deal on the strain of lager yeast and the temperatures at which it was fermented but lager yeasts generally produce greater quantities of hydrogen sulfide and other sulfur compounds.

Many bottom-fermenting yeasts are divided into powdery and flocculent yeasts. In the case of powdery yeast, the cells are finely divided in the fermentation medium and sink slowly to the bottom only at the end of fermentation. Because they do not clump well, they remain in suspension longer, which results in a higher degree of attenuation. The cells of flocculent yeast clump together after a short time to form large flocs that then settle rapidly. Consequently, flocculent yeast produces a clear but not a fully attenuated beer.

In lager brewing, unlike ale brewing, it is standard practice to use a pure strain of yeast, cultured from a single cell. However, in some cases two strains are used with different flocculation/sedimentation characteristics. The more weakly flocculating strain is often carried forward to secondary maturation. The two strains are propagated separately as pure cultures, used in separate fermenters, and blended on the way to the lagering cellar. (In this way it is possible to prevent one strain from taking over the other, as could happen if both strains were used for fermentation together.) This two-yeast system is still used in Europe; but it is rarely found in North America, where usually, one highly, attenuative powdery yeast is used, along with different beer conditioning processes.

Yeast Life Cycle

The life cycle of yeast is activated from dormancy when it is added (pitched) to the wort. Yeast growth follows four phases, which are: 1) the lag period, 2) the growth phase, 3) the fermentation phase, and 4) the sedimentation phase. These phases are somewhat arbitrary because they may overlap in time.

The Lag Phase

The lag phase starts when the yeast is pitched into the aerated wort. This stage is marked by a drop in pH because of the utilization of phosphate and a reduction in oxygen. Glycogen, an intracellular carbohydrate reserve, is essential as an energy source for cell activity since wort sugars are not assimilated early in the lag phase. Stored glycogen is broken down into glucose, which is utilized by the yeast cell for reproduction—the cell's first concern. Low glycogen levels produce abnormal levels of vicinal diketones (especially diacetyl) and result in longer fermentations.

The glycogen level can be estimated directly using a simple iodine test. A very dark-brown reaction indicates high glycogen levels; and as the sample becomes progressively more yellow, the glycogen level decreases. It should be noted that glycogen will react with other starchy material.

From the brewer's point of view, the lag phase of fermentation should be as short as possible—normally 8 to 24 hours, depending on yeast health, pitching rate, and wort temperature. The lag phase is generally longer when yeast has not been cultured in a solution similar to the wort for pitching.

The Growth Phase

The growth phase, often referred to as the respiration phase, follows the lag phase once sufficient glycogen reserves are built up within the yeast. This phase is evident by the covering of foam on the wort surface due to the liberated carbon dioxide.

In this phase, the yeast cells use the oxygen in the wort to oxidize a variety of acid compounds, resulting in a significant drop in pH. In this connection, some yeast strains will result in a much greater fall in pH than others within the same fermenting wort. For ale yeast the pH drop during respiration is dramatic, while for lager yeast it is much less precipitous. In general, ales have a slightly lower pH than lagers; typical values for ales are 4.0 to 4.5 and for lagers 4.4 to 4.7.

During this phase, yeast cells are growing logarithmically. Growth of yeast cells (sexual reproduction) is by budding, which is a form of cell division. The absolute amount of yeast growth is roughly proportional to the nutrients available in the wort. Commercially, one to three doublings in the cell number will occur. It is during the growth phase that most of the beer flavor, alcohol, carbon dioxide, and heat are produced.

The Fermentation Phase

The fermentation phase quickly follows the growth phase when the oxygen supply has been depleted. Fermentation is an anaerobic process. In fact, any remaining oxygen in the wort is "scrubbed," i.e. stripped out of solution by the carbon dioxide bubbles produced by the yeast. This phase is characterized by reduction of wort gravity and the production of carbon dioxide, ethanol, and byproducts which contribute to beer flavor. During this phase, yeast is mostly in suspension converting beer fermentables in the beer wort. Most beer yeasts will remain in suspension from 3 to 7 days, after which flocculation and sedimentation will commence.

The Sedimentation Phase

The sedimentation phase is the process through which yeast flocculates and settles to the bottom of the fermenter following fermentation. The yeast begins to undergo a process that will preserve its life as it readies itself for dormancy, by producing a substance called glycogen. Glycogen is necessary for cell maintenance during dormancy and, as mentioned, is an energy source during the lag phase of fermentation.

Nutritional Requirements

To grow successfully, yeast requires an adequate supply of nutrients—fermentable carbohydrates, nitrogen sources, vitamins, and minerals—for healthy fermentation. These nutrients are naturally present in malted barley or developed by enzymes during the malting and mashing process.

Inadequate nutrition can result in poor beer stability, mutation of yeasts, poor sedimentation, off-flavors, and sluggish or "stuck" fermentations. Stuck fermentations are characterized by a long lag phase accompanied by no or very little fermentation activity. Nutrition requirements will vary with yeast strains. Additional nutrients (proprietary blended "yeast foods") are often used either to avoid fermentation problems or simply to ensure consistent fermentations. This is especially the case as wort gravities increase and/or brewers use larger fermentation vessels (57).

Carbohydrates

Carbohydrates are available for yeast growth in wort as low-molecular-weight sugars such as the mono-, di- and oligosaccharides are available for yeast growth. Polysaccharides are not used by the yeast. The sugars are, in order of concentration, maltose, maltotriose, glucose, sucrose, and fructose, which together constitute 75 to 85% of the total extract. The other 15 to 20% consists of non-fermentable products such as dextrins, beta-glucans, pentosans, and oligosaccharides. Regardless of concentration, fermentable carbohydrates are usually assimilated by yeast in the following order: sucrose, glucose, and fructose are consumed most rapidly (24–49 hrs.); followed by maltose (70–72 hrs.); then maltotriose (after 72 hrs.) [26]. Some overlap in assimilation does occur. A majority of the yeast strains leave maltotetraose and dextrins unfermented (63).

The order of uptake reflects the fact that sucrose, glucose, and fructose are present in low concentrations and are utilized rapidly. Although maltose is the major sugar in all malt wort, the permease enzyme controlling its diffusion into the yeast cell is inhibited by glucose. But as the glucose concentration diminishes, the enzyme systems required for assimilating maltose are synthesized, and the yeast begins to utilize maltose and maltotriose, the major sugars in brewer's wort (63). Sucrose is split into fructose and glucose by the yeast-produced enzyme invertase.

Substantial addition of sucrose and glucose (e.g., when using glucose syrup adjuncts) will delay maltose and maltotriose fermentation, which may affect flavor. In fact, "stuck" fermentations may occur in worts with high glucose concentrations relative to other sugars from a condition sometimes referred to as "glucose repression" (26). In this event, the presence of high concentrations of glucose prevents the yeast from synthesizing the enzymes needed to assimilate the other sugars. A similar condition may arise if a high concentration of fructose is present, causing "fructose block." These situations are more likely in worts for low-calorie beers because the adjunct may be solely glucose.

Nitrogen

Nitrogen is available for yeast growth in wort as amino acids, peptides, and ammonium salts. Yeast prefers to use ammonium salts, but these are present in wort only in very small amounts (33). Amino acids and peptides are therefore the most important wort constituents. Amino acids collectively referred to as "free amino nitrogen (FAN)," are the principal nitrogen source in wort and are an essential component of yeast nutrition (45). It is the amino acids that the yeast cells use to synthesize more amino acids and, in turn, to synthesize proteins. Also, FAN is an important factor in yeast viability and vitality, fermentation rate, ethanol tolerance, carbohydrate uptake, and attenuation (14).

The optimum level of nitrogen FAN for yeast growth depends in part on the gravity of the wort. Noble reports that normal levels of between 110 and 140 mg/l (or ppm) are sufficient for yeast growth for a wort at 10°P (SG = 1.040) [47]. Fix considers FAN levels in the range of 300 to 325 mg/l ideal for an all-malt wort at 12°P (SG = 1.048) [19]. Excessive FAN levels can result excessive yeast growth, a higher beer pH, and susceptibility to bacterial attack while low FAN levels can lead to a sluggish fermentation.

The nitrogenous compounds in wort depend on the choice of raw barley, the use of adjuncts, the malting process, and the mashing program. Nitrogen deficiencies may occur if large percentages of adjunct (usually more than 40%) or wheat malt (more than 70%) are used by diluting the amino acids derived from the malt. The brewer can increase the FAN levels by using a thicker mash, mashing at a pH < 5.5, and including a mash protein rest to increase the protein levels (14). In the case of beers made with higher adjunct levels, protease enzymes can be added (57). If FAN levels are too high, the brewer can choose malt less then 11.5% protein that has an S/T ratio between 39 and 44% (14).

Vitamins

Vitamins such as biotin, panthotenic acid, thiamin, and inositol are essential for enzyme function and yeast growth. Biotin is obtained from malt during mashing and is involved in carboxylation of pyruvic acid, nucleic synthesis, protein synthesis, and synthesis of fatty acids. Biotin deficiencies will result in yeast with high death rates. Panthotenic acid is required by many strains of fermentation yeast and is an essential factor in carbohydrate and lipid metabolism and in cell membrane function. Panthotenic acid deficiencies can lead to the accumulation of hydrogen sulfide. Thiamine is essential in oxo-acid decarboxylation. Inositol is required for cell division; deficiencies will decrease the rate of carbohydrate metabolism.

Minerals

Yeasts are unable to grow unless provided with a source of a number of minerals. These include phosphate, potassium, calcium, magnesium, sulfur, and trace elements. Phosphate is involved in energy conservation, is necessary for rapid yeast growth, and is part of many organic compounds in the yeast cell. Potassium ions are necessary for the uptake of phosphate. Calcium improves the flocculation properties of yeast and should be present in a concentration greater than 50 mg/l (25). Magnesium is required for yeast growth and acts as an enzyme activator. Yeast requires sulfur for the synthesis of methionine and for cycteine, which is incorporated into protein, glutathione, coenzyme A, and thiamin. The elements zinc, copper, and manganese are required in trace amounts.

Zinc

The most important trace element is zinc, and at least 0.10 to 0.15 mg/l should be present in the wort (33). Zinc assists in protein synthesis in yeast cells and controls their nucleic acid and carbohydrate metabolism. Fermentations are accelerated by adding zinc chloride (0.2–0.3 mg/l) to the wort (45). Zinc, as well as boron and manganese, is usually available in small amounts from malt, hops, or water. However, brewers who use a high fraction of unmalted cereals in standard gravity batches typically experience zinc deficiencies (20). Zinc deficiencies have also been experienced when changing from copper vessels and galvanized pipes to stainless steel.

If a deficiency has been documented by laboratory analysis, the brewer can add yeast nutrient powders to the kettle or whirlpool. Brewers sometimes add zinc chloride or zinc sulfate to the brew kettle before knockout to remedy the deficiency.

Excessive levels of zinc ions can promote rapid fermentation and off-flavors and may even cause premature yeast flocculation. If too high, the zinc can even be toxic to the yeast.

Yeast Byproducts

The flavor and aroma of beer is very complex, being derived from a vast array of components that arise from a number of sources. Not only do malt, hops, and water has an impact on flavor, so does the synthesis of yeast, which forms byproducts during fermentation and conditioning. The most notable of these byproducts are, of course, ethanol and carbon dioxide; but in addition, a large number of other flavor compounds such as esters, higher alcohols, and acids are produced, all of which contribute to the taste, aroma, and other characteristics of the beer.

The formation of byproducts is influenced by wort composition and the choice of yeast strain. For example, non-flocculent yeasts tend to produce more volatiles than do flocculent strains. Lager yeasts produce more fatty acids and sulfur byproducts than do ale yeasts. Mutant strains of yeast have been known to produce markedly different levels of vicinal diketones.

Esters

Esters are considered the most important aroma compounds in beer. They make up the largest family of beer aroma compounds and in general impart a "fruity" character to beer. Esters are more desirable in most ales, and in some dark or amber lagers, lower levels are preferred in pale lagers.

Quantitatively and qualitatively, the two most important esters produced during fermentation are ethyl acetate and iso-amyl acetate (38). Ethyl acetate is a combination of ethanol and acetic acid and has a flavor threshold of 25 to 35 mg/l, with normal concentrations in beer averaging about 15 mg/l (12). It gives a fruity tone that sometimes can take on a solvent-like character. Iso-amyl acetate is called the banana ester and has a threshold level of 3.0 mg/l (18). It is a combination of amyl alcohols and acetic acid. The strain of yeast used plays an important role in these flavor components.

Most of the esters are formed during primary fermentation, and some ester formation occurs during conditioning. However, the level of esters could double with a long secondary fermentation (33).

Kunze reports that ester production is increased by 1) high fermentation temperatures, 2) restricting wort aeration, 3) increasing the attenuation limit, and 4) increasing the wort concentration to above 13% P (33).

Diacetyl and 2,3-Pentanedione

Diacetyl and 2,3-pentanedione, which are classified as ketones, are important contributions to beer flavor and aroma. Often these two ketones are grouped and reported as the vicinal diketone (VDK) content of beer, which is the primary flavor in differentiating aged beer from green beer. Of the two, diacetyl is more significant because it is produced in larger amounts and has a higher flavor impact than 2,3-pentanedione. A "buttery" or "butterscotch" flavor usually indicates the presence of diacetyl, while 2,3-pentanedione has more of a "honey" flavor. The flavor of diacetyl can initially be confused with that of caramel malts, but within time it is easy to distinguish the two: diacetyl usually tends to be unstable in most beers, imparting a raunchy flavor; the flavoring imparted by caramel malts, on the other hand, tends to be stable. Diacetyl is detectable more readily in lagers (especially in lighter-bodied lagers), where it is almost universally regarded as a defect.

Many heavier beers, such as European beers, have VDK at higher levels than the lighter beers typically produced in the United States and they still maintain good flavor. VDK is considered a desirable flavor in typical heavily hopped British ales (63). The taste threshold of diacetyl in ales is in the order of 0.1 to 0.14 mg/l and is somewhat lower for lagers (63). The threshold for Pilsners can be as low as 0.05 mg/l. Although diacetyl and 2,3-pentanedione are usually present in beers at levels below their taste threshold, their presence still makes a significant contribution to the overall flavor and balance of a beer.

The bacteria that directly promote diacetyl production consist of the strains *Pediococci* and *Lactobacillus*. In most cases infections, when they occur, happen through pitching yeast and not necessarily through unsanitary equipment. Air pickup during transfer from the primary fermenter to the aging tanks will cause the yeast to produce more diacetyl.

Acetaldehydes

There are many flavor-active acetaldehydes present in beer that are formed at various stages in the brewing process. They are produced by oxidation of alcohols and various fatty substances. Acetaldehyde levels peak during the early to mid-stages of primary fermentation or immediately after kraeusening, then decrease in concentration (55). By the end of the primary fermentation, aldehyde is reduced to ethanol. If oxygen is introduced back into the process, the ethanol is oxidized back into acetaldehyde.

Quantitatively acetaldehydes are commonly present at levels ranging from 2 to 20 mg/l and with a flavor threshold ranging from 5 to 50 mg/l, depending on the type of beer (38). Excess levels of acetaldehydes in beer impart a flavor variously described as "grassy," "bruised/raw apples," or "cidery."

The principal causes of high acetaldehyde concentrations in beer are the use of poor quality pitching yeast, high pitching rates, excessive wort oxygenation, and high fermentation temperatures (8). Other causes include rapid fermentation, pressure application during primary fermentation, and infected worts, especially those infected by strains of *Zymomonas anaerobia* (35). As in the case of diacetyl, the presence of active yeast in the maturation phase is a prerequisite for satisfactorily low acetaldehyde levels in the final product (25). Removal of acetaldehyde is favored by a warmer maturation stage.

Organic and Inorganic Sulfur Volatiles

Sulfur-containing compounds in beer arise from organic sulfur-containing compounds such as some amino acids and vitamins. They are also formed from inorganic wort constituents such as hydrogen sulfide, dimethyl sulfide, sulfur dioxide, and thiols make significant contributions to beer flavor. When present in small concentrations, sulfur compounds may be acceptable or even desirable (for example Burton ales), but in excess they give rise to unpleasant off-flavors, e.g., rotten-egg flavors.

Of the sulfur compounds, hydrogen sulfide is found in the greatest concentrations. Knudsen has reported its threshold: 10 μg/liter in lagers and 30 μg/liter in ale (30). Hydrogen sulfide is very volatile; it can be reduced during fermentation either by applying carbon dioxide counter-pressure or by ascending carbon dioxide gas during prolonged maturation times. Normally, ascending carbon dioxide gas purges most of the hydrogen sulfide formed during fermentation. Low maturation temperatures in the presence of yeast will reduce hydrogen sulfide, too. Since hydrogen sulfide production is a function of the yeast strain, it is often easier to switch yeasts if a given strain continually produces excessive levels. Poor-quality yeast can also result in sulfury flavors.

Sulfur dioxide is also present in beer, although usually in concentrations well below 10 mg/l, at which level it does not have a flavor impact in most beer (45). Top-fermentation beers generally contain less sulfur dioxide than bottom-fermentation beers (42).

Besides yeast metabolism the main sources of sulfur compounds in beer are raw materials (malt and hops) and spoilage organisms—in particular *Zymomonas anaerobia*, *Enterobacter aerogenes* and *Hafnia protea*.

Dimethyl Sulfide

Another major compound responsible for sulfury flavors in beer is dimethyl sulfide (DMS), which is a desirable flavor component in lager beer but not in ales. Dimethyl sulfide in lagers it will lead to a malty/sulfury note. The taste threshold for DMS is considered to be from 50–60 μg/liter (33). If the concentrations are too high, it has a relatively objectionable taste and aroma of cooked sweet corn.

Most British ales have DMS levels that range from 0 to 13 µg/l, which is well below the threshold (38). The DMS levels of British and continental lagers are from 18 to 27 µg/l and 44 to 114 µg/l, respectively (38). In fact, the DMS levels for medium- to high-gravity European lagers can run as high as 75–100 µg/l (38). At these levels the net effect is a malty, slightly sulfury tone. Beers produced in the United States rarely have high DMS levels, though regional specialty beers have DMS levels in the range of 50–75 µg/l because of high SMM levels in the malt.

Malting procedures have a much greater effect on beer DMS levels than fermentation conditions (24). During malting, DMS is released through the breakdown of S-methylmethionine (SMM), a malt precursor that is produced during the kilning process through protein breakdown (5). Thus high protein levels increase the potential for DMS formation in beer. Six-row malts contain higher levels of SMM, presumably because of their higher protein content. Pale malts generally have higher levels of SMM than do darker, highly kilned malts. Lower fermentation temperatures and high gravity worts favor DMS production too.

Fusel Alcohols

Fusel alcohols are a group of byproducts that are sometimes called "higher alcohols." They contribute directly to beer flavor but are also important because of their involvement in ester formation. Fusel alcohols have strong flavors, producing an "alcoholic" or "solvent-like" aroma. They are known to have a warming effect on the palate. About 80% of fusel alcohols are formed during primary fermentation (34). The yeast strain is very important, with some being able to produce up to three times as much fusel alcohols as others (24). Ale strains generally produce more fusel alcohols than lager strains (8).

The most important fusel alcohols are n-propanol, iso-butanol, 2-methyl-1-butanol, and 3-methyl-l-butanol (iso-amyl alcohol), with flavor threshold values in beer of 600, l00, 50, and 50 mg/1, respectively (38). Kunze reports that the concentration of fusel alcohols should not exceed 100 mg/l for ale beers and should fall within the range of 60 and 90 mg/l for lager beers (33). The cold-temperature fermentations that lagers undergo limit the production of fusel alcohols.

Other factors that increase the production of fusel alcohols are high fermentation temperatures, mutant yeasts, high wort gravities above 13% P, intensive aeration of the pitching wort, and low amino acid concentration in wort. The production of fusel alcohols has been linked to amino acid biosynthesis (41). Therefore, the balance of fusel alcohols depends on the balance of amino acids. Reportedly, tall slender fermentation vessels promote production of fusel alcohols (14).

Organic Acids

The most important organic acids found in beer are acetic, citric, lactic, malic, pyruvic and succinic. They confer a "sour" or "salty" taste to beers. Some of these organic acids are derived from malt and are present at low levels in wort, with their concentrations increasing during fermentation. Other acids are produced solely as a result of yeast metabolism. Organic acids can directly affect the flavor of beer by lowering its pH.

Fatty Acids

Fatty acids are minor constituents of wort and increase in concentration during fermentation and maturation. They give rise to "goaty", "soapy", or fatty flavors and can cause a decrease in beer foam stability. They are recognized as common flavor characteristics in both lagers and ales but are more prevalent with lager yeast strains (38). Warm or excessive aging of beer can result in elevated levels of fatty acids (14).

Nitrogen Compounds

Yeast also excretes some nitrogen compounds during fermentation and maturation as amino acids and lower peptides, which contribute to the rounding of the taste and an increase in palate fullness. Harvesting of the yeast too soon can therefore produce empty, dry beers even when they are subsequently lagered for a long time (34). The beginning of yeast autolysis can be detected by an excessive increase in the amino acid content.

Yeast Strain Selection

Selection of a yeast strain with the required brewing characteristics is vital from both a product quality and economic standpoint. The criteria for yeast selection will vary according to the requirements of the brewing equipment and the beer style, but they are likely to include the following:

- Rapid fermentation
- Yeast stress tolerance
- Flocculation
- Rate of attenuation at the desired temperature
- Flavor of the final product
- Good yeast storage characteristics
- Stability against mutation
- Stability against degeneration

Rapid Fermentation

Rapid fermentation, without excessive yeast growth, is important in producing a beer with the maximum attainable ethanol content consistent with the overall flavor balance of the product.

Yeast Stress Tolerance

The yeast strain should be tolerant to alcohol, osmotic shock, and temperature. Another stress point for yeast can be the collection, separation (centrifuging/pressing), and transfer (pumping) throughout the plant.

Flocculation

The flocculation characteristics of yeast are of great importance. The term "flocculation" refers to the tendency to form clumps of yeast called flocs. The flocs (yeast cells) descend to the bottom in the case of bottom-fermenting yeasts or rise with carbon dioxide bubbles to the surface in the case of top-fermenting yeasts. The flocculation characteristics need to be matched to the type of fermentation vessel used—a strongly cropping strain will be ideal for skimming from an open fermenter but unsuitable for a cylindroconical fermenter.

The degree and type of flocculation varies for different yeasts. Some are very flocculent, giving rise to a more "gravelly" consistency, whereas others are non-flocculent and are referred to as "powdery." Most yeast strains clump and flocculate to a moderate degree. As might be expected, the best balance is neither totally flocculent nor totally powdery, but somewhere near the middle of the range. Yeasts which flocculate prematurely fall out of suspension and fail to complete fermentation, whereas non-flocculent yeasts are difficult to crop and need removal by either finings or centrifugation.

Flocculation is genetically determined and depends on the yeast, but it is also affected by environmental factors. Falling pH and increasing alcohol content enhance flocculation (12). As the pH drops during fermentation, flocculation occurs more readily because the overall negative charge on the yeast cell becomes neutralized. Hop substances and calcium promote floc formation, too. In addition, artificial cooling at the end of fermentation promotes flocculation by reducing the effects of carbon dioxide evolution.

The flocculating power of the yeast determines the degree of attenuation. Strongly flocculating yeasts give rise to sweeter, less-fermented beers, while less-flocculent strains remain in the beer during aging and cause yeasty flavors (55). Highly flocculent strains are suitable for those beers where high diacetyl is desirable, such as certain British ales. The shorter contact time between yeast and attenuated wort (sugars) has the affect of not reducing the diacetyl precursors. For most styles of beer, attenuative yeast is definitely required. For dry stout, a super-attenuative strain is even better. In a few cases, such as for sweet stout and some brown ales, unattenuative yeast is preferable.

Rate of Attenuation

Attenuation refers to the percentage of sugars converted to alcohol and carbon dioxide, as measured by specific gravity. Most yeasts ferment the sugars glucose, sucrose, maltose, and fructose. To achieve efficient conversion of sugars to ethanol (good attenuation) requires the yeast to be capable of completely utilizing the maltose and maltotriose. Brewing yeasts vary significantly in the rate and extent to which they use these sugars. Lager strains are often better at utilizing maltotriose than their ale counterparts. The degree of attenuation obtainable exerts a great influence on the organoleptic properties of the resultant beer and, consequently, is one of the determinant factors in the process of yeast selection.

Beer Flavor

The selection of the yeast strain itself is perhaps one of the most important contributors to beer flavor. Different strains will vary markedly in the byproducts they produce: esters, higher alcohols, fatty acids, hydrogen sulfide, and dimethyl sulfide. The yeast strain must also be capable of reproducible flavor production.

Good Storage Characteristics

The storage characteristics of yeast are very important for maintaining viability during storage between fermentations and rapid attenuation when re-pitched.

Mutation of Yeast

Yeast mutations are a common occurrence in breweries, but their presence may never be detected. Usually the mutant has no adverse effect since it cannot compete with normal yeast and generally disappears rapidly. In some cases, though, mutant yeast will overcome the normal brewing yeast and may express itself in many different ways. For example, a mutation could affect the fermentation of maltotriose, or there could be a continuous variation in the fermentation rate. Reportedly, lager yeast mutates more rapidly than ale yeast (40). Most commonly, mutations are due to poor handling of brewing yeast.

Three types of yeast mutations are common in practice. There are those yeast strains that have a tendency to mutate from flocculence to non-flocculence; those that lose the ability to ferment maltotriose; and those yeast mutants that are respiratory-deficient (55).

A change from flocculence to non-flocculence can give the mutant yeast a selective advantage in cropping, which can cause a gradual change in brewing yeast performance. However, if the yeast crop is collected by skimming or from the bottom of the lager fermenter, the non-flocculent mutants will be eliminated. If the crop is collected by centrifuging, then the proportion of non-flocculent cells may increase in successive brews.

A mutant that has lost its ability to ferment maltotriose will likely take longer to reach the attenuation limit. Maltotriose non-fermenting yeasts have probably caused many reported stuck fermentations, which is why lager brewers regularly introduce a new yeast culture after a comparatively small number of fermentations (50).

The most frequently identified mutant found in brewing is respiratory-deficient yeast (55). Such yeasts tend to display slower fermentation rates and higher dead-cell counts. These mutants produce elevated levels of diacetyl, esters, and higher alcohols because of the absence of oxygen (17). At levels as low as 1% of the pitching yeast, these mutants can produce diacetyl at levels well above the threshold (18).

Degeneration of Yeast

Yeast degeneration refers to the gradual deterioration in performance of the brewing yeast. Yeast degeneration has a harmful effect on the course of brewing fermentations. It is characterized by some of the following symptoms: sluggish fermentations, premature cessation of fermentation (resulting in high residual fermentable levels in beer), gradual lengthening of fermentation times, and poor foam or yeast head formation (50). Some brewers have noticed that the flavor of beer becomes increasingly "dry" as a result of yeast degeneration.

It is believed that insufficient oxygen in the wort is the chief cause of yeast degeneration. It can also occur because of a shortage of calcium, zinc, and phosphate ions in wort and exposure of the yeast to excessive amounts of metal ions, notably iron and copper. Other factors leading to yeast degeneration include excessive quantities of fine trub in the wort, higher than normal fermentation temperatures, bacterial infection, long and intensive washing of yeast and excessively long or warm yeast storage.

Lager brewers are usually more concerned with yeast degeneration than ale brewers. Bottom yeasts, with their lower multiplication rate (because of the low temperatures used), are usually more susceptible to contamination, degeneration, and autolysis than are ale yeasts.

Autolysis is an irreversible degradation (self-digestion) of the cell material by the yeast's own internal enzymes. The enzymes that once acted to support life, start to hydrolyze the cell contents, releasing numerous low-molecular-weight substances such as peptides, amino acids, vitamins, and nucleic acid compounds.

> **Note | Obtaining Yeast Cultures**
>
> Pure yeast cultures are obtained from a number of sources. Most often, the yeast is already in use in the brewery, but it can also be obtained from other breweries, commercial distributors, or culture collections.

The effects of autolysis are numerous, but some of the most common are a substantial worsening of taste, with a tendency toward a yeasty, creosote-like taste; an increase in pH as a result of the excretion of amino acids; a reduction in biological and colloidal stability; and a reduction in flavor stability. Also, bitterness becomes broader and more persistent, and diacetyl concentration often increases. Changes in beer color (autolyzed yeasts darken and take on a red or orange tinge) are likely to occur, as well as an increase in beer infections.

Pure Culture Maintenance

Once yeast has been selected, accepted, and fully proven for use in brewing, it is essential that a pure culture (working master culture) is maintained in the laboratory yeast bank for prolonged periods. Some small breweries do not maintain a master culture but rather purchase fresh slants for each in-house propagation cycle or hold stock cultures at independent third party institutions. Some of the more common methods brewers can use for maintaining the purity and characteristics of their yeast strains are sub-culturing, desiccation, lyophilization, and freezing in liquid nitrogen.

Sub-culturing

Sub-culturing on Agar Slants

A popular method of sub-culturing involves maintaining the cultures on a medium suitable for yeast growth. Yeast cultures are best kept on agar slopes in 28 ml screw-capped McCartney bottles. Aluminum caps with rubber liners are preferred since bottles fitted with plastic caps suffer from poor survival rates (13). The medium can be either brewery wort solidified with 2% agar or MYGP medium, which contains 3 g of malt extract, 3 g of yeast extract, 5 g of peptone, 10 g of glucose, and 20 g of agar, all in 1 liter of water (22). The slope is inoculated by means of a sterile platinum loop from a previous slant culture and is incubated between 20 and 30°C for 1 to 3 days, until visible growth is observed. At the end of the incubation, the screw caps are tightened and held at 2 to 4°C for up to 6 months (8). Cultures must not be exposed to temperatures below freezing.

Subcultures will normally survive for several years but can lead to unacceptable levels of strain degeneration. Regular sub-culturing at intervals between 3 and 6 months is necessary to preserve culture viability (26). More frequent sub-culturing could increase the possibility of population takeover by mutant strains and of introducing contaminating organisms into the master culture. Consequently, it is advisable to keep at least two cultures of each strain. This technique is inexpensive and versatile, and it produces consistently good results.

> **Note | Alternatives to Sub-culturing Yeast**
>
> Some small-scale brewers would rather purchase a liquid culture for fermentation rather than run the risk of contamination inherent in starting from a slant.

Sub-culturing on Agar Slants with an Oil Overlay

Alternatively, if the agar slants are overlayed with sterile mineral oil, the known shelf life increases to at least two years. After inoculation and incubation for 3 days at 25°C, the culture is overlaid with a layer of sterile B.P. mineral oil (29). The screw-capped bottle is then closed and held at 2 to 4°C. The oil overlay keeps the sample from drying out and serves as an oxygen barrier. According to Graham, use of this method will increase the maximum time between transfers to about six months (23).

Desiccation

This method uses purified silica gel as a desiccant, or squares of Whatman No. 4 filter paper (29). It is more suitable for strains used by collection curators or for research rather than for industrial strains. The shelf life can be up to 3 to 6 years.

Lyophilization

Lyophilization or freeze-drying is another popular technique among research laboratories and culture collections. Cultures are rapidly frozen followed by drying under vacuum. This method is superior since the sub-culturing and growth required by maintenance techniques increase the opportunity for contamination and selection of mutant strains. Lyophilization differs from desiccation in that water is removed from the frozen material, and is recommended where large numbers of strains must be maintained. Although this method is widely used it results in an overall reduction in viability, up to 95% of the original cells die (8).

Freezing in Liquid Nitrogen

With this method pure cultures are kept in vials and submerged in liquid nitrogen (−196°C), thereby maintaining viability and genetic integrity for tens of years. Unlike the other techniques for maintaining cultures, no reported changes in genetic migrations have been reported (8).

Yeast Propagation and Scale-up

The objective of propagation is to produce large quantities of yeast with known characteristics for the primary role of fermentation, in as short a time as possible. Most brewers use a simple batch system of propagation, starting with a few milliliters of stock culture and scaling up until there is enough yeast to pitch a commercial brew. Scale-up introduces actively growing cells to a fresh supply of nutrients in order to produce a crop of yeast in the optimum physiological state.

One principal disadvantage of yeast propagation is that the potential exists to stop the natural selection of yeast. Starting new propagations from identical laboratory stocks rather than harvesting yeast from fermentations can interfere with the yeasts' ability to mutate to prevailing conditions or production requirements. A possible solution is to encourage selective harvesting of the best yeast in the brewery and to isolate yeast cultures from time to time.

Propagation of yeast is divided into two separate steps: the laboratory phase and the plant phase.

Laboratory Phase

The process initially begins in the laboratory when cultures are taken from the "working" master culture and grown in a progression of fermentations of increasing size until enough yeast is produced to transfer to the propagation plant. The number of transfer steps in the laboratory varies according to the final weight of yeast required for the propagation plant. Of course, the more transfers, the greater the risk of infection. Most yeast culturing is done in a laminar flow hood as shown in Figure 4.1.

There are many variations in the techniques used for propagating yeast. Typically, a culture from the master stock is used to inoculate a 50 ml Erlenmeyer flask containing 10 ml of sterile wort for 24 hours at 25°C. This yeast culture is then scaled up to approximately 100 ml in a 250 ml flask. After approximately 2 days' incubation at 25°C, the culture is used to inoculate a 2 liter flask containing 1 liter of sterile wort. Scale-up steps are kept small at the early stages to ensure good growth. The 2 liter flask is then incubated for 2 days at 25°C. The next scale-up is 5 liters in a 10 liter flask, incubated for 2 days at 25°C, and kept at 4°C, to form the stock culture for the brewery.

The last step involves transferring the micro-culture into a mini propagation unit termed the Carlsberg flask (see Figure 4.2)—named after the brewery that developed the basic process. This first step involves filling the flask with wort to approximately 80% of the total volume. The cold wort is then aerated through the membrane sample valve connected to the aeration lance. When the wort is well-aerated, yeast culture is introduced either aseptically through the membrane fitting or transferred directly into the flask. Actually it's not a flask in the conventional sense, but a cylindrical container capable of producing the necessary start-up batch of 20 liters, as well as other capacities.

Figure 4.1
Laminar Flow Hood
Courtesy of Streamline Laboratory Products

When the yeast has reached the necessary cell concentration, the next step involves transferring the yeast culture from the Carlsberg flask to a single vessel propagation plant.

Briggs *et al.* recommends starting out with a slope culture and scaling up, usually at round 10:1, to 20 liters using a general purpose yeast medium such as yeast, extract, peptone glucose (YPEG) for the initial scale-ups, and a sterile wort for the terminal phase (8). Terminal yeast counts should be within the range of 150–200 x 10^6 cells per ml at a viability of greater than 98% (8).

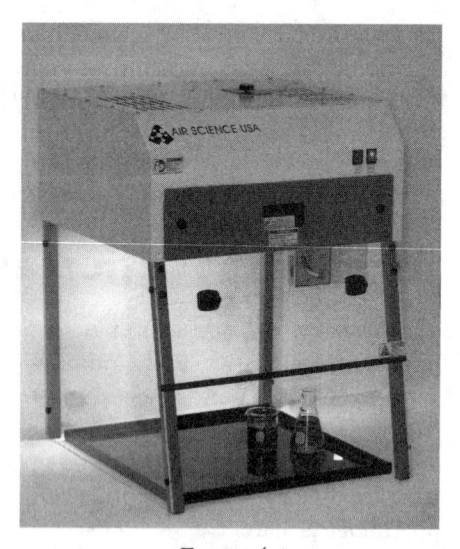

Figure 4.2
Carlsberg Flask
Courtesy of Alfa Laval Inc.

Plant Phase

After rigorously cleaning the yeast-propagation vessel (in the case of smaller breweries, a production fermenter) it is then filled with hot or cold wort and aerated with sterile air. Preferably the wort should be of the same quality as that used in fermentation. During propagation, temperature is maintained at a set level and the propagating yeast is intermittently aerated. When the yeast has reached the required cell concentration, it is pitched into an intermediate fermenter or directly to a production fermenter. As in the laboratory, yeast is grown in a progression of fermentations until there is enough to pitch a commercial-size brew.

Aeration

As mentioned, air or oxygen is passed continuously into the vessel through an efficient gas sterilizer to encourage yeast growth. Oxygen is preferable since it is sterile, whereas an air supply may contain impurities that must be removed before the air enters the vessel. The optimum rate of oxygenation for a system must be found by experiment, as the rate will affect the total crop produced.

Various methods of aeration have gained acceptance in practical operation. A few examples of practiced aeration techniques are indicated in the following to illustrate the wide variety of possibilities (65):

- Aeration every five minutes for one minute.
- Intermittent aeration for 1 minute every 15 minutes during the first 24-hour period and for 1 minute every 5 minutes during the second 24-hour period.
- Continuous aeration throughout the propagation process.
- Some suggest a circulation should be performed six times per hour. The amount of air can be variably adjusted up to 120 liters per minute for an oxygen content in the wort of 10–15 mg/l.

Optimum Propagation Temperatures

There is a wide variety of recommendations in this instance as well. Some brewers prefer to propagate their yeast at temperatures identical to those employed during fermentation in order to prevent temperature shock to the yeast (26). Others will start out at slightly higher temperatures with intermittent aeration to stimulate growth, dropping the temperature at each step until normal fermenting temperatures are reached at plant scale (40). The procedure for propagation of pure ale yeast culture is almost identical to that used for lager yeast, with the exception of the higher temperatures that are normal to ale fermentations.

Propagation Plants

The propagation plant usually consists of anywhere from one, two or more closed stainless steel vessels of increasing volume, which are usually situated in a separate room to minimize contamination of risk. Figure 4.3 shows two propagation lines with a three stage propagation system (three vessels) at the Grolsch Brewery in Holland. The air in the room should be filtered and

Figure 4.3
Yeast Propagation Plants
Courtesy of GEA Brewery Systems

the atmosphere maintained under a slight positive pressure. The vessels are equipped with attemperation jackets, control instruments, sight glasses, and a venting system that prevents contamination of the contents by external air.

One-tank System – Small breweries typically use single-vessel propagation plants. A single-vessel propagation plant utilizes one vessel to sterilize, aerate, and cool the wort, followed by inoculation with the laboratory culture in the same vessel. The one-tank system is set apart from the two-tank system above all in that after the yeast delivery, a certain amount of yeast remains in the system to serve as an inoculum for the following yeast batch.

The one-tank system is characterized primarily by its simple system design which means less equipment. Fast yeast growth is possible within a minimum amount of time. The constant conditions maintained by the system control results in a consistent pure culture yeast. However, since an additional wort sterilization is not possible, a possible biological risk is involved in this process. Considering that a certain amount of yeast always remains in the pure culture tank as inoculum, there is a possibility that the yeast could lose certain characteristic properties due to mutation.

Two-tank System – A two-vessel plant utilizes one vessel for wort sterilization and cooling and second vessel for the actual propagation. The two-tank process is a self-sufficient, self-contained system. The biological safety is ensured by possible wort sterilization. The automatic control guarantees flexible and specific adjustment of temperatures, aeration, and times. This means consistent beer quality. The advantages of the two-tank process previously mentioned, however, involve additional equipment in conjunction with increased cleaning and operating effort that also leads to increased acquisition costs.

Reportedly, the two-vessel plant is better for continuous propagation with one yeast strain while the single-vessel plant is better suited for specialty brewers using multiple strains of yeast (14).

Culture Contamination

It frequently happens that brewing yeasts carry a persistent low level of contaminants such as *Obesumbacterium proteus*, acetic acid bacteria, and slow-growing Torula-type yeasts. These organisms are generally regarded as harmless because their numbers never reach a point where they are likely to have adverse effects on the beer. On the other hand, *L. pastorianus*, *Z. anaerobia*, and *S. carlsbergensis* are strains considered harmful at low levels. Lager fermentations seem to be more susceptible to bacterial contamination than do ale fermentations primarily because the pH drops more slowly for lagers than for ales and bacteria are suppressed by falling in pH.

To deal with this problem, there is often no alternative but to replace the yeast from propagated stock or to periodically wash the yeast cultures to destroy bacterial contaminants. However, washing yeast will not destroy wild yeast; only reculturing can do so.

Microscopic Examination

Microscopic examination of the yeast culture can be useful in assessing the overall health of the population. Abnormal-looking or irregularly shaped yeast cells are signs of cell stress, possibly indicating potential problems with wort composition, aeration, poor yeast handling, or fermentation conditions. Microscopic examination is also useful in detecting extraneous particles such as diatomaceous earth, trub, grain particles, etc. that may interfere with proper yeast performance. However, microscopic examination is not particularly sensitive in detecting common brewery contaminants, such as *Pediococcus* spp., and *Lactobacillus* spp. and wild yeasts such as *Hansenula*, *Dekkera*, *Brettanomyces*, *Candida*, and *Pichia* (see Chapter 19). In this case microscopic examination must be supplemented by other tests, such as contrast-enhancing optics, staining methods, or plating on differential media. However, many small brewers have no way of assessing the bacterial load of brewery yeasts, and they may not detect the symptoms of infection until the beer has been in the bottle for some time.

Wild yeasts are usually very difficult to detect since they are morphologically similar to brewing yeasts. Wild yeasts resemble lactic bacteria in that they require the same conditions as brewer's yeast, and they are therefore hard to eradicate. Elimination of wild yeasts, or at least holding their numbers down, is principally a matter of using pitching yeast free of wild yeasts and keeping equipment sterile. Some brewers report that the critical level of wild yeasts is around 4 to 10 yeast cells per bottle of beer.

Yeast Washing

Pitching yeasts collected from brewery fermentations are never absolutely free of microbiological infection. In spite of whatever care and sanitary precautions are taken, some bacteria and wild yeast will contaminate the pitching yeast. The pitching yeast can contain healthy yeast cells and trub (dead yeast cells and organic residues) and may contain 5 to 15% dry solids (35). To minimize microbiological infection, brewery yeast can be washed using the following procedures:

1. Distilled or sterile wash
2. Acid wash
3. Acid wash with ammonium persulfate
4. Chlorine dioxide

Distilled or Sterile Water Wash

In the first method, the yeast slurry and cold, distilled or sterile water are mixed thoroughly in a decantation tank. The yeast is allowed to settle and the supernatant water is decanted, taking with it dead cells, trub, grain, and hop particles. The flocculent nature of the yeast makes yeast flocs denser than contaminants. The process may be repeated several times. This will physically remove a high percentage of contaminating microorganisms as well as old, degenerated, and dead yeast cells. An alternative to decantation is centrifugation, but this may increase chances of contamination. The washing should be also done with the least amount of water possible. This old method is still widely used with very good results.

Acid Wash

The second method is to wash the yeast with acids, e.g., tartaric, citric, sodium metabisulfite, sulfuric, or phosphoric acid which is typically the most commonly used acid. Acid washing lowers the pH of the yeast slurry to the point at which bacteria and weak yeast cells are killed off, but it does not harm the healthy yeast cells. Acid washing is very effective at killing "wort spoiler" bacteria (*Hafnia protea* and others), and it also kills *Flavobacterium proteus* and *Lactobacillus*. Pediococci, on the other hand, may survive the treatment as well as wild yeasts.

In this method, the yeast slurry is acidified with diluted acid to a pH of 2.0 to 2.3 with intensive mixing throughout the acid addition period. Good agitation is vital to the yeast slurry; otherwise, pH gradients will exist inside the yeast tank. Yeast exposed to pH values lower than planned will be damaged, as yeast exposed to higher pH values will minimize the effect of yeast washing. To avoid these problems acid washing tanks can be fitted with pH probes to monitor the pH levels of the yeast slurry. The yeast slurry is usually allowed to stand for a minimum of 2 hours, and during treatment the temperature should be maintained at 0 to 4°C (10). After the yeast has completely settled, decant off the liquid from the yeast, rinse several times, and cover with sterile wort. In many breweries, yeast is routinely treated in this way prior to each pitching.

Tartaric acid may tend to make yeast more flocculent, while phosphoric acid—though very popular with craftbrewers—may tend to make it more powdery, with corresponding effects on degree of attenuation. Tartaric acid is the safest of the acids to use, but all of these acids are equally effective.

Reportedly there are several problems with acid washing. For one thing, though it reduces the populations of most wort-spoiling bacteria, it is less successful with beer-spoilers such as lactic acid bacteria and is generally not effective on wild yeast and molds (28). Yeast containing these organisms should be discarded. Further, acid washing is controversial concerning its several alleged effects. These include reduced cell viability and vitality; changes in flocculation; reduced rates of fermentation, fining, and yeast crop size; and excretion of cell components. In spite of these claims, Simpson has reported that acid washing is only a problem if carried out incorrectly (62). Of course, the real solution to the problem requiring yeast washing is to trace the contamination to its source and eradicate it by a thorough cleaning and sanitation program and using good yeast-handling practices.

Acid Wash with Ammonium Persulfate

Some brewers use an acid-persulfate combination rather than just acid claiming that it is a more effective treatment than treatment with acid alone. Briggs *et al.* recommend the addition of a strong oxidizing agent as ammonium persulphate (0.75% w/v) with phosphoric acid (8). It is added to the diluted yeast slurry (2 parts water: 1 part yeast), then the slurry is acidified with phosphoric acid to pH 2.2 with a contact time of 2 hours (50). Alternatively, it is acidified to pH 2.8 and allowed to rest between 12 and 24 hours before pitching (50). However, the latter modification reportedly may result in a sluggish fermentation. This treatment is claimed to give at least 99% kill without impairing the viability of yeasts.

Despite the claims made, it does not seem absolutely certain that specific properties of the yeast (e.g., flocculation, fining behavior, etc.) are not affected by the treatment. Nor can it be assumed that the yeast is totally free of bacteria or that the original equilibrium will not eventually be restored.

Chlorine Dioxide

Chlorine dioxide, an alternative to distilled water or acid washing, is relatively new to the brewing industry and is gaining acceptance as a method for washing yeast. It kills via microbes by reacting chemically with sulfur-containing amino acids, the building blocks of protein which are used to form cell membranes. When these proteins are destroyed, the organism's cell membrane ruptures, killing the organism. Brewers who use chlorine dioxide to wash yeast begin by pouring it into a small amount of acidified water to activate the solution (28). The activated solution is then immediately added to the yeast slurry and mixed well. After 5 to 10 minutes of washing, depending on the contaminant levels, the yeast can be pitched or refrigerated until needed.

Yeast Viability and Replacement

Viability

Viability is a measure of yeast's ability to ferment—a property not possessed by dead cells. Yeast viability is determined by selective staining, by the standard-slide culture method, or by more advanced methods such as the Slide Viability Method, and fermentation tests.

Selective Staining

A more objective result of yeast viability is obtained by selective staining using buffered methylene blue or methylene violet. These stains show dead cells as blue or pink, respectively, on microscopic examination. Various other dyes have been recommended from time to time, such as methyl green, acridine orange, neutral red, and erythrosine.

The use of methylene blue as a viability stain in determining the health of a yeast strain before pitching is the method of choice of the American Society of Brewing Chemists. A yeast culture is applied to a microscope slide covered by a thin layer of nutrient media. After a sterile slipcover is place on top of the yeast and media, the slide is incubated for no longer than 18 hours at room temperature. Then the slide is examined at a magnification of 200x. Yeast cells giving rise to micro-colonies are considered viable, and dead cells can readily be distinguished, as they are stained blue. The percentage of viable cells is called the budding index. Typically, a healthy culture should contain at least 95% viable cells, and many brewers believe it would be inadvisable to use cultures with viabilities less than 85% (36).

Although still widely used, this method can give rise to inaccurate viability readings, particularly with weak or stressed cells. Other concerns in using this method are the possibility of causing cell damage by the dye itself and the fact that this method relies on the skill of the technician performing the test.

Standard-Slide Culture Method

A more accurate method in determining the viability of yeast is the standard slide-culture method that consists of three steps: perform a hemacytometer count on a suspension of cells, plate a measured quantity on a wort gelatin medium, and then incubate and count the resultant colonies (22). However, this method has its problems too due to cell clumping and the death of cells during preparation.

> **Note | Hemacytometer**
>
> A hemacytometer is a graduated counting chamber that can be viewed under a microscope to determine the concentration of yeast cells in suspension. It consists of a heavy glass slide with two counting chambers, each of which is divided into nine large 1 mm squares, on an etched and silvered surface separated by a trough. The engraved grid on the surface of the counting chamber ensures that the number of yeast in a defined volume of liquid is counted. The hemacytometer is placed on the microscope stage and the cell suspension is counted.

Slide Viability Method

In the Slide Viability Method, cells are suspended in a growth medium containing 6% gelatin, and the suspension is placed on a hemacytometer slide (22). The edges of the cover slip are sealed with a melted petroleum jelly to prevent desiccation of the cells. The slide is incubated for 20 hours at 18°C under conditions of high humidity and then re-examined for dividing cells. The percentage of dividing cells compared with the number of cells seen before incubation is taken as the percentage viability. Flocculence is tested by re-suspending a known amount of yeast cells in a measuring cylinder of beer and measuring the volume of sediment formed in a set time. The volume of sediment can be used as an indicator when the culture should be replaced.

Fermentation Tests

Although the above methods can assess whether a cell is alive or viable, they cannot readily assess the vigor of the yeast and whether they can produce a normal fermentation. In this case the brewer can perform a pilot fermentation tests using the rate and pH drop to assess the overall vitality of the culture (14).

Replacement

Most brewers discard yeast after a successive number of fermentations because it may be intermixed with other yeasts, contaminated with wild yeast and bacteria, or mutated to less desirable strains. The decision when to replace a yeast culture is also determined by the strain (ale yeast is typically more robust), fermentation performance, yeast-handling procedures, flocculation performance, and flavor and aroma characteristics. Some brewers use their yeast in production for less than three generations, while others only discard their yeast after 5 to 10 successive fermentations. However, there are exceptions to the rule, with some brewers routinely discarding yeast after 30 brewery fermentations (50). Those brewers who employ high pitching rates will probably have to replace their yeast more often because continuous use and high pitching rates tend to increase the average age of the yeast, thus reducing their vigor (25).

Some brewers are reluctant to propagate frequently because of the belief the newly propagated yeast does not perform in a satisfactory manner. According to Briggs *et al.*, the suggestion that newly propagated yeast performs poorly is unproven and more likely reflects poor propagation practices (8).

The decision to introduce a new culture should be based on microbiological and performance testing of the existing yeast. In order to keep track of the age and condition of the pitching yeast, it is important to keep records on each batch. A commonly adopted method is to give each new culture an identifying code and to give each subsequent yeast crop of that culture a sequential number. The tracking sheet should, at a minimum, include the batch number, the date the yeast was pitched, the quantity of yeast pitched, and the viability of the yeast. The brewer should also include visual observations (e.g., whether powdery, degree of liquidity, etc.), taste observations (off-flavors, etc.), fermenter from which the yeast was harvested, and whether and how the yeast was washed (e.g., with sterile water or acid).

Yeast Storage

In most breweries, yeast is stored during the period between cropping and re-pitching. Pitching yeast may be stored within the brewery as slurry in a yeast collection vessel, or as slurry stored under a layer of water or beer, or as pressed cake.

Yeast Collection Vessels

Most modern breweries store their yeast in sophisticated collection vessels under filtered sterile air or inert sterile gas pressure with external cooling and equipped with low shear stirring devices. A positive pressure is maintained to prevent the entry of contaminants. The temperature is kept at 2 to 4°C (to slow metabolic activity), and the slurry is agitated to ensure sufficient heat transfer (22). According to some brewers, a gentle agitation in the holding tank is necessary so that metabolic activity does not accelerate cell death and autolysis, while others recommend stirring the yeast only within 2 hours before use (25). Some brewers advocate oxygenation at this stage while others believe oxygen should be excluded to avoid yeast autolysis (6).

Slurried Storage Systems

Slurried storage systems are usually self-contained, thereby reducing the risks of contamination. Slurry yeast has the advantage over pressed yeast in that it gives more vigorous fermentations, requires lower pitching rates, and can be stored for longer periods (usually fewer than 4 weeks) without affecting viability. However, as a rule of thumb, it is not advisable to store slurried yeast in the holding tank for longer than 5 days, and preferably for 24 hours or less (31). Slurry yeast has the disadvantage that it contains more fermented beer, and slurry yeast handling systems are expensive to install. They are probably only justified in larger brewing operations.

Pressed Cake

Alternatively, yeast can be stored as pressed cake. The yeast recovered is stored at 0°C before re-suspension in wort for pitching (42). The principal disadvantages with pressed cake are that proper temperature control is difficult to achieve, there is an increased likelihood of contamination, and there is a limited storage time (usually fewer than 10 days) at which optimum viabilities are maintained (38).

References

1. Ahlquist, Eric. 1986. "Process and Quality Management of Yeast." *Brewers' Guardian* 115 (8).
2. Allen, Fal. 1994. "The Microbrewery Laboratory Manual – A Practical Guide to Laboratory Techniques and Quality Control Procedures for Small-Scale Brewers, Part 1: Yeast Management." *Brewing Techniques* 2 (4).
3. American Society of Brewing Chemists (ASBC). 1992. Yeast-6 Yeast Viability by Slide Culture. *Methods of Analyses*, 8th ed. The Society, St. Paul, Mn.
4. Awford, Bruce B. 1966. "Practical Aspects of Fermentation Control in a British Brewery." *MBAA Technical Quarterly* 14 (3).
5. Bamforth, C.W. and A. H. P. Barclay. 1993. "Malting Technology and the Use of Malt." *Barley: Chemistry and Technology*, edited by A. W. MacGregor and R. S. Bhatty. St. Paul, Minnesota: American Association of Cereal Chemists.

6. Bamforth, Charles W. 2006. *Scientific Principles of Malting and Brewing.* St. Paul, Minnesota: American Society of Brewing Chemists.

7. Bindesboll Nielson, N. and F. Schmidt. 1985. "The Fate of Carbohydrates During Fermentation of Low Calorie Beer." *Carlsberg Res. Commun.* 50(6).

8. Briggs, Dennis E., Boulton, Chris A., Brookes, Peter A., and Stevens, Roger. 2004. *Brewing: Science and Practice.* Cambridge, England: Woodhead Publishing Limited.

9. Boughton, Richard. 1987. "Practically Managing Yeast." *MBAA Technical Quarterly* 24 (4).

10. Boulton, Chris. 1991. "Yeast Management and the Control of Brewery Fermentations." *Brewers' Guardian* 120 (3).

11. Busch, Jim. 1995. "Some Finer Points in Preparing Wort and Yeast for Fermentation." *Brewing Techniques* 3 (5).

12. Campbell, I. 1987. "Microbiology of Brewing: Beer and Lager." *Essays in Agricultural and Food Microbiology*, edited by J.R. Norris and G.L. Pettipher. Chichester [West Sussex], England: John Wiley & Sons Ltd.

13. Campbell, I. 1988. "Culture, Storage, Isolation and Identification of Yeasts." *Yeast – A Practical Approach*, edited by I. Campbell and J. H. Duffus. Oxford, United Kingdom: IRL Press Limited.

14. Carey, Daniel and Grossman, Ken. 2006. "Fermentation and Cellar Operations." *Fermentation, Cellaring, and Packaging Operations.* Volume 2, edited by Karl Ockert. St. Paul, Minnesota: Master Brewers Association of the Americas, 2006.

15. Dowhanick, Terrance M. 1999. "Yeast – Strains and Handling Techniques." *The Practical Brewer*, edited by John T. McCabe. Wauwatosa Wisconsin: Master Brewers Association of the Americas.

16. Emeis, C.C. 1966. "Performance and Stability of Culture Yeast." *MBAA Technical Quarterly* 3 (265).

17. Ernandes, J.R., J. W. Williams, I. Russel, and G. G. Stewart. 1993. "Respiratory Deficiency in Brewing Yeast Strains – Effects on Fermentation, Flocculation, and Beer Flavor Components." *J. Am. Soc. Brew. Chem.* 51.

18. Fix, George J. 1989. *Principles of Brewing Science.* Boulder, Colorado: Brewers Publications.

19. Fix, George J. 1993. "Diacetyl: Formation, Reduction, and Control." *Brewing Techniques* 1 (2).

20. Fix, George and Laurie Fix. 1997. *An Analysis of Brewing Techniques.* Boulder, Colorado: Brewers Publications.

21. Gilliland, R.B. 1971. "Classification and Selection of Yeast Strains." *Modern Brewing Technology*, edited by W P.K. Findlay. Cleveland, Ohio: The Macmillan Press.

22. Gilliland, R. B. 1981. "Brewing Yeast." *Brewing Science*. Volume 2, edited by J.R.A. Pollock. London, England: Academic Press.

23. Graham, Michael D. 1997. "A Simple, Practical Method for Long-Term Storage of Yeast." *Brewing Techniques* 5 (2).

24. Hammond, John, MA. 1986. "The Contribution of Yeast to Beer Flavor." *Brewers' Guardian* 115 (9).

25. Henson, M.G. and D. M. Reid. 1987. "Practical Management of Lager Yeast." *The Brewer* 74 (879).

26. Hough, J.S., et al. 1982. *Malting and Brewing Science*. Volume 2. London, England: Chapman-Hall, Ltd.

27. Jolda, Deb, ed. 1996. "The Yeast Directory – The Complete Guide to Commercially Available Yeast Strains." *The 1996 Brewer's Market Guide*. Eugene, Oregon: New Wine Press.

28. Johnson, Dana and Katie Kunz. 1998. "Coming Clean – A New Method of Washing Yeast Using Chlorine Dioxide." *The New Brewer* 15 (4).

29. Kirsop, B.E. "Maintenance of Yeasts." 1991. *Maintenance of Microorganisms and Cultured Cells*, edited by B.E. Kirsop and A. Doyle. San Diego, California: Academic Press Limited.

30. Knudsen, Finn B. "Fermentation Principles and Practice." 1977. *The Practical Brewer*, edited by H. M. Broderick. Madison, Wisconsin: Master Brewers Association of the Americas.

31. Knudsen, Finn B. 1985. "Fermentation Symposium. Part II. Fermentation Variables and Their Control." *MBAA Technical Quarterly* 22 (4).

32. Kruger, Lyn. 1996. "The Inhibitory Effects of Carbon Dioxide on Yeast Metabolism and Fermentation." *The New Brewer* 13 (6).

33. Kunze, Wolfgang. 1996. *Technology Brewing and Malting*, translated by Dr. Trevor Wainwright. Berlin, Germany: VLB Berlin.

34. Leistad, Roger. 1983. *Yeast Culturing for the Home Brewer*. Ann Arbor, Michigan: G. W. Kent, Inc.

35. Lewis, Michael J. and Tom W. Young. 1995. *Brewing*. London, England: Chapman & Hall.

36. Lewis, M. J. and Young, T. W. 2002. *Brewing*. New York, New York: Kluwer Academic/Plenum Publishers.

37. Logsdon, David. 1999. Personal Communication. Wyeast Laboratories, Inc.

38. MacDonald, J., et al. 1984. "Current Approaches to Brewery Fermentations." *Progress in Industrial Microbiology*, edited by M. E. Bushell. Guildford, United Kingdom: Elsevier.

39. Mallet, John. 1992. "Pitching Control." *The New Brewer* 9 (4).

40. Maule, D.R. 1980. "Propagation and Handling of Pitching Yeast." *Brewers Digest* 55 (2).

41. Middlekauff, James E. 1995. "Microbiological Aspects." *Handbook of Brewing*, edited by William A. Hardwick. New York, New York: Marcel Dekker, Inc.

42. Moll, Manfred, ed. 1995. *Beers and Coolers*. New York, New York: Marcel Dekker, Inc.

43. Molzahn, Stuart. 1977. "The Propagation of Brewing Yeast." *Brewers' Guardian* 106 (7).

44. Morris, Rodney L. 1994. "Simple Detection of Wild Yeast and Yeast Stability." *Brewing Techniques* 2 (3).

45. Munroe, James H. 1995. "Fermentation." *Handbook of Brewing*, edited by William A. Hardwick. New York, New York: Marcel Dekker, Inc.

46. Munroe, James H. 1995. "Aging and Finishing." *Handbook of Brewing*, edited by William A. Hardwick. New York, New York: Marcel Dekker, Inc.

47. Noble, Stuart. 1997. "Practical Aspects of Fermentation Management." *The Brewer* 83 (991).

48. Palmer, F. 1969. "The Determination of Pitching Yeast Concentration," *MBAA Technical Quarterly* 6.

49. Palmer, John. 1997. "Conditioning – Fermentation's Grand Finale." *Brewing Techniques* 5 (3).

50. Piesley, John G. and Tomas Lom. 1977. "Yeast – Strains and Handling Techniques." *The Practical Brewer*, edited by H. M. Broderick. Madison, Wisconsin: Master Brewers Association of the Americas.

51. Rajotte, Pierre. 1994. *First Steps in Yeast Culture: Part One*. Montreal, Quebec: Alliage Éditeur.

52. Reed, G. and Peppler, H. 1973. *Yeast Technology*. Westport, Connecticut: AVI Publishing Company.

53. Riess, S. 1986. "Automatic Control of the Addition of Pitching Yeast." *MBAA Technical Quarterly* 23 (1).

54. Robson, F. O. 1971. "Techniques of Yeast Separation." *Modern Brewing Technology*, edited by W.P.K. Findlay. Cleveland, Ohio: The Macmillan Press.

55. Russell, Inge. 1995. "Yeast." *Handbook of Brewing*, edited by William A. Hardwick. New York, New York: Marcel Dekker, Inc.

56. Russell, Inge. 2006. "Yeast." *Handbook of Brewing*, edited by Fergus G. Priest and Graham G. Stewart. Boca Raton, Florida: CRC Press, Taylor & Francis Group.

57. Ryder, David S. and Joseph Power. 1995. "Miscellaneous Ingredients in Aid of the Process." *Handbook of Brewing*, edited by William A. Hardwick. New York, New York: Marcel Dekker, Inc.

58. Ryder, David S. and Joseph Power. 2006. "Miscellaneous Ingredients in Aid of the Process." *Handbook of Brewing*, edited by Fergus G. Priest and Graham G. Stewart. Boca Raton, Florida: CRC Press, Taylor & Francis Group.

59. Scheer, Fred M. 1991. "Ask the Brewmaster: Chlorine Removal, Polyphenols." *The New Brewer* 8 (1).

60. Scheer, Fred M. 1996. "The Danger of Using Two Different Yeast Cultures in Fermentation." *The New Brewer* 13 (6).

61. Simpson, W.J. 1987. "Kinetic Studies of the Decontamination of Yeast Slurries with Phosphoric Acid and Acidified Ammonium Persulfate and a Method for the Detection of Surviving Bacteria Involving Solid Medium Repair and the Presence of Catalase." *J. Inst. of Brewing* 93.

62. Simpson, W.J. and Hammond, J. R. M. 1989. "The Response of Brewing Yeast to Acid Washing." *J. Inst. of Brewing* 95.

63. Stewart, G.G. and I. Russell. 1985. "Modern Brewing Biotechnology." *Food and Beverage Products*, edited by Murray Moo-Young. Oxford, England: Pergamon.

64. Thompson, Donald, Karl Ockert, and Vince Cottone. 1999. "Craft Brewing." *The Practical Brewer*, edited by John T. McCabe. Wauwatosa Wisconsin: Master Brewers Association of the Americas.

65. Westner, Hans. 2000. "One-Tank or Two-Tank Process?" *Brewing and Beverage Industry International.* (12) 2.

Chapter Five

Brewing Water

High consumption of good-quality water is characteristic of beer brewing. More than 90% of beer is water and an efficient brewery will typically use between 4 and 6 liters of water to produce one liter of beer. Some breweries use much more water, especially small breweries. In addition to water used for beer production—mashing, boiling, sparging, filtration, and packaging—breweries also use water for heating and cooling as well as cleaning and sanitation of equipment and process areas. Each uses requires a somewhat different water quality too.

Evaluation of Brewing Water

The mineral content of brewing water has long been recognized as making an important contribution to the flavor of beer. This is especially important since water composes more than 90% of the beer. Brewers interested in brewing a particular beer style first need to evaluate whether or not the their water is suitable by comparing it to the water analyses of a flagship brewery or to the water used to produce the beer style in the regions of its origin. For example, the water of Dublin for stouts, the water of Burton-on-Trent for dry, hoppy pale ales and so on.

Historically, different regions have become famous for their classic beer styles as defined by the waters available for brewing. For example, the famous brewing waters from the deep wells at Burton-on-Trent are known for their excellent qualities in brewing full-flavored pale ales. Burton water is high in permanent hardness because of the high calcium and sulfate content, but it also has a lot of temporary hardness from a high level of bicarbonate. Munich water is poor in sulfates and chloride but contains carbonates, which are not very desirable for pale beers but ideal for producing darker, mellower lagers. The carbonates raise the mash pH, producing wort with a higher dextrin to maltose ratio. The water from Vienna is more mineralized than that of

Munich. Pilsen, renowned for its pale lagers, has very soft water, with very little mineral content and produces beers famous for their pale color and hop flavor. The water of Dortmund contains appreciable amounts of both carbonate and chloride that aid in the production of full-flavored lagers and pale ales. Higher concentrations of chlorides are suitable for some mild ales and stouts, as are certain alkaline waters containing magnesium sulfate and sodium bicarbonate. The mineral (ion) content of classic brewing waters can be found in Table 5.1.

However, since most brewery operations today are dependent on municipal water supplies and that they produce a wider range of beer styles some mineral salt adjustment is required. See Section on Water Treatment for mineral salt adjustment.

Table 5.1

Mineral Content of Classic Brewing Waters (mg/l or ppm)

Mineral (ion)	Pilsen	Munich	Dublin	Dortmund	Burton-on-Trent	Vienna
Calcium	7	75	115–200	250–260	265–350	160–200
Sulfates	5–7	10	55	120–280	450–820	125–215
Magnesium	2–8	18–20	4	20–40	25-60	60–65
Sodium	2–30	2–10	12	60–70	25–55	8
Chloride	5	2	20	60–105	15–35	10–40
Biocarbonate	10–35	150–200	125–320	180–550	140–320	120–240
Hardness	30	250–265	300	750	850–875	750
TDS	35	275	350	1,000	1,100	850

pH, Alkalinity, and Water Hardness

pH

When water molecules are ionized, they produce hydrogen (H⁺) and hydroxyl (OH⁻) ions, which carry an electrical charge. These ions in the water determine its fundamental character—whether it is acid (excess H⁺) or alkaline (excess OH⁻). The term "pH" refers to the hydrogen cation (H⁺) concentration in water and is defined as the logarithm of the reciprocal of the hydrogen-ion concentration. The scale ranges from 0 to 14 with a pH of 7 being neutral. Higher pH values denote alkalinity, lower pH values acidity. Because it is a logarithmic scale, a solution at pH 5 is 10 times more acidic than a solution at pH 6, 100 times more acidic than a solution at pH 7, and 1,000 times more acidic than a solution at pH 8.

Alkalinity

Alkalinity is a measure of the buffering capacity of the bicarbonate ions and, to some extent, the carbonate and hydroxide ions of water. These three ions all react with hydrogen ions to reduce acidity and raise pH. Alkalinity is normally given in mg/l as calcium carbonate ($CaCO_3$) for all three ions. As a rule, water with more than 100 mg/l of calcium carbonate is considered alkaline and should be treated. Water with less than 100 mg/l calcium carbonate is considered "soft" or mildly alkaline.

The pH increasing effect of carbonate hardness opposes the pH lowering effect of the other calcium and magnesium ions that exist as calcium chloride, calcium sulfate, magnesium chloride, and magnesium sulfate. The bicarbonate ion is a very strong buffer. Thus, water with a high buffering capacity or alkalinity tends to have a pH that is very stable, even when bases or acids are added to it.

Residual Alkalinity

The result of the competition between the pH increasing and lowering properties of water is determined by the residual alkalinity (RA). The residual alkalinity is the difference between carbonate (carbonate + bicarbonate) hardness and non-carbonate hardness, which is shown by the following relationship:

$$\text{Residual alkalinity} = (\text{bicarbonate}/3.5) - (\text{magnesium}/7.0)$$

A positive residual alkalinity indicates the water will increase the pH of the mash, a residual alkalinity of zero indicates that the water will not change the mash pH, and a negative value indicates that the mash pH will be decreased when the water is used (21). Residual alkalinity is usually expressed as milliequivalents per liter (meq/l) of sodium carbonate, or on some analysis reports as calcium carbonate.

Water Hardness

Total water hardness is the measure of the bicarbonate, calcium, and magnesium ions present in the water. Total hardness is expressed as mg/l of calcium carbonate ($CaCO_3$), which determines the degree of softness or hardness. Generally, a measurement of fewer than 50 mg/l is considered very soft water, 50 to 100 mg/l is considered soft water, 100 to 200 mg/l is considered medium-soft water, 200 to 400 mg/l is considered moderately hard water, 400 to 600 mg/l is considered hard water, and greater than 600 mg/l is considered very hard water. Where calcium and magnesium measurements are not given in an analysis, dividing the hardness reading by 1.5 and by

3.0 roughly indicates, respectively, the calcium and magnesium in solution (assuming a 2/1 Ca/Mg ratio).

The hardness is made up of two parts: temporary (carbonate) and permanent (non-carbonate) hardness. Temporary hardness is always strongly alkaline, and permanent hardness is usually only slightly acidic.

Temporary Hardness

In the United States, temporary hardness results from calcium, magnesium, and sodium salts of carbonates and bicarbonates in the water. Most of the hardness is attributed to calcium bicarbonate. The hardness that bicarbonate ions contribute is temporary because they are easily precipitated and removed when water is boiled or treated with certain acids.

A measure of bicarbonates (temporary hardness) greater than 150 mg/l is considered undesirable because it will inhibit the proper acid pH balance necessary during mashing, resulting in inadequate conversion of starch to sugars. Some brewers believe that temporary hardness should be fewer than 100 mg/l for light beers. Brewers will often add dark roasted malts to maintain the correct acidity given a high concentration of bicarbonates. However, the strong flavoring properties of malts prohibit the use of large quantities of dark roasted malts for some beers.

Permanent Hardness

Permanent hardness is that portion of total hardness remaining after the water has been boiled. Permanent hardness results from calcium and magnesium salts of sulfates and chlorides remaining in the water. Generally, permanent hardness will increase the acidity (lower the pH) of water. A certain amount of permanent hardness is desirable in the brewing of all-grain beers, as it favors enzyme conversion (starch to sugars). Most beer styles require some form of hardness in the water.

Minerals in Brewing Water

The principal ions most likely to be found in brewing water are those of calcium, magnesium, sodium, potassium, sulfate, phosphate, chloride, carbonate, and nitrate/nitrite. The minor ions of iron, copper, zinc, and magnesium may be found in trace amounts.

Principal Ions

Calcium (Ca^{2+})

The calcium ion is by far the most influential mineral in the brewing process. Calcium reacts with phosphates, forming precipitates that involve the release of hydrogen ions, in turn lowering the pH of the mash. This lowering of the pH is critical in that it provides an environment for alpha-amylase, beta-amylase, and proteolytic enzymes. This in turn will enhance proteolysis and saccharification, leading to an increase in extract yield and a rise in soluble nitrogen (5). The lowering of the pH may sufficiently reduce wort viscosity for faster drainage. Also, polyphenols (tannins) will be extracted to a lesser degree at a lower pH, resulting in less astringent beers with less color. During boiling, calcium lowers the pH, which in turn reduces the rate of extraction and isomerization of the alpha acids. The wort will therefore be less bitter (5). The presence of calcium ions in the boil is essential to good break formation, as it aids in the precipitation of proteins and other materials, which might otherwise cause beer haze formation (23). Calcium enhances yeast flocculation and sedimentation during fermentation. During aging, calcium improves clarification, stability, and flavor of the finished beer.

Too much calcium, however, may cause too much phosphate to precipitate, thus robbing the wort of a vital yeast nutrient. High calcium concentrations also tend to diminish the extraction of hop resins and delay their isomerization (14). Finally, high calcium concentrations can result in a beer with a harsh, thin flavor.

Magnesium (Mg^{2+})

Magnesium ions react similarly to calcium ions, but since magnesium salts are much more soluble, the effect on wort pH is not as great. Magnesium is most important for its benefit to yeast metabolism during fermentation. Magnesium carbonate reportedly gives a more astringent bitterness than does calcium carbonate (23). Normally, malt contains sufficient magnesium to supply the required amount; but if a high adjunct ratio is employed, addition of small amounts of magnesium salts to the brewing water may be advisable (1). Some Flanders browns are brewed with water relatively high in magnesium.

Sodium (Na^+)

Sodium has no chemical effect; it contributes to the perceived flavor of beer by enhancing its sweetness. Levels from 75 to 150 mg/l give a round smoothness and accentuate sweetness, which is most pleasant when paired with chloride ions than when associated with sulfate ions. In the presence of sulfate, sodium creates an unpleasant harshness, so the rule of thumb is that the more sulfate in the water, the less sodium there should be (and vice versa). In any case, 150 mg/l of sodium sulfate is considered the upper limit for brewing water. Some brewers prefer to add potassium chloride in place of sodium chloride, as potassium is free from the slightly sour flavor of the sodium ion (5). As a bicarbonate salt, sodium raises the alkalinity and the wort pH.

Some waters with appreciable amounts of sodium salts must be restricted to the production of the sweeter dark beers, since it is virtually impossible to reduce this alkalinity without adversely affecting other desirable characteristics of the water.

Potassium (K⁺)

Like sodium, potassium can create a "salty" flavor effect. It is required for yeast growth and inhibits certain mash enzymes at concentrations above 10 mg/l (21). However, it is rarely present in high enough concentration to have any effect on beer flavor.

Sulfate (SO$_4^{2-}$)

Sulfates, positively affects protein and starch degradation, which favors mash filtration and trub sedimentation. However, its use may result in poor hop utilization (bitterness will not easily be extracted) if the levels are too high. It can lend a dry, crisp palate to the finished beer; but if used in excess, the finished beer will have a harsh, salty, and laxative character. The effect of sulfate is magnified and worsened by potassium and sodium. Light ales, bitters, and even mild ales benefit from the presence of sulfate in brewing water. Lagers cannot tolerate too many sulfates, especially those of magnesium and sodium salts. The sulfate/chloride ratio should be higher in bitter beers than in mild ales (5). Acceptable sulfate concentrations in the brewing water are in the range of 10–250 mg/l (3).

Phosphate (HPO$_3^{2-}$)

Phosphates are important pH buffers in brewing and useful for reducing the pH in mashing and during the hop-boil (3).

Chloride (Cl^{1-})

Calcium and magnesium chlorides give body, palate fullness, and soft-sweet flavor to beer. The certain roundness on the palate given by sodium chloride (common table salt) makes this salt eminently suited for all types of sweet beers—for both dark beers and stouts. Pale beers with concentrations of sodium chloride above 400 mg/l acquire a pasty flavor. At concentrations above 500 mg/l chloride has a harmful effect on yeast activity, and leads to poor clarification and a flat taste (14).

Carbonate (CO$_3^{2-}$)

The presence of carbonate ions and their effect in raising pH can result in less fermentable worts (a higher dextrin/maltose ratio), unacceptable wort color values, difficulties in wort filtration, and less efficient separation of protein and protein-tannin elements during the hot and cold breaks. High carbonate waters can affect hop flavor, too; e.g., hop bitterness becomes increasingly harsher.

Requirements for carbonate water vary considerably. Brewers report that carbonate concentrations exceeding 50 mg/l can ruin light colored beers by causing harsh, bitter flavors. Dark beers, on the other hand, especially those containing a high proportion of roasted malts like some alkalinity derived from the carbonates to balance the extra acidity in the grist. Carbonate in excess of 200 mg/l is tolerable only when dark-roasted malt is used to buffer its excessive acidity.

Nitrates (NO_3^{1-}) and Nitrites (NO_2^{1-})

Nitrate, in and of itself, is not a problem; it has no effect on beer flavor or brewing reactions. However, high nitrite levels may reduce the fermentation rate, dampen the rate of pH reduction, and give rise to higher levels of vicinal diketones (24). An additional effect is the reduction in the size of the yeast crop. Taylor reports that if nitrate remains in the beer, it also causes gushing when nitrite reacts with amino acids (24).

Minor Ions

Iron (Fe^{2+})

Iron in large amounts can give a metallic taste to beer. Iron salts have a negative action at concentrations above 0.2 mg/l during wort production, preventing complete saccharification, resulting in hazy worts, and hampering yeast activity (14). At concentrations above 1 mg/l, iron weakens yeast and increases haze and oxidation of tannins (16). In addition, iron can confer dark colors to worts and beers by interacting with phenolic substances from the malt and hops (3).

Copper (Cu^{2+})

Copper, in concentrations as low as 0.1 mg/l can act as catalysts of oxidants thus leading to irreversible beer haze. At levels more than 10 mg/l, copper is toxic to yeast (14). Brewing water should contain fewer than 0.1 mg/l of copper (3). Fortunately, copper is rarely a problem.

Zinc (Zn^{2+})

Zinc plays an important role in fermentation and has a positive action on protein synthesis and yeast growth. It also impacts flocculation and stabilizes foam (promotes lacing) (1). Moll recommends zinc levels between 0.08 and 0.20 mg/l and reports zinc content above 0.6 mg/l adversely affects fermentation and colloidal stability (16). Zinc is toxic to yeast and inhibits enzymes at levels above 1 mg/l. Zinc gives a metallic taste if the levels are too high.

Manganese (Mn^{2+})

Manganese is important for proper enzyme action and has a positive action on protein solubilization and yeast. If the concentration of manganese is too high it can inhibit yeast growth and negatively affect colloidal stability; and in appreciable amounts, it can impart an unpleasant taste. Ideal manganese levels in the brewing water should be fewer then 0.02 mg/l or better yet less than 0.05 mg/l (3).

> **Note | Chemicals Used for Water Treatment**
>
> Chemicals used for brewing in the United States must meet appropriate purity specifications in order to be approved by the Alcohol and Tobacco Tax and Trade Bureau. Chemicals meeting these specifications may be labeled FCC (Food Chemicals Codex).

Water Treatment

Most brewers find it necessary to treat the water before using it for brewing. There are a number of water treatments but generally they fall into the following categories:

1. Removal of particulate matter
2. Reduction in alkalinity
3. Mineral salt adjustment
4. Dechlorination
5. Microbiological control

The type of water treatment(s) needed will depend on a number of factors, including the style of beer produced and the quality and consistency of the incoming water supply.

Removal of Particulate Matter

If your brewery uses municipal water, the water must meet certain mandated conditions; however, the water picks up contaminants in the pipes from the water treatment plant to the brewery. These contaminants usually consist of scale from the pipes, sediment, and rust particles. Consequently, incoming water can be filtered using either a depth filter, which can exclude 98% of the particles more than 2 microns (μm) in diameter or a cartridge membrane filter, which can exclude particles more than 5 μm (3).

Reduction in Alkalinity

Boiling

While "temporary hardness" can be removed by boiling the water, it is seldom used on a commercial scale because of the costs associated with heating the water.

Calcium Hydroxide

The addition of calcium hydroxide—slaked lime or hydrated lime can be used to reduce the level of calcium, magnesium, and sodium salts of carbonates and bicarbonates in the water. Where calcium and magnesium are primarily in chloride or sulfate compounds, this treatment is noticeably less effective. In addition to reducing temporary hardness, the lime softening process has some efficacy as a bactericide and partially removes other inorganic contaminants such as iron, manganese, and other heavy metals. Other advantages of adding lime include its simplicity and its low chemical costs. Some of the disadvantages include the disposal of the calcium carbonate precipitate and the constant adjustment of lime based on the incoming water composition (25).

Food Grade Acids

Alkalinity can be reduced by additions of food grade acids (sulfuric, hydrochloric, phosphoric), but sometimes citric and lactic acids can be used too. In theory, sulfuric and hydrochloric acids can be used, but most brewers do not want to add excess sulfate or chloride ions to their mashes because of potential flavor effects. However, if the water is low in chlorides, the addition of hydrochloric acid is valuable. Comrie recommends sulfuric acid because phosphoric and lactic acids produce buffer salts, and these tend to keep the pH of the beer above its best range unless the water already contains a large amount of sulfate (5). However, many brewers believe that pure, food-grade phosphoric acid is the best choice for brewing: it poses no flavor problems, and the ion it adds to the wort is phosphate is a natural malt component and a vital yeast nutrient. Citric acid is used as well for adjusting pH. Lactic acid is excellent for brewing too. It does not taste like sour milk as some brewers think; in fact, its chemical structure is very similar to that of alcohol. Its flavor blends very well with beer, and it is possibly the mildest of all the acids. Lactic acid is unstable in crystalline form and is usually sold as a liquid at 88% concentration. Acidification is popular with many large commercial and craft breweries.

Carbon Dioxide

In Germany and in other countries where mineral acids (i.e., food grade acids) are prohibited, pH reduction can be achieved by using carbon dioxide. Carbon dioxide is introduced into the high pH water by means of a diffuser, which is typically installed in an existing pressurized pipe or at the bottom of a tank. Small carbon dioxide bubbles are then released by the diffusers into the high pH water, which once dissolved into water produces as weak acid—carbonic acid. This substance reacts immediately with alkalis turning them in to neutral carbonates and bicarbonate salts. A pH probe is installed downstream of the gas injection point, thus measuring pH in the water after absorption and carbon dioxide reaction.

Mineral Salt Adjustment

Historically, breweries were located on sites with established, consistent water supplies having characteristic mineral compositions. This led to the emergence of regional beer characteristics in locations such as Burton-on-Trent, Dortmund, Pilsen, and Vienna. Mineral salt adjustment was held to a minimum and, often, recipes were adapted to the shortcomings of the brewing water. For example, dark malts often were used because their natural acidity neutralized the excess alkalinity of high carbonate waters. Water treatments only involved the addition of calcium salts to the brewing water or the wort to precipitate bicarbonate ions in order to reduce alkalinity.

Today it's more common for brewers to adjust the composition of the water since most receive water from different sources and the fact they brew many different beer styles. The amount of mineral salt, i.e., ions to be added, is based on the style of beer being brewed. In the case of craft breweries, brewing water usually receives minimal in house treatment.

Some of the more common brewing "salts" (minerals) that are added to brewing water are calcium and magnesium salts of sulfates; and sodium, calcium, and potassium salts of chlorides. Calcium and magnesium salts are used to counteract alkalinity too.

Sulfate ions usually are associated with pale ales, whereas chlorides are associated with mild ales. As a rule, a sulfate/chloride ratio of 3:2 is suggested for pale ales and a ratio of 2:3 for mild ales (5). Stouts require little, if any sulfate, and their calcium will be present mainly as chloride.

Calcium Sulfate

Calcium sulfate is often used as a source of calcium ions and is generally used in brewing British pale ales and bitters. Calcium sulfate treatment is sometimes referred to as "Burtonization" (after Burton-on-Trent in England) because Burton-on-Trent waters are rich in gypsum and this area is world-famous for its pale ales. Calcium sulfate gives a dry edge that accentuates the hop bitterness and enhances the flavor and fullness of the beer. Calcium sulfate requires a good mixing because of its limited solubility in water. Ryder *et al.* recommend at least 90% of the calcium sulfate should pass through a U.S. No. 100 sieve to encourage a quick dissolution (20).

Magnesium Sulfate

Magnesium sulfate is similar to calcium sulfate but is not as effective as calcium in reducing the pH of the mash as demonstrated by the calculation for residual alkalinity. Magnesium sulfate accentuates the bitterness in the beer; used in excess, it produces an unpleasant bitter flavor. It is generally avoided in pale lagers.

Sodium and Calcium Chlorides

Sodium chloride is used to increase sodium and chloride content. Like calcium sulfate, it accentuates bitterness and enhances the flavor and fullness of the beer. It is generally used in brewing Dortmund and Burton pale ales. Treatment with sodium chloride should not exceed 150 mg/l (20). Calcium chloride adds calcium without adding sulfate or carbonate and generally is associated with a fuller, mellower taste. Potassium chloride is used as a substitute for part of a sodium chloride treatment.

Dechlorination

Water used in brewing should be free of chlorine since it hampers yeast growth, causes off-flavors, and, if in high concentrations, can corrode stainless steel. Chlorinated municipal water supplies will usually require treatment before use. There are several methods available to the brewer for removing chlorinated compounds, which are discussed below.

Heating the Water

Scheer reports that heating water in the holding tank at a temperature of 78°C overnight is sufficient to eliminate all chlorine, whether bound or free (22). However, this technique will not work if the water has been treated by a process called "chloramination." Chloramination involves adding ammonia and chlorine to the water to produce monochloramines and dichloramines—substances that can produce medicinal odors and flavors—and requires the brewer to resort to carbon filtration.

Carbon Filtration

Activated carbon filtration is a common method used to remove chlorinated compounds, including THMs and organic pollutants such as pesticides.

Water filtration begins by soaking the activated carbon in clean water for at least 12 to 24 hours in the filter vessel. It is important to take into account bed expansion, 20 to 30% by volume, when sizing a tank. This initial step will remove the air that occupies the pores of the carbon. The next step is to backwash until the effluent is clear, removing carbon fines and dust from the surface of the carbon bed. The filter now is ready to filter the water. Most brewers prefer between 10 and 15 minutes of contact time with a hydraulic loading of 4 to 6 gallons per minute per square foot. The activated carbon should be certified under the National Sanitary Foundation for use in drinking water applications.

Rather than using filter vessels to remove chlorine, craft brewers use activated carbon contained in cartridge filters with a 5 micron rating. Some of the advantages in using cartridge filters is minimal downtime since they can be changed in a matter of minutes and require no presoaking, sterilization, or backwashing. The disadvantages of using cartridge filters are the cost (per unit) of treated water and the low flow rates compared to those of bed filters.

Activated carbon has a finite capacity to adsorb chemicals, and will require replacement or regeneration, depending on the level of contamination and the capacity of the filter used. There are a number of ways that brewers can clean and disinfect activated carbon. The first step is to backwash the activated carbon in the vessel. Tubbs recommends doing this once a week or when there is a 15 psi pressure drop across the bed due to the accumulation of suspended solids (28). After cleaning the activated carbon, the next step is to sterilize the carbon with steam from the bottom of the bed to the top for 15 minutes until the top of the bed reaches 60°C. After steaming, the bed is backwashed again with ambient water to remove any material loosened from the cleaning. Instead of steam, the carbon can be sterilized using a 2% by weight solution of sodium hydroxide or a 2 to 3% solution of sodium carbonate (28).

Potassium Metabisulfite

Excess chlorine can also be removed by rousing 1 to 2 mg/l potassium metabisulfite into the water (21). The bisulfite at this concentration acts as a harmless bacteriostat.

Microbiological Control

The microbiological safety of water is measured by testing the water for coliform bacteria. This test serves purely as an indicator, as it is based on the assumption that water meeting the requirements for coliform bacteria is also free from other potential pathogenic bacteria.

In general, most of the water used for brewing will be heated prior to mashing and boiled in the kettle, thereby ensuring it is free from microorganisms. However, water used for diluting worts, beer, or rinsing equipment may not be microbiologically sterile. Water can be sterilized using physical sterilants such as ultraviolet light and filtration or by using chemical sterilants such as chlorine, chlorine dioxide, and ozone.

Physical Treatment

Ultraviolet Radiation – Sterilization with ultraviolet radiation (UV) uses mercury vapor lamps at wavelengths between 200 and 300 nm to destroy organisms (15). Organisms vary in their susceptibility to UV. Viruses and bacteria generally are more easily killed than some yeasts, fungi, and spores. The method is widely accepted because of its nonchemical nature: no taints or off-flavors arise, and no residues remain in the water. It is also relatively inexpensive in respect both to initial capital and maintenance costs. However, one disadvantage is that no residual "kill-potential" remains in the water, and it is limited to the point of application. Thus, it is important to avoid re-infection of the water after UV treatment.

Sterile Filtration – Bacteria and fungi can also be removed by using a cartridge membrane filter with a 0.45 µm pore size.

Chemical Treatment

An alternative approach in disinfecting water is to treat it with chemical sterilants. This technique has the advantages of providing a degree of post-treatment microbiological protection and facilitating a centralized point of treatment. A possible limitation is that contact times in excess of 1 hour are usually required in order to achieve the necessary biocidal effect. Disinfection treatments of interest include:

1. Chlorine
2. Chlorine dioxide
3. Ozone

Chlorine – Where chlorination is practiced in breweries, it is usually in the form of dosing with sodium, potassium, or calcium hypochlorites. Calcium hypochlorite exists in a solid form that releases chlorine on contact with water. Handling the solid, however, requires more labor time such as opening bags and pouring than the use of gas or liquids which are more easily automated. Liquid sodium hypochlorite is both inexpensive and safer than the use of gas or solid chlorine. All forms of chlorine are widely used despite their respective drawbacks. Briggs *et al.* recommend that the level of chlorine should initially be 5 mg/l and held for at least 30 minutes to allow the sterilant to act (3).

One advantage in using chlorine is its ability to remain in solution long enough to have a useful "residual" sterilizing effect. One drawback is that chlorine from any source reacts with natural organic compounds in the water to form potentially harmful chemical by-products (e.g., trihalomethanes (THMs) and haloacetic acids). These byproducts are both carcinogenic in large quantities and regulated by the United States Environmental Protection Agency. The formation of THMs and haloacetic acids may be minimized by effective removal of as many organics from the water as possible prior to chlorine addition. Although chlorine is effective in killing bacteria, it has limited effectiveness against *Giardi*, cysts of other protozoa and some viruses (3). A further potential problem may arise in the formation of scale where hard water is treated.

Any residual chlorine remaining in the water can be removed by aeration, evaporation, by filtration through active carbon, or by adding bisulphate or sulfite to the water (3).

Chlorine Dioxide – As an alternative to sodium hypochlorite is chlorine dioxide which is generated on-site immediately before use. Unlike sodium hypochlorite, it does not form trihalomethanes (THMs) or chlorophenols, compounds that are extremely detrimental to beer. It is quite effective against a wide variety of beer spoilage organisms, including bacteria, yeast, and molds.

Chlorine dioxide also removes other undesirable compounds from water, including cyanides, sulfides, aldehydes, and mercaptans. Chlorine dioxide is less effective at preventing re-infection than chlorine because it has a shorter "residual" period. Chlorine dioxide is used at a maximum level of 0.5 mg/l to treat final rinse water after disinfection (18).

Ozone – As a water treatment, use of ozone has proved successful in brewery applications in removing most water born organisms. To use ozone as a disinfectant, it must be created on-site by passing dry air or oxygen through an electrical generator and added to the water by bubble contact.

Some of the advantages of ozone include the production of relatively fewer dangerous by-products (in comparison to chlorination) and the actual removal of taints and odors. It is also said to be more effective against *Giardia*, cysts of other protozoa and some viruses and bacteria than chlorine or chlorine dioxide (3). Although fewer by-products are formed by ozonation, it has been discovered that the use of ozone produces a small amount of the suspected carcinogen Bromate. Another one of the main disadvantages of ozone is that it leaves no disinfectant residual in the water. Ozone-treated water can be corrosive to a variety of materials, especially when it is heated. The capital cost of an ozone generating plant can be high too. In addition, depending upon the quality of the water to be treated, post-filtration with sand may be required to remove oxides, and/or activated carbon may be needed to remove other oxidation products. The U.S. Food and Drug Administration has accepted ozone as being safe; and it is applied as an anti-microbiological agent for the treatment, storage, and processing of foods.

Contact time should last from 3 to 15 minutes at an initial rate of $1-3 \text{ g/m}^3$ with higher doses of ozone used if iron and manganese ions are to be oxidized (3).

References

1. Bamforth, Charles W. 2006. *Scientific Principles of Malting and Brewing.* St. Paul, Minnesota: American Society of Brewing Chemists.

2. Bernstein, Leo and I. C. Willox. 1977. "Water." T*he Practical Brewer*, edited by H. M. Broderick. Madison, Wisconsin: Masters Brewers Association of the Americas.

3. Briggs, Dennis E., Boulton, Chris A., Brookes, Peter A., and Stevens, Roger. 2004. *Brewing: Science and Practice*. Cambridge, England: Woodhead Publishing Limited.

4. Carey, Daniel J. 1991. "Microbrewery Design and Performance." *MBAA Technical Quarterly* 28 (1).

5. Comrie, A. A. D. 1967. "Brewing Water." *Brewer's Digest* 42 (7).

6. De Clerck, Jean. 1957. *A Textbook of Brewing*. Volume 1, translated by K. Barton-Wright. London: Chapman-Hall Ltd.

7. Fix, George J. 1989. *Principles of Brewing Science*. Boulder, Colorado: Brewers Publications.

8. Fix, George J. 1995. "The Role of pH in Brewing." *The New Brewer* 12 (6).

9. Fix, George and Laurie Fix. 1997. *An Analysis of Brewing Techniques*. Boulder, Colorado: Brewers Publications.

10. Hough, J. S., et al. 1982. *Malting and Brewing Science*. Volume1. London, England: Chapman-Hall, Ltd.

11. Hough, J. S. 1991. *The Biotechnology of Brewing*. London, England: Cambridge University Press.

12. Kemmer, Frank N., ed. 1988. *The NALCO Water Handbook*. New York, New York: McGraw-Hill Book Company.

13. Kunze, Wolfgang. 1996. *Technology Brewing and Malting*, translated by Dr. Trevor Wainwright. Berlin, Germany: VLB Berlin.

14. Moll, M.M. 1979. "Water in Malting and Brewing." *Brewing Science*. Volume 1, edited by J.R.A. Pollock. London, United Kingdom: Acadmic Press.

15. Moll, Manfred, ed. 1995. *Beers and Coolers*. New York, New York: Marcel Dekker, Inc.

16. Moll, M.M. 1995. "Water." *Handbook of Brewing*, edited by William A. Hardwick. New York, New York: Marcel Dekker, Inc.

17. O'Rourke, Timothy. 1994. "Making the Most of Your Water." *The New Brewer* 11 (1).

18. O'Rourke, Timothy. 1998. "The Treatment and Use of Water in Brewing." *Brewers' Guardian* 127 (12).

19. Ryder, David S. and Joseph Power. 1995. "Miscellaneous Ingredients in Aid of the Process." *Handbook of Brewing*, edited by William A. Hardwick. New York, New York: Marcel Dekker, Inc.

20. Ryder, David S. and Joseph Power. 2006. "Miscellaneous Ingredients in Aid of the Process." *Handbook of Brewing*, edited by Fergus G. Priest and Graham G. Stewart. Boca Raton, Florida: CRC Press, Taylor & Francis Group.

21. Sanchez, Gil W. 1999. "Water." *The Practical Brewer*, edited by John T. McCabe. Wauwatosa Wisconsin: Master Brewers Association of the Americas.

22. Scheer, Fred. 1991. "Ask the Brewmaster: Chlorine Removal, Polyphenols." *The New Brewer* 8 (1).

23. Stewart, G.G. and I. Russell. 1985. "Modern Brewing Biotechnology." *Food and Beverage Products*, edited by Murray Moo-Young. Oxford, England: Pergamon.

24. Taylor, T.G. 1988. "The Impact of Water Quality on Beer Quality." *The Brewer* 74 (890).

25. Taylor, David G. 2006. "Water." *Handbook of Brewing*, edited by Fergus G. Priest and Graham G. Stewart. Boca Raton, Florida: CRC Press, Taylor & Francis Group.

26. Theaker, P.D. 1988. "Maintaining Water Quality Standards." *Brewers' Guardian* 117 (3).

27. Thompson, Donald, Karl Ockert, and Vince Cottone. 1999. "Craft Brewing." *The Practical Brewer*, edited by John T. McCabe. Wauwatosa Wisconsin: Master Brewers Association of the Americas.

28. Tubbs, Jerry. 1996. "Incorporating Activated Carbon Filtration." *The New Brewer* 13 (6).

29. Tubbs, Jerry. 1998. "The Basics of Cartridge Filtration." *The New Brewer* 15 (1).

30. Tubbs, Jerry. 1998. "The Basics of Cartridge Filtration – Part II." *The New Brewer* 15 (3).

Chapter Six

Brewing Adjuncts

Adjuncts are nothing more than unmalted grains such as corn, rice, rye, oats, barley, and wheat. Adjuncts are used mainly because they provide extract at a lower cost (a cheaper form of carbohydrate) than is available from malted barley or to modify the flavor of the beer. Adjuncts are used to produce light-tasting, light-colored beers that have the alcoholic strength of most beers.

Uses of Adjuncts

Adjunct use results in beers with enhanced physical stability, superior chill-proof qualities, and greater brilliancy. The greater physical stability has to do with the fact that adjuncts contribute very little proteinaceous material to wort and beer, which is advantageous in terms of colloidal stability. Rice and corn adjuncts contribute little or no soluble protein to the wort, while other adjunct materials, such as wheat and barley, have higher levels of soluble protein. Except for barley, adjuncts also contribute little or no polyphenolic substances.

Adjuncts can be used to adjust fermentability of a wort. Many brewers add sugar and/or syrup directly to the kettle as an effective way of adjusting fermentability, rather than trying to alter mash rest times and temperatures.

Adjuncts are often used for their flavor contribution. For example, rice has a very neutral aroma and taste, while corn tends to impart a fuller flavor to beer. Wheat tends to impart a dryness to beer. Semi-refined sugars add flavor to ales that has been described as imparting a luscious character. Adjuncts will also alter the carbohydrate and nitrogen ratio of the wort, thereby affecting for formation of byproducts, such as esters and higher alcohols.

Adjuncts are used for color adjustment, as in the case with dark sugars. On the other hand, adjuncts such as rice and pure starches and sugars are used to dilute malt colors to produce lighter colored beers.

Some adjuncts are used for their chemical properties—raw barley and wheat, which contribute glyco-proteins to enhance foam stability (26). Other adjuncts, low in protein, are used to improve colloidal stability since they will dilute the amount of potential haze-forming proteins.

Finally, the use of adjuncts can result in increased brewing capacity, reduced labor costs, improved hot and cold breaks, and shorter brewing cycles. For example, brewery capacity can be increased without adding brewhouse vessels by using adjuncts. Adjunct use increases the gravity of the wort in the kettle allowing the brewer to expand barrelage of the beer followed by a water adjustment later in the brewing process.

As is so often the case, benefits in one area are often offset by problems in another. If the level of adjuncts used is too high, the brewer runs the risk of producing wort with insufficient insoluble nitrogen for yeast growth.

The proportion of adjuncts used varies from 10 to 30% in Europe, to 40 to 50% (e.g., malt liquors) for some U.S. brewers, to as high as 50 to 75% in certain African countries (22). Although not usually practiced by brewers, adjunct levels up to 100% of total grist composition can be used, but will require the addition of exogenous enzymes (26). However, in certain countries, for example Germany, malt is the only permitted source of fermentable extract because of the German purity law or "Reinheitsgebot."

Types of Adjuncts

Cereal adjuncts can either be added to the cereal cooker, the mash tun or directly to the brew kettle.

Cooker Mash Adjuncts

Cooker mash adjuncts consist of non-gelatinized cereal products (meal, grits, flour, or dry starch) whose starches are in their native forms. A non-gelatinized adjunct needs to be heated in a separate cereal cooker to complete liquefaction since the starch gelatinization temperature of the adjunct is higher than that used for the malt saccharification (starch hydrolysis) temperature. The cooked adjunct is then added directly to the mash in the mash tun. The malt enzymes from the malt mash can be used to hydrolyze the starch from the adjunct, converting it to sugars ready for fermentation. In some cases, starch hydrolysis can be augmented or replaced by adding commercially-available exogenous enzymes (bacterial alpha-amylase) directly to the mash in either the mash tun or mash conversion vessel. Table 6.1 shows typical gelatinization temperatures for various adjuncts.

Table 6.1
Gelatinization Temperatures of Various Cereals

Cereal (Source of Starch)	Gelatinization Temperature °C
Corn	68–74
Sorghum	68–80
Rice	68–75
Wheat	52-65
Rye	58–70
Malted Barley	61–65
Raw Barley	53–58

Mash Tun Adjuncts

The adjunct can be mashed directly with the malt in the mash tun in two ways: 1) when the starch gelatinization temperature of the adjunct is lower than the malt saccharification temperature required for mashing or 2) when the adjunct has been pre-gelatinized (e.g., for flakes, torrified cereals, and refined starches). The amount of cereal adjunct that can be converted will depend on the diastatic power of the malt. Again exogenous enzymes can be added to allow the use of higher levels of raw grains.

Kettle Adjuncts

Kettle adjuncts consist of syrups and sucrose sugar. These are sometimes called "wort extenders" because of the extract they readily contain. British brewers commonly use syrups and sugars, whereas the rest of the world are more likely to use non- and pre-gelatinized adjuncts cereal adjuncts such as corn, rice, and wheat.

Classification of Cereal Adjuncts

Most of the brewers' adjuncts are based on a limited range of cereal grains. The non-malt brewing materials used in greatest quantity today are those derived from corn and rice, although barley, wheat, and sorghum grain are sometimes used.

Types of Milled Products

Flours

Flours are produced as by-products, during the manufacture of corn and rice. Flours must be cooked before being mixed in with the malt mash.

Grits

Grits consist of uncooked fragments of starchy endosperm derived from cereal grains. The starch of these adjunct products is in its native form, and is not readily attacked by the malt diastase enzymes during mashing. Consequently, these adjuncts must be processed by boiling in a cereal cooker to bring about solubilization and gelatinization of the starch granules to render them susceptible to diastatic enzyme attack. Unlike in America and Australia, grits are rarely used in British brewing, as cereal cookers are not found in most traditional British breweries.

Flakes

There are two different manufacturing processes used to produce brewing flakes. In the traditional process, corn and rice grits or whole barley are steam-cooked to soften the endosperm, which is then rolled flat and dried. Gelatinization occurs during the steaming process. Another process involves "micronizing" these materials prior to flaking by subjecting the grain to internal heating by infrared heat. Flakes can be added directly to the mash tun since the starch granules have been gelatinized. Flaked cereals have been used for many years in the United Kingdom and Belgium, being valued for the soft and subtle grain-like sweetness they impart.

Torrified Cereals

Torrified cereals are produced by heating the grain, which makes the endosperm expand and pop, thus rendering the starch pre-gelatinized and easily milled. Torrified cereals can be added directly to the mash tun since the starch granules have been gelatinized. Most of the nitrogen is denatured in the kernel and is not solubilized, thus contributing little or no nitrogen to the mash. Fat content is slightly higher than for other adjuncts, but this is usually not an issue when using a higher adjunct level (6).

Torrified cereals and malt can be ground simultaneously and mashed-in together. However, higher yields are found by cooking the torrified product separately between 71 and 77°C prior to adding it to the malt mash (32). Proper mill settings are critical: large particle size leads to poor yield and too fine a grind leads to problems with runoff (6).

Torrified cereals absorb more water because of their expanded nature than other adjuncts, so a higher ratio of water to cereal must be used, especially for torrified barley (8). The use of torrified cereals leads to increased lauter grain bed depth and to a slower runoff (32). Coors reports that the flavor of beer produced with torrified adjuncts is unchanged (8).

Refined Starches

Refined starches can be prepared from many cereal grains. In commercial practice, refined wheat starch, potato starch, and corn starch have been used in breweries; corn starches, in particular, are used in the preparation of glucose syrups. Wheat starch has been employed in breweries in Australia and Canada, where local conditions make it economical to use. However, the most important source of refined starch is corn.

Refined starch is the "purest" mash adjunct available to the brewer. One of its drawbacks is its strict handling requirements, as it is extremely fine and must be contained in well-grounded lines and tanks to prevent explosions resulting from static electricity. The starches are usually handled as slurry. Refined starches contain very little nitrogenous material, and on a dry-weight basis, they have extracts of 103 to 105% because of hydrolysis gains. Since starches are converted wholly to soluble materials, they do not cause runoff problems. Furthermore, pure starches do not contribute to beer flavor.

Types of Cereal Adjuncts

Corn

Corn products have traditionally been the adjunct of choice among brewers. They are extremely consistent in terms of quality, composition, and availability and produce a spectrum of fermentable sugars and dextrins similar to that produced by malt upon enzymatic conversion. All corn-derived adjuncts contain a certain amount of oil, though generally this does not seem to exercise any harmful effect upon the finished beer.

Corn has a sweet, smooth flavor that is compatible with many styles of beer. It is the most popular adjunct used in American breweries. It lowers the protein and polyphenol content of beers, thereby lightening the body and reducing haze potential. Corn will provide a somewhat neutral flavor to the finished beer. A "corn" taste may be apparent, making it generally more suited to the sweeter dark beers and lagers than to the drier pale ales. It is, however, one of the best adjuncts to use for full-bodied bitters. Some brewmasters claim that the use of corn (10–20%) will help stabilize the flavor of beer.

The following adjunct products are processed from corn: corn grits and meal, corn flakes, and refined starch.

Corn Grits – Corn grits (also known as "brewers' grits") are used extensively in the United States, Canada, France, Belgium, and Italy, though normally only for economic reasons. Corn grits produce a slightly lower extract than other unprocessed adjuncts and contains higher levels of protein and fat (32). The gelatinization temperature range for corn grits (62–74°C) is slightly lower than that for rice grits (64–78°C).

Corn Flakes – Corn flakes are similar to corn grits except for a lower moisture content and a slightly higher extract value. Flaked corn has all the advantages of flaked rice and will contribute flavor.

Refined Corn Starch – Refined corn starch has not found widespread use because its price is higher than those of corn grits and brewers rice (32). There are no runoff problems, and fermentations tend to attenuate better with corn starch, while colloidal stability is unaffected (32). Beer flavor is not affected, but the beer is slightly thinner due to higher attenuation limits (6). Refined corn starch can be used as a total adjunct, even though gelatinization and liquefaction of the starch proceeds at lower temperatures; or it can be mixed with brewer's rice or corn grits (8). When it is used as a total adjunct, care should be taken to prevent sticking at the bottom of the cooker.

Rice

Rice is currently the second most widely used adjunct material in the United States in the production of light-colored lager beers (32). Rice has hardly any taste of its own, which is regarded as a positive characteristic since rice will not interfere with the basic malt character of the beer. It promotes dry, crisp, and snappy flavors and is employed in several premium brands, including Budweiser. Some brewers prefer rice because its lower oil content compared to corn grits (6). One disadvantage in using rice is the need to use an additional cooking vessel because its gelatinization temperature is too high for adequate starch breakdown during normal mashing.

Different types of rice vary widely in their suitability for use in brewing. Short-grain rice is preferred because medium-and long-grain varieties can lead to viscosity problems. In milling rice, a certain proportion of the rice kernels are chipped and broken, rendering them unsuited for table use because of their impaired physical appearance. It is this portion of the broken rice that is designated as "brewer's rice."

Rice Grits – The advantages of rice grits are that they are relatively easy to handle and there is little associated starch dust. The gelatinization temperature range for rice grits is between 64 and 78°C, although it is common practice to mash-in and hold at 33 to 38°C as a protein rest (6). Usually 5 to 10% of malt is added to the cooker because the malt enzymes are essential for partial liquefaction necessary to pump the fluid (6). Yields with rice grits are better than those of corn grits, and rice is usually economically competitive.

Rice Flakes – Rice flakes are processed in a manner similar to corn flakes to pre-gelatinize the starch. They have the advantage that they do not require cooking in the brewery but can be added directly to the cereal cooker.

Barley

Unmalted barley gives a rich, smooth, "grainy" flavor to beer. Unlike the other adjuncts, unmalted barley will contribute foam (head) retention to the finished beer because of lower levels of proteolysis (30). However, the nitrogenous and complex proteins that contribute to head retention also contribute to chill haze problems. Clarity problems make unmalted barley inappropriate for light beers, which is one reason why corn and rice adjuncts are preferred. It is essential in dry stout, e.g., Guinness Stout.

Unmalted barley can be employed for as much as 50% of the total grist, but it usually makes up no more than 10 to 15% as an adjunct. High levels of unmalted barley can lead to a slightly harsh taste in the beer. Incomplete degradation of beta-glucans could be a problem which can increase wort viscosity and runoff times; thus affecting the stability of the finished product.

Barley may be de-husked before use, to increase the extract yields, but doing so may lead to runoff difficulties since the husk provides material for filter bed formation. To minimize the level of fine material, the raw grain is usually hammer-milled rather than being milled with a conventional roller mill; this can result in problems in lautering (32).

Flaked Barley – Flaked barley is the easiest to use and definitely is preferred by brewers. Barley flakes impart a dry, grainy flavor to the beer. In the darker beers, especially stouts, where haze problems are no worry, flaked barley is unsurpassed as an adjunct for flavor and head retention. It is recommended that the flaked barley be included with malt in a protein rest before starch conversion.

Wheat

Unmalted wheat often is used as an adjunct by brewers who wish to enhance head retention and foam stability. It also contributes to the body or "palate fullness" of the beer. Its high content of proteins greatly enhances foam stability. Beers made from significant amounts of wheat adjuncts are likely to be light in flavor and smooth in taste qualities. Wheat adjuncts are used in the same manner as barley adjuncts; but unlike with barley, there is almost no husk in wheat. Thus, tannins are not much of a problem. The gelatinization temperature range for wheat is between 52 and 64°C.

Refined Wheat Starch – Refined wheat starch is used very little in the United States because of its high price compared to more readily available adjuncts (32). Chemically, wheat starch is very similar to refined corn starch. Its gelatinization is similar to that of malt gelatinization, and it can be added directly to the malt mash; however, higher yields can be achieved by using a cereal cooker (6). Runoff problems with wheat starch appear to be caused by the formation of impermeable layers of "fines" in the grist. Mixing in 10% malt and holding between 44 and 48°C for 30 minutes in the mashing cycle will give beta-glucanase time to break down the beta-glucans prior to the temperature being raised to 96°C (32). This procedure should result in few or no runoff problems. Beer brewed with refined wheat starch is quite comparable to that brewed with corn grits in analysis and flavor, but the lautering times for wheat starch are much longer than those for corn grits (8).

Wheat Flour – Wheat flour is unusual in that it is not pre-gelatinized, nor is it subjected to cooking at the brewery. Wheat flour can be used directly in infusion mashes, but higher extracts may be obtained if the flour is pre-soaked or is pre-cooked (4). Mash tuns can accommodate more than 10% of wheat flour in the grist, whereas as much as 40% can be handled in some lauter tuns. Wheat flour often is processed into pellets for easier handling.

Oats

The high protein, fat, and oil content of oats are theoretically a deterrent to their use in brewing. However, oats have been used in the brewing process, particularly in brewing oatmeal stout.

Syrup and Sugar Adjuncts

Syrups and sugars are adjuncts that are used in brewing too, especially with British and Belgium brewers. Syrups and sugars can be added to the wort either at the boiling stage or as primings during racking. If added during boiling, syrups can be used to extend brewhouse capacity or to improve beer stability. Syrups that are added directly to the kettle to supplement fermentable carbohydrates effectively extend the capacity of the brewhouse to produce more fermentable wort. Syrups and sugars also allow for shorter boiling times and high-gravity brewing. Syrups and sugars can improve beer stability by diluting the non-starch constituents of the wort, such as proteins and polyphenols that contribute to haze. Syrups and sugars that are added as primings to the beer may be used for sweetening, body, and color, as well as to provide fermentable carbohydrate for secondary fermentation and conditioning in the cask (27).

Syrups

The two major syrups used in brewing are sucrose- and starch-based. The sucrose-based syrups have been refined from natural sources such as sugar cane or beets. The starch-based syrups are produced from cereals by hydrolysis using acid, exogenous enzymes, or a combination of the two to produce a range of syrups with different fermentabilities. Starch-based syrups contain both carbohydrates and nitrogenous materials, minerals, and yeast growth factors that make these materials equivalent to adding concentrated wort to the beer production stream (4).

In recent years, there has been a great development in the range of starch-based syrups produced from corn and wheat. In the United States, these adjuncts are produced exclusively from yellow corn; while in Europe, they are produced from corn and wheat. The starch-based syrups are commonly referred to as "glucose" syrups. This name is misleading, however, since the syrups contain a large range of sugars, depending on the method of manufacture—dextrose, maltose, maltotriose, maltotetraose, and larger dextrins (27).

Sugars

Sucrose

Sucrose (table sugar) is made from cane or beet sugar. Granulated sugar, the normal end product of the refining process, may be added directly to the kettle, but usually is dissolved in a solution before being added. Granulated sugar, as a disaccharide, is not completely fermented by yeast. It has to be hydrolyzed to change into glucose and fructose, which are fermentable. Some brewer's preparations contain both sucrose and invert sugar (4).

Sucrose imparts a cidery taste due to "spillover" byproducts created during its fermentation. On the other hand, partially refined forms of sucrose, such as light and dark-brown sugar, give a rum-like flavor, which some people like very much in ales. The strong flavors of these brews seem to mask the cidery note.

Candi Sugar – A few brewers specify candi sugar. This sugar is made by allowing a supersaturated solution of sucrose to cool slowly. Thin rods or string may be inserted into the solution to act as nuclei on which the large crystals grow over a period of from 2 to 3 weeks. The crystals are chipped off the rods, but string candy usually is sold on the string. It is claimed that candi contributes to good head retention in a high-gravity, lightly hopped beer. Candi is also quite often used in liquid form and is commonly used by Belgium brewers.

Dextrose

Dextrose is also known as corn sugar and is available in the trade in the purified form as a spray dry or as a crystalline powder. Dextrose sugar is added directly to the brew kettle during boiling. Its addition alters the sugar composition of the resulting wort and thus affects fermentation, leading to a more aromatic and flavorful product (1). Dextrose is commonly used in the production of ales, and leaves no taste of its own in the beer.

Malto-Dextrin

Malto-dextrin is the most complex fraction of the products of starch conversion. It is tasteless, gummy, and hard to dissolve. It is often said to add body (palate fullness) to beer, increase wort viscosity, and add smoothness to the palate of low-malt beers. However, it is easy to increase the dextrin content of grain beers by changing the mash schedule or using dextrin malt. Malto-dextrin is of interest mainly as a supplement to extract brews.

Caramel

Caramel is used in brewing as a flavor and/or coloring agent. For example, many milds and sweet stouts contain caramel for both flavor and color. Caramel may be used either in the kettle or in primings to make minor adjustments to the color of the beer, but the choice of malt grist and the grade of adjuncts added to the kettle will determine the fundamental color of the beer.

Invert Sugar

Sucrose can be split into its two component sugars (glucose and fructose). This process is called inversion, and the product is called invert sugar. Commercial invert sugar is a liquid product that contains equal amounts of glucose and fructose.

References

1. Bradee, Lawrence H. 1977. "Adjuncts." *The Practical Brewer*, edited by H. M. Broderick. Madison, Wisconsin: Master Brewers Association of the Americas.
2. Bradee, Lawrence, Will Duensing, Scott Halstad, Ray Klimovitz, and Andy Laidlaw. 1999. "Adjuncts." *The Practical Brewer*, edited by John T. McCabe. Wauwatosa Wisconsin: Master Brewers Association of the Americas.
3. Brenner, M.W., M. J. Arthurs, and E. E. Stewart. 1968. "Yellow Corn Grits vs. Brewers Corn Syrups as Adjuncts for Lager Beer." *MBAA Technical Quarterly* 5 (1).
4. Briggs, Dennis E., Boulton, Chris A., Brookes, Peter A., and Stevens, Roger. 2004. *Brewing: Science and Practice*. Cambridge, England: Woodhead Publishing Limited.
5. Burton, A. H. and J. R. Palmer. 1974. "Production Scale Brewing Using High Proportions of Barley." *Journal Institute of Brewing* 80.
6. Canales, A.M. 1979. "Unmalted Grains in Brewing." *Brewing Science*. Volume 1, edited by J.R.A. Pollock. Reading, United Kingdom: Academic Press.
7. Chantler, John. 1990. "Third Generation Brewers Adjunct and Beyond." *MBAA Technical Quarterly* 27 (3).
8. Coors, Jeffery. 1976. "Practical Experience with Different Adjuncts." *MBAA Technical Quarterly* 13 (2).
9. Dougherty, Joseph J. 1977. "Wort Production." *The Practical Brewer*, edited by H. M. Broderick. Madison, Wisconsin: Master Brewers Association of the Americas.

10. Enari, T.M. and M. Linko. 1967. "Unmalted Barley in Brewing." *MBAA Technical Quarterly* 4, (3).

11. Fix, George and Laurie Fix. 1997. *An Analysis of Brewing Techniques.* Boulder, Colorado: Brewers Publications.

12. Geiger, K. 1972. "Brewing with Wheat Starch."*MBAA Technical Quarterly* 9 (4).

13. Hahn, R.R. 1965. "Brewing with Sorghum Brewers Grits as the Adjunct Cereals." *MBAA Technical Quarterly* 2 (3).

14. Hobson, Derek. 1980. "Influence of Brewhouse and Raw Materials." *Brewers' Guardian* 109 (10).

15. Hough, J.S., et al. 1982. *Malting and Brewing Science.* Volume 1. London, England: Chapman-Hall, Ltd.

16. Hough, J.S. 1991. *The Biotechnology of Brewing.* London, England: Cambridge University Press.

17. Hudson, J.F. 1985. "Adjuncts and Their Use in Brewing." *Brewers' Guardian* 114 (7).

18. Knarr, Skip. 2006. "Adjuncts and Other Ingredients." *Raw Materials and Brewhouse Operations.* Volume 1, edited by Karl Ockert. St. Paul, Minnesota: Master Brewers Association of the Americas.

19. Kunze, Wolfgang. 1996. *Technology Brewing and Malting*, translated by Dr. Trevor Wainwright. Berlin, Germany: VLB Berlin.

20. Lloyd, W.J.W. 1988. "Brewers' Solid Adjuncts." *Brewers' Guardian* 117 (5).

21. Marchbanks, C. 1987. "A Review of Carbohydrate Sources for Brewing." *Brewing & Distilling International* 17 (1).

22. Martin, P.A. 1978. "Cereal and Sugar Adjuncts." *Brewers' Guardian* 108 (8).

23. Mathes, F.E. 1957. "Normal Brewing Practices with the Use of Liquid Adjuncts." *Technical Proceedings, 70th Annual Convention, MBAA Technical Quarterly* 2 (11).

24. Moll, Manfred, ed. 1995. *Beers and Coolers.* New York, New York: Marcel Dekker, Inc.

25. O'Rourke, Timothy and David Pierpoint. 1994. "Developments in the Use of Brewing Adjuncts in Britain." *Brewers' Guardian* 123 (7).

26. O'Rourke, Timothy. 1999. "Adjuncts and Their Use in the Brewing Process." *Brewers' Guardian* 128 (3).

27. Richards, P.J. 1988. "Brewer's Liquid Adjuncts." *Brewers' Guardian* 117 (4).

28. Rooney, L.W. 1969. "Properties of Sorghum Grain and New Developments of Possible Significance to the Brewing Industry." *MBAA Technical Quarterly* 6 (4).

29. Ryder, David S. and Joseph Power. 1995. "Miscellaneous Ingredients in Aid of the Process." *Handbook of Brewing*, edited by William A. Hardwick. New York, New York: Marcel Dekker, Inc.

30. Sfat, M.R. 1973. "Trends in Brewing Materials." *MBAA Technical Quarterly* 10 (4).

31. Stewart, G.G. and I. Russell. 1985. "Modern Brewing Biotechnology." *Food and Beverage Products*, edited by Murray Moo-Young. Oxford, England: Pergamon.

32. Stewart, Graham G. 1995. "Adjuncts." *Handbook of Brewing*, edited by William A. Hardwick. New York, New York: Marcel Dekker, Inc.

33. Stewart, Graham G. 2006. "Adjuncts." *Handbook of Brewing*, edited by Fergus G. Priest and Graham G. Stewart. Boca Raton, Florida: CRC Press, Taylor & Francis Group.

34. Vogel, E.H., Jr. "Rice as a Brewing Adjunct." *Brewers Digest* 25 (4), 1950.

35. Wainwright, T. 1971. "Biochemistry of Brewing." *Modern Brewing Technology*, edited by W.P.K. Findlay. Cleveland, Ohio: The Macmillan Press.

36. Wainwright, Trevor. 1993. "Using Sugar Adjuncts in Brewing." *Brewers' Guardian* 122 (12).

37. Wilson J. 1990. "Brewing Sugars – The Versatile Brewing Adjuncts." *The Brewer* 76 (906).

Chapter Seven

Brewery Cleaning and Sanitation

Cleaning and sanitation are an integral part of a brewery and should be taken into consideration at every phase of the beer brewing process. Cleaning proceeds sanitation and prepares the way for sanitation treatment by removing organic/inorganic residues and microorganisms from the brewery equipment. Sanitation reduces the surface population of viable microorganisms after cleaning and prevents microbial growth on the brewery equipment.

Careful selection of chemicals for cleaning and sanitizing should consider:

- Effectiveness and suitability for the job
- The best application method for the chemical (i.e., foaming or non-foaming)
- Exposure time, temperature and concentration specifications
- Whether it is approved for surfaces coming into contact with food
- Toxicity and the operator safety requirements
- Impact on the environment and how it will affect the pollutant load of wastewater
- Corrosiveness
- Stability and ease of storage, and the possible impacts of any spillage

Cleaning Detergents

There are two types of cleaning detergents: alkaline-based or acid-based detergents that are often formulated with surfactants, chelating agents, and emulsifiers to enhance the effectiveness of the detergents. A detergent must be capable of wetting surface(s) to allow it to penetrate the soil deposits in order to act more quickly and efficiently. The detergent must have the capacity to break the soil into fine particles and to hold them in suspension so that they do not redeposit on the cleaned surface. Detergents also must have good sequestering power to keep calcium and magnesium salts (beerstone) in solution.

There are literally hundreds of detergent mixtures available that are specifically tailored to the needs of the brewer. The most effective detergents in the brewery today are formulated with alkaline solutions that have chelators and surfactants.

Alkaline-Based Detergents

Alkaline detergents are most effective in removing organic soils, i.e., oils, fats, proteins, starches, and carbohydrates encountered in brewing. Alkaline detergents work by hydrolyzing peptide bonds and breaking down large, insoluble proteins into small, more easily soluble polypeptides. Alkaline detergents will not remove calcium oxalate and other inorganic compounds that lead to a buildup of beerstone.

Alkali detergents, while having good properties for brewhouse cleaning, have two major disadvantages in cleaning brewery equipment. First, alkali detergents are not suitable for cleaning vessels with high residual carbon dioxide levels or where carbon dioxide top pressure is maintained during the cleaning cycle. The presence of carbon dioxide has a significant impact on the effectiveness of alkali-based cleaning solutions. For example, sodium hydroxide in the presence of excess carbon dioxide is reduced to sodium carbonate, a less effective cleaning agent—thus, resulting in higher cleaning costs. In some cases, there is the threat of a tank implosion that is brought about by the neutralization of alkaline detergents by carbon dioxide. This can be prevented by leaving the manway doors open when cleaning and by equipping the tank with pressure relief and vacuum relief valves. Alternatively, acid-based detergents can be used where carbon dioxide is a problem, which makes them more suitable for cleaning-in-place (CIP) systems. Secondly, alkaline based detergents react with the salts to form insoluble precipitates that accumulate on the surface to form scale or more commonly known as beer stone (calcium carbonate/oxalate).

Some of the more common alkalis are sodium hydroxide, sodium percarbonate, sodium hydrogen carbonate, sodium orthosilicate, and sodium metasilicate.

Sodium Hydroxide

Of the cleaning agents, sodium hydroxide (NaOH), otherwise known as caustic soda, is widely used in breweries worldwide. Its effectiveness in dissolving proteinaceous soils and fatty oils by saphonification is virtually unsurpassed. This makes it a natural choice for cleaning sludge off the bottoms of boilers and for cleaning beer kegs. Sodium hydroxide is an acutely excellent emulsifier too. It is unrivaled in its ability to dissolve protein and organic matter if used in conjunction with chlorine, surfactants, and chelating agents.

The solution can be used manually or in an automated cleaning-in-place (CIP) system. Caution must be taken if it is used manually because it is extremely corrosive, particularly on the skin and eyes. If it is used in a CIP system, a low-foaming or defoaming surfactant may be incorporated to enhance penetration and hold down foam generated by the soil as it is removed. High foam can cause pumps to cavitate and render their action ineffective in a re-circulating operation.

Traditionally, the brewing industry has used solutions of sodium hydroxide in the concentration range of from 1 to 5% w/v NaOH (33). For most CIP purposes, a level of 2% w/v NaOH is more than adequate; but in practice, higher concentrations may be necessary to obtain a satisfactory level of sequestration (40). The concentration usually is listed on the label, so it can be diluted with water. For example, diluting a 20% solution with one-part concentrate to nine parts water (1:9) will give the required 2%.

Sodium Hydroxide/Hypochlorite Solutions

Caustic/hypochlorite mixtures are particularly effective in removing tannin deposits, but are used for a great variety of cleaning tasks. These mixtures can be used in CIP systems for occasional purge treatments or to brighten stainless steel. A temperature of 22°C is an industry standard. The solution most commonly used consists of approximately 20 to 30% sodium hydroxide and 3 to 5% w/v available chlorine. Hypochlorite is incompatible with common sequestrants such as EDTA and sodium gluconate.

Acid-Based Detergents

Acid detergents are often used in a two-step sequential cleaning regime with alkaline detergents. Heavy soils, tannins, hop oils, resins, and glucans are unaffected by acid detergents. Acid detergents are also used for the prevention or removal of beerstone, water scale (calcium and magnesium carbonates), and aluminum oxide. Acid detergents are more effective against bacteria than are alkaline detergents (37).

Acid detergents are effective at cold temperatures and do not absorb carbon dioxide, thereby saving energy and avoiding vacuum collapse. In addition, water consumption generally is lower for breweries using acid detergents because acid detergents are rinsed more easily than alkaline detergents (38). Acid detergents produce a low pH detergent, which local regulations may require. Acid detergents are non-corrosive and completely safe for use on the stainless steel normally used in breweries (Types 304 and 316).

The most widely used acids in the brewing industry are phosphoric, nitric, sulfamic, and glycolic (hydroxyacetic). Because inorganic acid detergents are more aggressive, they are better cleaners, more corrosive, and more economical, whereas organic acid detergents are safer and less aggressive but more expensive to use. Acid detergents are used alone or in combination, and for best results, they are often formulated with corrosion inhibitors and surfactants.

> **Note | Water Quality and Chemical Effectiveness**
>
> Impurities in water can significantly reduce the effectiveness of the cleaning and sanitizing chemicals. For example, hard water containing substantial amounts of calcium, magnesium and iron hardness can result in scale, affecting the ability of detergents and sanitizers to make contact with the surface that requires cleaning. In such cases, the addition of buffering agents to highly alkaline (above 8) or highly acidic (below 5) water may help to improve the effectiveness of chemicals while also reducing mineral buildup.

Phosphoric Acid

Phosphoric acid is used widely in the removal of beerstone and similar deposits on surfaces such as protein material resins and yeasts. Its performance is greatly enhanced by adding an acid-stable surfactant, which promotes penetration of surface deposits and also assists in the process of rinsing at the end of the cleaning process. Phosphoric acid is not effective in removing beerstone until it reaches 16°C. However, some brewers believe it lacks the strength to remove pre-existing beerstone and is better at removing organic deposits. It often is used at a concentration between 2 and 3% w/v phosphoric acid for cleaning. Small quantities of complex organic acids are often added to enhance its effectiveness.

Phosphoric acid does not corrode stainless steel even at temperatures up to 80°C, and in combination with a corrosion inhibitor it can be used on aluminum at temperatures above 80°C. This combination makes phosphoric acid extremely suitable for keg cleaning (38). Phosphoric acid, in combination with effective corrosion inhibitors, also is suitable for use in installations composed of mixed materials such as stainless steel/copper, aluminum/copper, and aluminum/copper/brass.

Nitric Acid

Nitric acid not only is used to remove beerstone and other inorganic deposits, it also has biocidal properties when used either as a pure acid or in more stable, less hazardous mixtures with phosphoric acid (2). In addition, nitric acid attacks protein. Nitric acid does not corrode aluminum or stainless steel, but it is corrosive to other metals. Nitric acid offers the added benefit of inducing a passivating effect on stainless steel by forming a protective layer of chromium oxide on the stainless steel surface.

Additives

Surfactants, chelating agents, and emulsifiers often are added to enhance the effectiveness of both alkaline and acid detergents.

Surfactants

Surfactants, also referred to as wetting agents, are a large group of compounds that, when dissolved in water, gives a product the ability to remove dirt from surfaces of equipment thus increasing the penetration of the cleaning solution.

In more technical terms:

- They enable the cleaning solution to fully wet the surface being cleaned so that dirt can be readily loosened and removed.
- They can clean greasy, oily, particulate-, protein-, and carbohydrate-based stains.
- They are instrumental in removing dirt and in keeping them emulsified, suspended, and dispersed so they don't settle back onto the surface being cleaned.

There are three main groups of surfactants: anionic, nonionic, and cationic. All three groups have both high- and low-foaming qualities. The anionic types are negatively charged surfactants that have excellent wetting and dispersing qualities. They are most commonly used in detergent formulations, the most important being sulfated alcohols and alkyl aryl sulphonates such as Teepol®. The nonionic agents are neutral charge surfactants that principally are used in high-pressure systems. Nonionic agents are the workhorses of the brewing industry and are well-suited for use in conditions where solutions are subjected to high pressure. Both anionic and nonionic detergents have very good wetting, dispersing, and rinsing powers; but they are of little use, unaided, against most brewery deposits. The cationic wetting agents are positively charged and have the poorest detergent qualities of the group, but they do demonstrate good bactericidal action. As a result, they are used far more widely as sanitizers than as wetting agents. Mixtures of the wetting agents are often used; however, cationic and anionic wetting agents cannot be mixed together since they will neutralize each other and thereby eliminate their desirable properties.

Chelating Agents

Chelating agents (sequestrants) are chemicals that are incorporated into the detergent formulation that prevent scale buildup, i.e., the precipitation of calcium and magnesium salts onto equipment surfaces. The chelating agent is purchased either as part of a proprietary formulation or as an additive for blending. Commonly used sequestrants are EDTA (ethylene diamine tetra-acetic acid), NTA (nitrilo-tri-acetic acid), ADPA (aceto diphosphonic acid), sodium gluconate, and sodium tripolyphosphate.

Chelating agents exhibit different chelating power, and so manufacturers of brewery detergents often formulate detergents for specific tasks. The makeup of the additives can also differ, depending on their physical form; powdered detergents tend to contain polyphosphates, and liquids often contain gluconic acid.

Chelating agents vary in their effectiveness under different pH conditions. In the neutral or moderately alkaline range, EDTA is considerably more effective than the gluconic acids, but as the pH is raised in the presence of free caustic soda, the sequestering action of the gluconates rapidly surpasses that of EDTA. According to Kretsch, sodium gluconate's functionality at high temperatures and high pHs makes it ideal for use in concentrated sodium hydroxide-based cleaners (16). Thus, sodium gluconates are particularly suited for use in solutions used in cleaning-in-place systems, for bottle washing, and for similar applications where the solution is recovered and reutilized (35). More specifically, gluconates work extremely well at high caustic concentrations, particularly those greater than 2% sodium hydroxide (7). On the other hand, EDTA performs better at lower caustic values, up to 2%.

Emulsifiers

Emulsification is a measure of a detergent's ability to break down fats and oils into smaller units that are removed more easily during rinsing. Polyphosphates function as emulsifying agents and are employed widely in the formulation of alkaline detergents for use in brewery processing applications. Phosphates are divided broadly into two classes: the orthophosphates and the condensed or complex phosphates.

Orthophosphates

The most widely used of the orthophosphates is trisodium phosphate (TSP), which is a good soil remover and emulsifier. It is very effective in softening water by precipitation, producing an easily rinsed non-adherent precipitate. TSP is hard on the skin, so rubber gloves should be worn while using it.

TSP also is available in combination with sodium hypochlorite (a source of chlorine). It is known commercially as chlorinated trisodium phosphate and is in powder form. Chlorinated trisodium phosphate contains 3 to 5% available chlorine, which adds some bactericidal capability. Chlorinated trisodium phosphate assists in the cleaning operation, where the oxidizing power is of value. It is considered a better manual hand cleaner than a sanitizer. The dry powder retains its activity over long periods and releases hypochlorous acid on contact with water. Since all the chlorine atoms can be hydrolyzed to produce an equivalent amount of hypochlorous acid, the available chlorine content is twice the total chlorine content. The formulated derivatives have different solubilities, available chlorine contents, and rates of dissolution.

> **Note | Removing Beerstone**
>
> Beerstone usually appears as a dull brownish to brownish-white film on the metal's surface and is most likely to occur on stainless steel rather than on copper surfaces. A dual-chemical approach is needed in removing beerstone: an alkaline detergent to "digest" the organic component (protein), followed by an acid detergent to dissolve the inorganic (oxalate) minerals. In this situation, cleaning with 0.5–1.0% nitric acid or sometimes phosphoric acid is effective (6).
>
> Alkaline detergents are also routinely supplemented with sequestrants such as EDTA that work well in removing beerstone. Kretsch recommends a 5 to 15% EDTA solution between 38 and 60°C from 4 to 24 hours, depending on deposits (16).
>
> If a strong acid-based detergent is used, there is no need for an extra acidic cleaning/descaling rinse, providing all the alkali is removed and the equipment is used within a short time.
>
> If not removed, beerstone affects flavor; worse, it can harbor microorganisms. Furthermore, the surface of the metal in contact with the beerstone can escape passivation, resulting in permanent tank damage. The metal underneath the deposit becomes oxygen-depleted via biological or chemical means, and corrosion occurs.

Complex Phosphates

The complex phosphates used most widely in the formulation of detergent mixtures including tetrasodium pyrophosphate, sodium tripolyphosphate, sodium tetraphosphate, and sodium hexametaphosphate. All the complex phosphates, particularly sodium tripolyphosphate, have good synergistic properties in aiding the detergent in its cleaning action. The four phosphates have good sequestering and dispersing powers, with sodium hexametaphosphate having the greatest power to sequester calcium ions, and with tetrasodium pyrophosphate being best for sequestering magnesium ions (34). Sodium tripolyphosphate doesn't have the sequestering power of sodium hexametaphosphate and tetrasodium pyrophosphate, but can sequester both calcium and magnesium ions making it suitable in providing a broad spectrum of activity. The complex phosphates also have dispersing powers, as they are able to break up large masses into small ones. Salisbury reports that complex phosphates revert to orthophosphates with a resulting loss of sequestering power at raised temperatures or in the presence of strong caustic soda (34).

Sanitizing Agents

Sanitizing agents (often called disinfectants) are used to reduce the number of microorganisms to acceptable levels in brewing. Sanitizing may be accomplished by physical methods or through the use chemical sanitizers. Physical methods include the use of either hot water or steam to kill bacteria. Chemical sanitizing generally involves either immersing the object in a sanitizing solution for a specific amount of time or spraying/wiping the object with the solution and allowing it to air-dry. Chemical sanitizers differ in their effectiveness on certain microorganisms and in the concentration, temperature and contact time required to kill bacteria. Common chemical sanitizers include chlorine compounds, quaternary ammonium compounds, hydrogen peroxide, peroxyacetic acid, anionic acids, and iodophores.

Physical Sanitation

Some brewers favor steam or hot water for brewery sanitation on the grounds that chemical sanitizers can taint the beer with objectionable odors. To be effective, steam must be wet (not superheated) and free from air. Such treatment is effective, providing it is long enough to maintain a temperature of 100°C for at least 15 to 20 minutes. However, heating equipment to above 60°C for 20 minutes will eliminate most microbes. Steam is not suitable for certain types of cleaning. Where it is used, the plant equipment must be clean; otherwise, soil will bake onto the surface and cause sterility problems later.

Physical versus Chemical Sanitation

The advantage of physical sanitation is the elimination of chemical sanitizing agents; however, its application is limited due to the energy required to produce steam or hot water, and its suitability and effectiveness for some applications is limited. Chemical sanitation is therefore predominantly used in the beer brewing industry.

Alkaline-Based Sanitizers

Chlorine

Chlorine based sanitizers are widely used in the beer brewing industry. Chlorine compounds are broad spectrum germicides which act on microbial membranes, inhibit cellular enzymes involved in glucose metabolism, have a lethal effect on DNA, and oxidize cellular protein. Chlorine has activity at low temperature, is relatively inexpensive, and leaves minimal residue or film on surfaces. In properly blended products, chlorine based sanitizers are relatively non-toxic, colorless, non-staining, and easy to prepare and apply.

The activity of chlorine is dramatically affected by such factors as pH, temperature, and organic load. It is better to use chlorine based sanitizers at a pH range of 6.0–7.5. The effectiveness of chlorine increases with increase in temperature. However, above 50°C the liberated chlorine is rapidly lost to the atmosphere, reducing the effectiveness of the solution. Chlorine also loses its effectiveness when large amounts of organic matter are present. Chlorine is less affected by water hardness when compared to other sanitizers (especially the quaternary ammonium compounds).

The major disadvantage to chlorine based compounds is corrosiveness to many metal surfaces (especially at higher temperatures). Chlorine contamination will cause the formation of chlorophenols and chloramines, a class of chemicals with horrible medicinal flavors. Some of these chlorine compounds can be tasted in concentrations as small as a few parts per billion. Health and safety concerns can occur due to skin irritation and mucous membrane damage in confined areas. In recent years, concerns have also been raised about the use of chlorine based compounds and the formation of potentially carcinogenic trihalomethanes (THMs) under certain conditions.

Sodium Hypochlorite – Among the chlorine-based disinfectants, sodium hypochorite has been traditionally employed in breweries because of its relatively low cost and very powerful germicidal properties. Sodium hypochlorite is commonly known as household chlorine bleach, Clorox®, or chlorinated soda. It's better to use sodium hypochorite at a neutral or slightly alkaline pH because it can rapidly corrode stainless steel in acid conditions. Lewis recommends 200 mg/l (or ppm) of chlorine for 10 seconds at room temperature in most brewer settings (22). If not rinsed well, chlorine can also form crystals of sodium hypochlorite that will dissolve later in wort or beer.

Hydrochlorous Acid – An effective form of chlorine is hydrochlorous acid that works well within a pH range of 4 to 6. Most brewers, however, prefer to use hydrochlorous acid between a pH of 7 and 8. Hydrochlorous acid is less effective at that pH range, but it is much safer to use. If the pH is too acidic, chlorine becomes corrosive to stainless steel and dangerous to people.

Chlorine Dioxide – Chlorine dioxide's properties make it an ideal choice for most brewery operations. It is an ideal alternative to other chlorine-based sanitizers, providing all of chlorine's benefits without any of its weaknesses and detriments. Chlorine dioxide is less reactive with organic compounds than other chlorine-based sanitizers and, therefore, less affected by incompletely cleaned surfaces. Other properties of chlorine dioxide that differentiate it from other oxidizing biocides/disinfectants are:

- Chlorine dioxide possesses broad spectrum anti-microbial capabilities.

- Chlorine dioxide is not sensitive to system pH and does not foam.
- It provides a residual disinfectant level.
- Chlorine dioxide is significantly less corrosive than chlorine.
- It is approved by the EPA for drinking water disinfection.
- Chlorine dioxide is 100–1,000 times more effective at removing/preventing biofilms on equipment than chlorine. Biofilms shelter bacteria from turbidity and attack from chlorine as well as providing a rich nutrient source for further recontamination.
- Holds a USDA D-2 approval as a terminal sanitizing rinse, not requiring a water flush.
- Chlorine dioxide is more effective than other chlorine based sanitizers, requiring a lower dose and resulting in a lower environmental impact.

Chlorine dioxide's primary disadvantages are worker safety and toxicity. Its highly concentrated gases can be explosive and exposure risk to workers is higher than that for chlorine. Its rapid decomposition in the presence of light or at temperatures greater than 50°C makes on site generation a recommended practice. Chlorine dioxide is generated typically from sodium chlorite solution and applied at rates from 5 to 25 mg/l.

Quaternary Ammonium

Quaternary ammonium compounds, commonly referred to as "quats" or "QACs," are used extensively in breweries because of their stability and non-corrosiveness. They have rapid bactericidal action at very low concentrations but selective biocidal activity. QACs are efficient against gram-positive bacteria but less effective against gram-negative bacteria (38). They are very effective against yeast and mold too.

QACs are active and stable over a broad temperature range. They are less affected by light soil than are other sanitizers. However, heavy soil dramatically decreases activity. QACs generally have higher activity at alkaline pH. Acidity decreases the efficiency of many QACS to such an extent that at pH 3 their germicidal activities almost disappear. While lack of tolerance to hard water is often listed as a major disadvantage of QACs when compared to chlorine, some QACs are fairly tolerant of hard water when formulated with other chemicals. Activity can be improved by the use of EDTA as a chelator.

QACs possess considerable wetting power and are well-suited for use in formulations of combined cleaner/sanitizing agents. They can be either high- or low-foaming. Concentrations of upward to 200 mg/l are required to adequately sterilize the beer plant where immunities are known to develop (2). However, some brewers recommend using a dilute solution containing 100 mg/l for approximately 15 minutes for adequate sterilization.

QACs tend to adsorb onto surfaces, giving a mottled appearance after rinsing with water. It is well known that QACs can affect the head retention of beer. Therefore, they should never be used to clean anything that is exposed to the beer.

Quaternaries commonly used in breweries in Britain are the alkyltrimethyl ammonium bromides (Vantoc® AL), benzalkonium chlorides (Marinol®), lauryldimethylbenzyl ammonium chloride (Vantoc® CL), alkyldimethylbenzyl ammonium chloride (Noramium® C.85), and didecyldimethyl ammonium bromide (Deciquam® 222). The benzyl derivatives appear to have a somewhat better killing effect on wild yeasts than the aliphatic derivatives.

Acid-Based Sanitizers

Hydrogen Peroxide

Hydrogen peroxide (HP) has a broad spectrum with slightly higher activity against gram-negative than gram-positive organisms. It is not detrimental to the environment as when hydrogen peroxide decomposes, hydrogen and oxygen are formed. Recommended temperature of use is 16°C at a rate of 100 mg/l.

Peroxyacetic Acid (PAA)

Peroxyacetic acid (peracetic, PAA) has been known for its germicidal properties for a long time. One of the advantages of peroxyacetic acid is that, once it is dosed into water, there are no vapor issues as with chlorine-based compounds. Its other advantages include the absence of phosphates and foam, and its biodegradability. PAA is relatively stable at use strengths of 100 to 200 mg/l (17). Other desirable properties include low corrosiveness and tolerance to hard water. PAA solutions have been shown to be useful in removing biofilms.

PAA is highly active against both gram-positive and gram-negative microorganisms. The germicidal activity of PAA is dramatically affected by pH. Any pH increase above 7–8 drastically reduces the activity.

PAA is the only sanitizer that carries a label indicating that it is an oxidizer and corrosive. Local fire departments should be notified before using this product, as some consider it a fire hazard. Solutions of 0.20 to 0.35% are non-corrosive to common materials (35).

Anionic Acids

Anionic acids are one of the fastest growing sanitizing groups in the craft brewing industry. They are chemicals composed of two functional groups—a lipophilic portion and a hydrophillic portion—which results in a negative charge. The negatively charged anionic acid sanitizers react with positively charged bacteria by attraction of opposite charges. Their low use pH, detergency, stability, low odor potential, and non-corrosiveness to stainless steel make them highly desirable in some applications.

Anionic acid sanitizers are usually formulated with phosphoric acid to lower the pH and thereby increase their killing power. Antimicrobial activity decreases with the increase in pH; above a pH of 6.0, anionics show only bacteriostatic properties—they inhibit growth of bacteria but do not kill them (35). Anionic acids are effective in removing or controlling the formation of mineral films too.

Anionic acids are effective against all microorganisms at any temperature and are not affected by organic load. Anionics are not affected by sunlight, as are iodine, chlorine, stabilized chlorine dioxide, peroxide, and peracetic acid. Because they do not impart a flavor or odor and are detergent-based, anionics are used in part's buckets and usually reassembly of cleaning-in-place (CIP) systems. They are effective in hard and soft water and eliminate the need for acid rinsing. They are either high- or low-foaming and are employed at concentrations between 50 and 200 mg/l.

Disadvantages include: relatively high cost, a closely defined pH range of activity (pH 2 to 3), low activity on molds and yeasts, and incompatibility with cationic surfactant detergents.

Iodophores

Iodophores have a wide biocidal spectrum, react directly with the cell, and are not subject to immune species of yeast, bacteria, or molds. These are iodine-containing formulations are usually composed of elemental iodine, a surfactant, and an acid such as phosphoric acid. The surfactant reduces the staining and corrosive properties of iodine, which is stated to be a more effective sanitizer than chlorine at comparable concentrations. Iodophores based on phosphoric acid are unaffected by carbon dioxide absorption, will not cause a vacuum in pressure cleaning, and, unlike alkaline sanitizers, show no reduction in sanitizing properties in the presence of high carbon dioxide levels. The major disadvantage with some acid-containing formulations is that they are high-foaming because of the high level of surfactants. However, there are low-foaming second-generation iodophores that do not require as much acid.

Iodophores are extremely effective cold sanitizers (recommended temperatures varying from 21 to 26°C). Iodophore use solution temperatures should not exceed 48°C or they will begin to 'gas-off'. Effective sanitation usually is accomplished at concentrations of about 25 mg/l of free iodine with contact times generally from 3 to 5 minutes. Sterilization can be accomplished at concentrations of 50 mg/l. Salisbury reports that iodine monochloride is effective at concentrations of 12 mg/l and has reduced foaming characteristics at that concentration (34).

The germicidal performance of different iodophore formulations may differ greatly. Products yielding the same pH and iodine concentration may yield vastly different germicidal activities at equivalent dilutions. Iodophores can be used in very hard water.

Iodophores are non-corrosive to Types 304 and 316, stainless steel. However, under certain conditions, iodophores are very corrosive to stainless steel, i.e., when CIP temperatures are over 27°C, pH solutions are above 6, or solutions are not allowed to completely drain from the tank.

Iodine is less likely to taint the beer, as it does not react (as chlorine does) with proteins. Consequently, many users find it unnecessary to post-rinse, provided that vessels are blown out with sterile air or carbon dioxide after use. Iodine can be as toxic as chlorine, so care should be taken in its handling.

Cleaning Methods

Nearly all brewery equipment including tanks, fermenters, brew kettles, and lauter tuns contain product reside that must be removed between batches or at routine intervals in continuous operations. There are two widely used methods for cleaning equipment:

- Manually
- Clean-In-Place systems

Manually

Manually cleaning is just exactly what it sounds like—a worker physically cleans the brewery equipment. Many craft brewers do not have the luxury of cleaning-in-place systems and have to manually clean and sanitize their equipment. They often have to use soft-bristled brushes, non-abrasive pads, cloths, and handheld spray hoses for cleaning. When cleaning manually, great care must be taken to assure that brushes and equipment are cleaned to avoid cross-contamination. The choice and concentration of detergents and sanitizers is limited in manual applications, given the risks to the user. Foaming detergents containing high amounts of wetting agents but of low alkalinity are normally used for manual scrubbing. In addition, the temperature of the water is usually limited between 48 and 50°C.

Manually cleaning often results in inconsistent cleaning and the use of more cleaning chemicals and water that is really necessary. In addition, there are safety concerns about tank entry and exit and exposure to toxic fumes from cleaning agents.

Clean-In-Place (CIP) Systems

Clean-in-place (CIP) systems were developed by the dairy industry as a means of reducing the amount of labor needed for cleaning and sanitizing operations.

One of the main advantages of CIP systems is that they can recirculate and allow the reuse of chemicals and rinse water, thereby reducing consumption by as much as 50%. CIP systems largely remove human contact with cleaning and sanitizing agents, thus reducing the risk of harmful exposure. They also assure a more consistent cleaning by removing some of the common sources of human error in cleaning.

This CIP unit (see Figure 7.1) includes a rinse/recirculation tank providing surge volume, a centrifugal pump for solution delivery, a heat exchanger to temper cleaning solutions, chemical feed equipment, control panel, associated control valves necessary to ensure system functionality, and tank spraying devices. CIP systems are usually equipped with in-line monitoring instrumentation that measures the concentration and temperature of chemicals. Correct chemical dosing is usually controlled by conductivity sensors, which should be well maintained and recalibrated to ensure that the correct amounts of chemical are being used for dosing.

Figure 7.1
Stationary CIP System
Courtesy of GEA Brewery Systems

CIP cycles are usually designed to clean a particular soil for a given set of processing equipment. A typical CIP cycle could include pre-rinse, caustic wash, rinse, acid rinse, sanitization, and post-rinse steps. See Table 7.1 for a typical CIP cleaning and sanitation program used in a brewery.

Table 7.1
CIP Cleaning and Sanitation Program

Action	Temperature	Duration
Pre-rinse	Cold or hot	5–10 min
Alkaline cleaning; sodium hydroxide (1.5–4%)	Cold or hot (60–65°C)	10–60 min
Intermediate rinse	Cold or hot	10–30 min
Acidic cleaning; phosphoric, nitric, sulfuric acid (1–2%)	Cold	10–30 min
Intermediate rinse	Cold	10–30 min
Disinfection - by disinfectant solution - by hot water	Cold (85–90°C)	 10–30 min 45–60 min
Final rinse	Cold	5–10 min
Source: Storgärds, 2000		

CIP systems can be installed permanently or configured as a mobile unit and easily moved form tank to tank. Permanent systems are inflexible, cost more than mobile systems, but generally are more uniform in cleaning and sanitizing the equipment. Dedicated systems also have the advantage in preventing cross-contamination of soils. For example, the same CIP system is used only for cold storage tanks.

Although individual pieces of equipment are equipped with CIP systems any pipework connected to this equipment or for that matter, detachable fittings such as hose adapters, valves, swing bends, components of fillers, flexible hoses, etc., will need to be cleaned manually followed by immersion in soak baths for disinfection. Too often these items are not subject to the same scrutiny as the equipment fitted with a CIP system and turn out to be a reservoir for taints or infection.

Spraying Devices

Spray devices are used to apply CIP fluids to the surface being cleaned. Two general types are available: static and dynamic.

Static spray devices are motionless heads with drilled or fixed nozzles. Popular versions include spray balls (see Figure 7.2), tubes and bubbles. Spray tubes and bubbles are essentially the given shape with drilled spray holes. Static spray balls are normally designed for 20 to 30 gpm and 20 to 30 psi pressure drop. The effective cleaning diameter of a static spray ball is about 8 feet. Static spray balls are typically used in small tanks, bright beer tanks, and water tanks where a combination of chemical strength, high temperature, and low mechanical action is all that is required for cleaning.

Dynamic spray devices have a moving spray head or body, which is driven by the cleaning media (see Figure 7.2) and/or mechanical means. Spray balls are normally left in the tank during processing. Dynamic spray-balls are used in tanks of larger diameter and tanks with heavier, more tenacious soils, where greater mechanical force and broader spay coverage is needed. Examples of vessels in which these devices are used include brew kettles, fermenters, lagering tanks, and large bright beer tanks.

Figure 7.2
From Left to Right Static and Dynamic Spray Balls
Courtesy of Alfa Laval, Inc.

CIP Single-Use or Recovery Systems

CIP units are available as single-use or recovery types, or as a combination of both.

Single-Use Type – In the single-use method, the cleaning solution is prepared and re-circulated through the system and discharged at the completion of the cleaning cycle. In some systems, the cleaning solution is used as a pre-rinse for the next cleaning cycle and then discharged. The main cleaning is done with a fresh solution. The advantages of this method are that the risk of cross-contamination is reduced and the absence of a cumulative soil load in the solution usually permits use of lower concentrations (34). The single-use method is suitable only in areas of heavy soil where cleanliness and microbiological hygiene are of paramount importance (e.g., propagation plant). In broad terms, a single-use unit is less expensive since it requires only one tank and one pump and needs less space. It is more flexible as it can use different strengths of cleaning solution for each plant, and the detergents are always fresh at the beginning of the cycle. This makes it suitable for cleaning a range of plants with different soil loads. Single-use CIP systems, because of their versatility in washing processing equipment, are suitable for craftbrewers. On the other hand, their consumption of water and chemicals usually is higher, so more effluent is produced.

Recovery-Type – In the recovery type, the solution is recovered at the completion of the cleaning cycle. These systems are used where the cleaning task is less burdensome, for example, bright beer tanks. A recovery unit, while more expensive to install can have lower running costs—reduced chemical and wastewater processing costs. This is especially true if hot cleaning is used, as both the heat and the chemical are recovered. If the unit has multiple pump sets, it can carry out more than one clean at the same time.

Cleaning and Sanitation Manual

The first step in any effective cleaning and sanitation program is the development of a detailed, up-to-date manual. This manual should establish a systematic procedure for cleaning and sanitizing each major piece of brewery equipment, listing the frequency, method, and materials to be used for cleaning and sanitizing. For each cleaning and sanitizing procedure discussed in the manual, the weight or volume of material used should be given relative to the amount of water used, along with the concentrations of each material involved. The manual should also include the schedule of routine microbiological assessments or surveys to evaluate the sanitary conditions and, hence, the effectiveness of the cleaning and sanitizing operation.

The manual should include an outline of safety precautions, along with any instructions regarding the use of protective clothing and protective equipment. Material Safety Data Sheets (MSDS) pertaining to chemicals used should be included in the manual. Finally, it is essential that the manual be kept up to date, by a designated person of the brewery.

OSHA requires MSDS for each hazardous material in use in the workplace. When purchasing chemicals, ask for an MSDS; OSHA requires suppliers to provide MSDS sheets. Listed on the MSDS is information on health risks; symptoms of overexposure, appearance, and smell of chemicals; and proper work practices and personal hygiene that can reduce exposure to chemicals. Also listed is what to do in case of a medical or other emergency.

Material Corrosion Resistance

Stainless Steel

Many types of stainless steel are used in brewing beer. The type of stainless steel used in brewing and fermentation equipment is the nonmagnetic 300 series, which includes several types. Those more common to brewing are 304 and 316L stainless steel. Types 304 and 316L have very good corrosion-resistant properties and are easily welded. Most brewery equipment is constructed from

Type 304 stainless steel. Type 316L, which has better corrosion resistance properties, is often used but the material cost is much higher. Other 300 series metals are to be avoided for brewery use, especially 303.

The corrosion inhibitor in stainless steel is the passive oxide layer that protects the surface. Beerstone (calcium oxalate) can cause corrosion if not removed, because the metal beneath the deposit becomes oxygen-depleted.

Chemical Agents

Acid Detergents – Some acids can be used for a variety of stainless steel cleaning and removing beerstone. Phosphoric and nitric acids are the preferred choice because they do not attack stainless steel, though nitric acid is considered best. Nitric acid should be blended with a phosphonic to keep it from fuming off. Excessive concentrations and contact times can damage a stainless steel surface finish. Hydrochloric (muratic) acid is very effective in cleaning and removing beerstone too. If muratic acid is used, it is imperative that the vessel be rinsed afterwards because chlorides can corrode stainless steel.

Alkaline Detergents – Sodium hydroxide, commonly used in the CIP systems of commercial breweries, is quite effective for removing organic deposits from stainless surfaces. However, although it is effective in dissolving organic carbon-based compounds, it tends to increase the occurrence of calcium-based deposits on stainless steel. In addition, caustic solutions must be used carefully, because high concentrations can damage the surface of the stainless steel if the contact time is too long.

Alkaline Disinfectants – Alkaline disinfectants, e.g., sodium hypochlorite, do not corrode stainless steel Type 316. However, when these disinfectants are exposed to Type 304 stainless steel, which are less resistant, temperatures should be restricted to 60°C and concentrations to 250 mg/l (13). Prolonged contact with bleach solutions (longer than 30 min.), especially with temperatures greater than 60°C, can pit the surfaces of stainless steel containers.

Passivation

As mentioned, stainless steel's resistance to corrosion and discoloration is in part due to a passive oxide layer that protects the metal. The oxide forms naturally on clean surfaces exposed to the atmosphere, but this formation can take up to two weeks, which is too long for breweries. A technique known as passivation, using acid mixtures containing oxidizing agents, can be used to enhance the formation of the passive oxide layer. The most frequently cited oxidizing acid solution used for passivation is nitric acid but given the dangers of nitric acid many equipment suppliers recommend citric acid or a blend of nitric and phosphoric acid (19). Typical solutions range from 4 to 10% citric acid by weight. Reportedly citric acid may lead to off-flavor problems of the first fill. The passive oxide layer can be damaged by improper or excessive use of chlorine-based sanitizers or contact with steel tools.

Copper

Copper generally is more acid-resistant than alkaline-resistant. Copper is usually resistant to non-oxidizing acids such as acetic, hydrochloric, and phosphoric, but is not resistant to oxidizing acids such as nitric and sulfuric or to non-oxidizing acid solutions that have oxygen dissolved in them (30). Alkaline detergents will blacken copper due to the formation of oxides. Commercial detergents usually contain buffering agents and inhibitors that prevent corrosion of copper.

Aluminum

Caustic cleaners react with aluminum, actually dissolving the metal and pitting the surface. The reaction with aluminum can produce a potentially dangerous situation, in that flammable hydrogen gas is produced. Proper ventilation is necessary under these conditions. The unsightly pitting that can occur can be a good harboring point for bacteria. For these reasons, aluminum is no longer used in the manufacture of brewing equipment and implements.

References

1. Ault, R.G. and R. Newton. 1991. "Brewing Hygiene and Biological Stability of Beers." *Modern Brewing Technology*, edited by W.P.K. Findlay. Cleveland, Ohio: The Macmillan Press.

2. Barrett, M.A. 1979. "Detergents and Sterilants in the Brewery." *Brewers' Guardian* 108 (8).

3. Barrett, M.A. 1991. "Detergents and Sterilants." *The Brewer* 77 (919).

4. Block, Seymour S. 1991. *Disinfection, Sterilization, and Preservation*. Philadelphia, Pennsylvania: Lea & Febiger.

5. Bousfield, K. 1993. "Practical Cleaning of Brewery Plant." *The Brewer* 79 (941).

6. Briggs, Dennis E., Boulton, Chris A., Brookes, Peter A., and Stevens, Roger. 2004. *Brewing: Science and Practice*. Cambridge, England: Woodhead Publishing Limited.

7. Docherty, R.A. 1984. "Stepwise Guide to Cleaning in the Brewery." *Brewing and Distilling International* 14 (2).

8. Dumpleton, C. 1985. "Detergents and Detergency." *Journal of Sterile Services Management* 155 (6).

9. Fix, George and Laurie Fix. 1997. *An Analysis of Brewing Techniques*. Boulder, Colorado: Brewers Publications.

10. Foss, Greg. 1997. "The Dirt on Brewery Cleaning – A Review of Procedures and Chemicals." *Brewing Techniques* 5 (2).

11. Giorgio, Greg. 1990. "QC: Fact of Fiction?" *The New Brewer* 7 (1).

12. Hamilton, Gordon. 1979. "Cleaning-in-Place Plant and Procedures." *Brewers' Guardian* 108 (7).

13. Hough, J.S., et al. 1982. *Malting and Brewing Science*. Volume 2. London, England: Chapman–Hall, Ltd.

14. Johnson, Dana. 1997. "Applications of Chlorine Dioxide – A Postrinse Sanitizer That Won't Leave a Bad Taste in Your Mouth." *Brewing Techniques* 5 (2).

15. Jones, Donald. 1988. "The Importance of Passivation." *The New Brewer* 5 (1).

16. Kretsch, John (Ace Chemical Products, Inc.). 1993. "Practical Considerations for Brewery Sanitation." *106th MBAA Anniversary Convention*. Anaheim, California: Master Brewers Association of the Americas.

17. Kunze, Wolfgang. 1996. *Technology Brewing and Malting*, translated by Dr. Trevor Wainwright. Berlin, Germany: VLB Berlin.

18. Landman, L. Charles, Jr. 1999. Personal Communication. National Chemicals, Inc., Winona, Minnesota.

19. Lewis, Ashton. 1999. "A Passivation Primer." *The New Brewer* 16 (4).

20. Lewis, Ashton. 2006. "Plant Sanitation." *Brewing and Engineering and Plant Operations*. Volume 3, edited by Karl Ockert. St. Paul, Minnesota: Master Brewers Association of the Americas.

21. Lewis, Michael. 1990. "Squeaky Clean." *The New Brewer* 7 (1).

22. Lewis, M.J. and Young, T. W. 2002. *Brewing*. New York, New York: Kluwer Academic/Plenum Publishers.

23. Lupa, Mark. 1992. "Laboratory Safety." *The New Brewer* 9 (5).

24. MacDonald, J., et al. 1984. "Current Approaches to Brewery Fermentations." *Progress in Industrial Microbiology*, edited by M.E. Bushell. Guildford, United Kingdom: Elsevier.

25. Mallet, John. 1993. "CO_2: Its Dynamics in Beer." *The New Brewer* 10 (2).

26. Middlekauff, James E. 1995. "Microbiological Aspects." *Handbook of Brewing*, edited by William A. Hardwick. New York, New York: Marcel Dekker, Inc.

27. Millspaw, Micah. 1994. "The Care and Feeding of Stainless Steel." *Brewing Techniques* 2 (4).

28. O'Shea, James M. III. 1999. Personal Communication. Five Star Products & Services, L.L.C., Commerce City, Colorado.

29. Palmer, John J. 1995. "Corrosion in the Brewery: How to Get the Upper Hand." *Brewing Techniques* 3 (2).

30. Parkes, Steve. 1998. "Beerstone – Tackling a Classic Brewers' Nemesis." *Brewing Techniques* 6 (4).

31. Pugsley, Alan. 1992. "It's A Way of Life." *The New Brewer* 9 (5).

32. Rosner, D. 1986. "Cleaning of Cylindroconical Fermenting and Maturation Vessels." *The Brewer* 72 (857).

33. Salisbury, Mike. 1983. "Cleaning Storage Vessels in CO_2 Atmospheres." *Brewers Digest* 58 (6).

34. Salisbury, Mike. 1999. "Brewery Cleaning and Sanitizing." *The Practical Brewer*, edited by John T. McCabe. Wauwatosa Wisconsin: Master Brewers Association of the Americas.

35. Scheer, Fred. 1990. "Filtering and Cleaning." *The New Brewer* 7 (1).

36. Schoof, Robert F. 1977. "Safety." *The Practical Brewer*, edited by H. M. Broderick. Madison, Wisconsin: Master Brewers Association of the Americas.

37. Schoorens, J. 1982. "Acid Cleaning and Related Disinfection in Breweries." *The Brewer* 68 (807).

38. Siegel-Sebolt, Cecil. 1992. "Compliance Please." *The New Brewer* 9 (5).

39. Stillman, C.G. 1983. "Detergents for CIP." *Brewers' Guardian* 112 (1).

40. Stillman, C.G. 1990. "Sanitizing Treatments for CIP Post-Rinses." *Brewing & Distilling International* 21 (3).

41. Stillman, C.G. 1991. "Primary and Secondary Treatments in Brewery CIP." *Brewers' Guardian* 120 (7).

42. Storgärds, E. 2000. *Process Hygiene Control in Beer Production and Dispensing*. Ph.D. thesis. Espoo, Finland: VTT Technical Research Centre of Finland.

43. Talley, Charlie. 1999. Personal Communication. Five Star Products & Services, L.L.C., Commerce City, Colorado.

44. Thompson, Donald, Karl Ockert, and Vince Cottone. 1999. "Craft Brewing." *The Practical Brewer*, edited by John T. McCabe. Wauwatosa Wisconsin: Master Brewers Association of the Americas.

45. Tuthill, Arthur H., Richard E. Avery, and Roger A. Covert. 1997. "Cleaning Stainless Steel Surfaces Prior to Sanitary Service." *Dairy, Food and Environmental Sanitation* 17 (11).

46. Warner, Eric. 1992. "Working with CO_2." *The New Brewer* 9 (5).

47. White, George Clifford. 1992. *Handbook of Chlorination and Alternative Disinfectants*. New York: Van Nostrand Reinhold.

48. Woods, Melinda. 1995. "Sanitation: A Common Sense Approach." *The New Brewer* 12 (2).

Chapter Eight

Malt Milling

The objective of milling is to reduce the malt to particles sizes, which will yield the most economic extract (wort) and will operate satisfactorily under brewhouse conditions and throughout the brewing process. The more extensive the malt is milled, the greater the extract production. However, the fine grind can lead to subsequent wort separation problems and a loss of extract in the spent grains during wort separation. As a result, the brewer needs to consider the equipment used in the brewhouse when determining the particle size when milling the malt. For example, mash tuns require comparatively coarse grists while lauter tuns can use finer grists and mash filters still finer grists.

Malt Handling and Storage

Malts can be purchased either in bags or in bulk shipments. Small craft breweries are more likely to purchase their malt in bags that weigh 50 to 55 lbs., which is particularly the case when purchasing specialty malt. Almost all other breweries purchase malt in bulk shipments transported either by truck or rail as bulk shipment offers the greatest savings in malt price.

Weighing and Cleaning

For larger breweries, a typical flow of malt upon arrival at the brewery begins by weighing the vehicle on the weigh-bridge and unloading the malt via a hopper. After unloading, the vehicle is tare-weighted, with the difference in weights being the weight of the malt. Alternatively, an in-line tipping conveyor can be used for recording weights. The malt is then run through a sieve cleaner to separate oversize malt and to separate magnetic impurities, i.e. iron and steel fragments, using a magnetic separator. Dust is separated by an aspirator by means of an intake fan and dust collector.

Silos

Once weighed and cleaned, the malt is conveyed to silos for storage. Silos are constructed of either corrugated galvanized steel, which is considerably cheaper or welded steel, which reportedly offers a higher level of protection from moisture. The minimum size for silos is usually 800 cubic feet, large enough to accommodate one truckload of malt.

Silos are filled pneumatically or mechanically by screw auger (available with a solid screw or a flex auger), horizontal conveyors, or bucket elevators. Pneumatic filling involves vacuuming or blowing malt from one location to another via sealed piping. It has the advantage of being able to lift the malt for several stories and to make several directional changes through gentle piping curves. Its disadvantages include high energy requirements and a tendency to damage malt if the system has not been properly designed.

The solid screw auger has greater capacity and handles malt more gently than the flex auger. Flex augers, as the name implies, have the ability to make several directional changes and are well-suited to installation after major components are in place. Flex augers perform best and damage malt least when operated at full capacity. Large breweries are more likely to use horizontal conveyors and bucket elevators because they offer more efficient movement of grain, lower power consumption, and minimal breakage, especially to very dry and friable malt. Horizontal conveyors are known as "chain and flight" conveyors. Bucket elevators are used to move malt to the top of the silo by means of buckets fitted to a rubber belt.

Hopper

Before milling, malt is removed from the silo and transferred to the feed hopper supported on a mechanical platform located at the grist case—a buffer containing the milled malt before the mash vessel. Load cells are employed to translate the mechanical load on the scale hopper in order to record and control the weight. Alternatively, a dump scale mounted atop the mill can be used for weighing the malt.

Grist Case

After milling, the milled malt is transferred to a temporary storage hopper, commonly called the grist case, that feeds the mash tun. Typically the grist case is located above the mash tun allowing the milled grain to be fed by gravity at the start of the brewing process, called mashing. Sometimes the grist case is set up with chutes or conveyors in order to serve more than one mashing vessel. A continuous read-out of the grist weight can be achieved by suspending the grist case on load cells.

Types of Malt Milling

Dry Milling

In breweries, dry milling is commonly performed by roller mills or hammer mills. If the wort separation involves using a mash tun or lauter tun, roller mills are employed. Hammer mills are largely used for the later generation of mash filters and continuous brewing systems.

Before milling the malt the brewer must consider the size distribution of malt kernels, their modification, moisture content, the mashing methods, and the wort separation method. For example, the degree of modification must be considered before milling the malt. Poorly modified malts require fine grinding and well-modified malts can be more coarsely ground. If wort separation is done with a mash filter, the malt can be ground finer than with a lauter tun. Mash tuns, with their deep grain bed, require a coarser grind of the malt compared to malt for lauter tuns.

Roller Mills

Roller mills are particularly suited for milling malt when the primary objective is to leave the malt husk intact. An intact husk helps wort separation and may reduce extraction of tannins and other undesirable components.

Roller mills are either two-, four-, or six-roll configuration. In general, the more rolls, the greater the flexibility and capacity of the mill. Multi-roll malt mills provide a degree of control that favors gentle treatment. The grind is controlled by the rate of the feed of the unground malt, the roll corrugations and size, the spacing between the rolls, and the speed at which the rolls are driven.

The roller mills are mounted horizontally in a rigid frame in such a way that one is fixed and the other adjusts to change the gap between the rolls. The rolls can be corrugated to match most any job requirement: coarse grooves for cracking, fine grooves for grinding and crimping, or smooth rolls for very fine grinding as in flour mills.

The relative speed between the rolls will determine what effect they have on the malt. If the speeds are the same, i.e., 1:1, the action is primarily compression, though some cutting will occur with corrugated rolls. As the relative speed between the rolls changes, they are said to operate with a differential. Differential speeds are described as the ratio of the fast roll to the slow roll, with the speed of slow roll expressed as one. For example, a mill operating at 1.5:1 differential would have the fast roll turning 1.5 times faster than the slow roll. A differential provides a shearing, grinding action between the rolls and is required to achieve maximum reduction and efficiency with a roll mill grinder.

A practice used by some commercial brewers is to double grind (8). The first grind uses wide-set rollers that separate the husk from the malt kernel, and this is followed by a light grind through a pair of more closely-set rollers. This practice allows for sufficient grinding of the kernel, yielding maximum extract while keeping the husks intact so that the grain bed remains sufficiently permeable to permit rapid wort runoff.

Once the malt is milled, it is collected in a grist case (or hopper) that may serve one or more mashing vessels. The grist case may also serve the purpose of mixing the ground malt with other solid adjuncts.

Two-Row Mills – Two-roll mills (see Figure 8.1) are single-pass mills commonly used by craft breweries and/or for well-modified malts. They are considerably less expensive than comparable four-roll, five-roll, and six-roll malt mills. However, the uniformity of milling fineness is not as good as with multi-roll mills. While two-roll mills are still in operation, they have largely been succeeded by multi-roll mills, which are of a four- or six-roll design.

Figure 8.1
Two-roll mill
Courtesy of Roskamp Champion

Multi-Row Mills – Multi-roll mills provide greater control of the rate of feed of the unground malt, the spacing between rolls, and the rate of speed, either uniform or differential, at which the rolls are driven. They can either be four, five, or six rolls with built-in vibrating screens to sort the various fractions that are produced from the malt—husks, coarse grits, fine grits, and flour. Unlike two-roll mills, these mills are more suited for milling under-modified malts.

Four-roll mills are equipped with two pairs of rolls with an intermediate separation stage allowing insufficiently crushed particles to be passed between rolls a second time. A sieving mechanism can be installed to provide a degree of selection for the brewery. The use of four-roll mills is widespread, but they are relatively inflexible in the quality of grist produced and are better used in mash tun operation for traditional ale brewing where coarse ground grists from more highly modified malts are employed (4). Using a four-roll mill to produce a finer grind required with less well-modified lager malts can lead to unacceptable husk damage.

The six-roll mill is the unit of choice in medium-to-large breweries and is the most flexible of all in handling malts of widely different qualities and in producing a wide range of grinds. Thus it can cater both to the lauter tun and mash tun. The best grist is obtained from six-roll malt mills with installed screens.

Hammer Mills

A hammer mill (see Figure 8.2) consists of a rotor made of two or more plates with pins to carry the hammers. Hammers are simply flat metal bars with a hole at one or both ends. They may have some type of edge preparation such as hard facing or carbide coating to provide better wear resistance. Malt reduction in a hammer mill is primarily a result of impact between the hammer and malt particles, or between particles and screen when the material accelerated by the rotor flies out into the screen. Hammer mills are used for grist preparation for high pressure filters and the Meura 2001 because in these devices the survival of large fragments of husk is irrelevant in collecting the wort.

Malt Conditioning

A refinement to dry milling employed by numerous breweries is conditioning of malt with steam or warm water. This practice minimizes the risk of fracturing the malt husks, thus the husks become tougher and more flexible due to absorbed moisture, while the endosperm remains dry and friable. The resulting grist is said to be characterized by a husk fraction containing a very high proportion of split husks with very few endosperm particles adhering. When properly operated, malt conditioning is claimed to increase the husk volume by 10 to 20% (4). The large husk particles are more resistant to damage during mashing and are said to arrive at the wort separation stage in better condition, resulting in reduced lautering time and higher mash/lauter tun capacity. Mash and lauter tuns can be loaded more deeply with conditioned malt as the bed density is reduced and its porosity increased.

Figure 8.2
Hammer mill
Courtesy of Roskamp Champion

Briggs *et al.* report malt conditioning gives a better yield of extract, better attenuation, and faster saccharification (2). It also said conditioning protects the enzymes of the mash against thermal shock.

When conditioning the malt, it passes through a screw conveyor in which the husks are exposed momentarily to a steam atmosphere or to a warm water spray. There is usually a buffer vessel between the screw conveyor and the dry roller-mill to ensure a constant flow of conditioned grain to the mill and to prevent steam or water from entering the mill.

Sampling

Dry milling and conditioned dry milling operations have the advantage that the crushed malt may be sampled by the brewer and assessed visually for uncrushed kernels, excessive tearing of the husks, and excessive flour. An intact husk aids in wort separation and may reduce extraction of phenolic compounds (15).

Wet Milling

Wet milling is very common in Africa and Asia, as it simplifies the grinding, but it is not common in the United States. In a wet milling operation, the whole uncrushed malt is pre-steeped in hot water to the point where the husks reach a water content of approximately 20% and the endosperm remains nearly dry, which results in a semiplastic, almost pasty consistency. The duration and temperature of steeping depends on the modification and the moisture content of the malt. The steeped malt is then passed through a two-roll mill in which the endosperm is squeezed out of the surrounding husk, leaving it fully available to the subsequent actions of mashing. The benefits of this process are the larger husk particles permit rapid-runoff during wort separation and the smaller endosperm particles allow higher extract recovery. There is, of course, no hazard relating to dust and explosions. There are some disadvantages to wet milling—high energy costs, difficulty in maintaining sanitary conditions in the mill, and increased maintenance and roller-wear, particularly when raw barley is crushed.

Sieve Analysis of Crushed Malt

Not only is the grist profile critical to achieving maximum conversion of available extract during mashing but it is also important for efficient wort filtration and sparging. Consequently, a complete screen analysis of the crushed malt when milling should be made on a regular schedule, with any adverse change in the mill's performance warranting an immediate correction.

Screen analysis consists in placing a sample of grist in the top set of horizontal sieves, which is shaken manually or mechanically for a set period of time. The particle size distribution can then be determined by measuring the percentage, by weight, of the grist retained on each, successively finer sieve.

Although a mechanical sieve-tester reduces labor, it does not appear to be a necessity in obtaining reproducible and repeatable results. However, it is critical that a manual procedure for screening the malt be documented and that this same procedure is followed by everyone in the brewery.

The most commonly used sieves for measuring particle distribution are those recommended by the American Society of Brewing Chemists (ASBC) and the European Brewing Convention (EBC) as shown in Table 8.1. Each sieve is a specific size with a given fraction name—husks, coarse grits, fine grits, flour, and fine flour. The names given to the fractions do not represent morphologically pure fractions since the composition of the fraction changes with the type of milling.

Each brewery will have to establish its own grist profile based on the wort quality, extraction efficiency, and other parameters appropriate for the brewery.

Mashing Systems

Typically, the malt is ground coarser for infusion systems than those used in more intensive mashing programs, such as temperature-controlled and

Table 8.1
Standard Sieve Sizes

ASBC			EBC			
Sieve Number	Mesh width	Fraction name	Sieve number	Wire thicknes in mm	Mesh width in mm	Fraction name
10	2.000	Husk	1	0.31	1.270	Husk
14	1.410	Husk	2	0.26	1.010	Coarse girts
18	1.000	Husk	3	0.15	0.547	Fine grits 1
30	0.590	Coarse Grits	4	0.07	0.253	Fine grits 2
60	0.250	Fine grits	5	0.04	0.152	Flour
100	0.149	Flour	Tray	--	--	Fine Flour
Pan	--	Fine Flour	--	--	--	--

decoction mashes. An infusion mash tun requires barley husks remain relatively undamaged, because they are needed to provide a filter medium for the removal of solid material from the wort. An infusion mash tun doesn't have the rakes and knives that a lauter tun (a vessel specifically designed for wort separation) has to cut and lift the grain bed.

Mash filters require a finer particle size distribution largely because of the shallow filter bed. Although the malt can be ground finer for a mash filter operation, it is still important to avoid an excess of fine flour (5). Excess fine flour in the grist tends to form dough, which can partially blind the pores of the cloths, adversely affecting filter pressure and filtration efficiency. Unlike mash tun and lauter tun wort separation systems, when mash filters are employed for wort separation, preservation of the husk is unnecessary. This is because compressibility of the spent grain bed is reduced, and a more extreme pulverization of the grist is required.

Safety

Grain handling systems should be well-ventilated to prevent the possibility of a grain-dust explosion. A tiny spark created by the metal rollers hitting a stone or staple in the malt can be enough to ignite an explosion. For this reason, grain cleaners are necessary. Grain rooms should be well-ventilated and clean to keep the concentration of dust to a minimum. Electrical outlets, switches, and motors should be dust proof to minimize the chance of a grain explosion. Also, dust arising from malt handling and processing is an important hazard in breweries because of its long-term effects upon the mucous membranes when inhaled.

References

1. Atkinson, B., Brown, P., Frisby, R., Heron, P., Hudson, J., Laws, D., Lloyd, W., Putnam, R., Reed, R., Tann, P., Tub, R., and Jackson, G. 1985. *Manual of Good Practice for Casking Conditioned Beer.* Nutfield, England: Brewer's Society and Brewing Research International.
2. Briggs, Dennis E., Boulton, Chris A., Brookes, Peter A., and Stevens, Roger. 2004. *Brewing: Science and Practice.* Cambridge, England: Woodhead Publishing Limited.
3. Carey, Daniel J. 1991. "Microbrewery Design and Performance." *MBAA Technical Quarterly* 28 (1).
4. European Brewery Convention (EBC).1999. *Milling, Manual of Good Practice.* Nürenberg, Germany: Getränke-Fachverlag Hans Carl.
5. Dougherty, Joseph J. 1977. "Wort Production." *The Practical Brewer*, edited by H. M. Broderick. Madison, Wisconsin: Master Brewers Association of the Americas.
6. Fahrendorf, Teri. 1994. "An Efficient Grain-Handling System." *The New Brewer* 11 (1).
7. Fahy, Ann and Jim Spencer. 1999. "Wort Production." *The Practical Brewer*, edited by John T. McCabe. Wauwatosa Wisconsin: Master Brewers Association of the Americas.
8. Foster, Terry. 1990. *Pale Ale.* Boulder, Colorado: Brewers Publications.
9. Haase, Nicolas. 1999. Personal Communication. Drinktec USA, Ltd./Kuenzel Maschinenbau GmbH.
10. Hough, J.S. et al.1982. *Malting and Brewing Science.* Volume 1. London, England: Chapman-Hall, Ltd.
11. Kruckenberg, Linda. 1999. Personal Communication. Roskamp Champion.
12. Kuenzel, Wilhelm. 1999. Personal Communication.
13. Kuhr, Jim. 1996. "Malt Handling in the Brewhouse." *The New Brewer* 13 (6).
14. Kunze, Wolfgang. 1996. *Technology Brewing and Malting*, translated by by Dr. Trevor Wainwright. Berlin, Germany: VLB Berlin.
15. Lewis, Michael J. and Tom W. Young. 1995. *Brewing.* London, England: Chapman & Hall.
16. Mallet, John. 1994. "Small Brewers and Malt." *The New Brewer* 11 (2).
17. Noonan, Gregory J. 1996. *New Brewing Lager Beer.* Boulder, Colorado: Brewers Publications.
18. Pfisterer, Egbert and Robert Gentry. 1999. "Improving Brewing Material Yield and Efficiency." *The New Brewer* 16 (1).

19. Rehberger, Arthur J. and Gary E. Luther. 1995. "Brewing." *Handbook of Brewing*, edited by William A. Hardwick. New York, New York: Marcel Dekker, Inc.

20. Schwarz, Paul and Richard Horsley. 1996. "A Comparison of North American Two-Row and Six-Row Malting Barley." *Brewing Techniques* 4 (6).

21. Thompson, Donald, Karl Ockert, and Vince Cottone. 1999. "Craft Brewing." *The Practical Brewer*, edited by John T. McCabe. Wauwatosa Wisconsin: Master Brewers Association of the Americas.

22. Wackerbauer, K., C. Zufall, and K. Hölsher. 1993. "The Influence of Grist from a Hammer-Mill on Wort and Beer Quality." *Brauwelt International* 2.

Chapter Nine

Mashing

Mashing involves mixing milled malt and solid adjuncts (if used) with water at a set temperature and volume to continue the biochemical changes initiated during the malting process. The malt and adjunct particles swell, starches gelatinize, soluble materials dissolve, and enzymes actively convert the starches to fermentable sugars. The end result is wort with a fixed gravity (OG), a set ratio of fermentable and non-fermentable sugars, and proteins (soluble and non soluble) that affect physical and biochemical changes during fermentation. The composition of the wort will vary according to the style of beer.

Chemical Changes at Mashing

The mashing process is conducted over a period of time at various temperatures in order to activate the enzymes responsible for the acidulation of the mash (traditionally for lagers) and the reduction in proteins and carbohydrates. Enzymes are biological catalysts responsible for initiating specific chemical reactions. Although there are numerous enzymes present in the mash, each with a specific role to play, this discussion is limited to the three principal groups and their respective processes.

1. Acidifying (phytase enzymes)
2. Protein-degrading (proteolytic enzymes)
3. Starch-degrading (carbohydrase enzymes)

Acidification

The acid rest is responsible for reducing the initial mash pH for traditional decoction mashing of lager beers. In recent years, because of the use of well-modified malts, the general trend has been to simplify and shorten the lager mash by eliminating the acid rest in mashing.

Acidification of the mash is primarily done by the enzyme phytase, which is active at 30 to 53°C and breaks down insoluble phytin, a complex organic phosphate containing both calcium and magnesium, to phytic acid. Phytic acid has a strong affinity for calcium ions, and it forms calcium phosphate and releases hydrogen ions in the process. Inorganic malt phosphates also react with calcium to release hydrogen ions, but the phytic acid reaction is more efficient. When mashing, phytase activity is greater with under-modified malt than with highly modified malt.

Protein Degradation

Although the term "protein rest" has long been adopted for low-temperature mashing systems, most of the protein degradation occurs during the malting process. The proteins that do dissolve during this phase are not significant and reportedly precipitation, not proteolysis, accounts for the disappearance of proteins during mashing (26). Protease enzymes comprise the group of enzymes that reduce high-molecular-weight proteins to simpler amino-acid constituents by breaking the peptide bonds between proteins. The enzymes proteinase and peptidase are two main enzymes of this group.

Proteinase

Proteinase is responsible for degrading albumins already dissolved in the wort, along with insoluble globulins, into simpler medium-sized proteins (peptones and polypeptides). The reduction of albumins and globulins is important for reducing haze. Haze is also caused by polyphenols (tannins) derived from the husk of the malt and hops. The medium-sized proteins (polypeptides and peptones) are not useful yeast nutrients, but are both important for foam stability and thus head retention, as well as for body, or palate fullness. Some brewers prefer to limit the duration of the protein rest to improve the foam quality of the beer. Proteinase has an optimum temperature range between 50 and 60°C and an optimum pH range of 4.2 to 5.3 (11).

Peptidase

The enzyme peptidase is responsible for degrading medium-sized proteins (peptones and polypeptides) into smaller proteins (peptides and amino acids). Unlike polypeptides and peptones, peptides and amino acids are essential for yeast growth. Peptidases are very active between 45 and 53°C and at a pH of 4.2 to 5.3.

Starch Degradation

By far the most important change brought about in mashing is the conversion of starch molecules into fermentable sugars and unfermentable dextrins. The principal enzymes responsible for starch conversion are alpha- and beta-amylase. Together, alpha- and beta-amylase are capable of converting only 60 to 80% of the available starch to fermentable sugars, which is mostly maltose and maltotriose (11).

Limit dextrinase, a de-branching enzyme that breaks down limit dextrins, has little effect at mashing temperatures since its optimum temperature is 40°C. Consequently, a large percentage of the dextrins pass to the finished beer, contributing to palate fullness, or body.

Alpha-amylase

Alpha-amylase very rapidly reduces insoluble and soluble starch by splitting complex starch molecules into many shorter chains (i.e., partially-fermentable polysaccharide fractions—dextrins and maltotriose) that can be attacked by beta-amylase. Given a long enough "rest," the alpha-amylase can dismantle most of the dextrins to maltose, glucose, and small, branched "limit dextrins."

Alpha-amylase is more resistant to high temperatures than beta-amylase and acts optimally between 72 and 75°C, but it is destroyed at 80°C (19). Calcium is commonly added to the mash as a co-factor and to provide better stability at high temperatures (31). Its optimum pH is 5.6 to 5.8.

Beta-amylase

Beta-amylase is the other mash enzyme capable of degrading starch. Beta-amylase is more selective than alpha-amylase since it breaks off two sugars at a time from the starch chain. The disaccharide it produces is maltose, the most common sugar in malt.

The optimum temperature of beta-amylase at mashing is between 60 and 65°C, and it is rapidly inactivated at 70°C (19). Parkes *et al.* report an optimum range at about 55 and 60°C, with beta-amylase being less effective at 65°C (31). Since the alpha- and beta-amylases have different optimum temperatures, many traditional mashing systems use two or more temperature rests for starch conversion (17). Its optimum pH is 5.4 to 5.5.

Stages of Starch Breakdown

Three distinct stages can be distinguished in the enzymatic breakdown of starch: gelatinization, liquefaction, and saccharification.

- Gelatinization is the formation of long-chain aqueous starch, which allows for more efficient alpha- and beta-amylase conversion. Gelatinization is best determined by following the changes in viscosity of the starch molecules. It occurs when the starch gels, resulting in a sharp rise in viscosity. The gelatinization temperature is different for each type of cereal. Malt and barley starch gelatinize in the presence of amylases at 60°C, whereas rice starch gelatinizes between 80 and 85°C (19). Adjuncts such as rice or corn require

boiling or "hot flaking" to achieve gelatinization since they have high gelatinization temperatures.
- Liquefication is the process of reducing the very large chains of starch molecules into shorter molecules, resulting in a reduction in viscosity.
- Saccharification covers the conversion of liquefied starch to dextrins and maltose and is brought about by the joint action of alpha- and beta-amylase. A pH of 5.2 to 5.5 should be the target of the saccharification rest for all mashes (28).

Beta-Glucans

During gelatinization, beta-glucans are hydrolyzed by beta-glucanases. These enzymes are most active between 35 and 40°C and are deactivated between 50 and 60°C (11). Beta-glucans increase wort viscosity, lengthening the

> **Note | Maltose/Dextrin Ratio**
>
> The temperatures that are chosen for starch conversion will determine the maltose/dextrin ratio of the wort. The higher the temperatures used at mashing, the greater will be the proportion of dextrins in the wort, producing more full-bodied, less-fermentable worts with greater mouthfeel. At the lower end of the temperature range, more maltose will be produced, resulting in a beer with more alcohol. This leaves a narrow range (often referred to as the "brewer's window") between these two temperature extremes that the brewer can influence the ratio of sugars that can be extracted into the wort. A low-temperature mash results in more fermentable wort by allowing the mash to stand at a temperature that favors the action of beta-amylase (i.e., maltose) but has a slightly lower yield. While a high-temperature mash results in less fermentable wort because more of the original starch remains as unfermentable dextrins, but with a higher yield.
>
> A prolonged stand between 60 and 65°C will yield highly fermentable wort that is rich in maltose and has a higher attenuation limit. Temperatures between 72 and 75°C will produce dextrin-rich wort with a low attenuation limit. A temperature rest at 65°C for 10 minutes, followed by a boost in temperature to 70°C held for an additional 10 to 15 minutes (or until conversion), will produce a medium-bodied beer. High conversion temperatures are used to brew long-maturing beers, such as traditional lagers that will be aged in cold storage. Most amber and golden lagers, and even the pale Pilsner/Dortmunder lagers, rely on a richer dextrin complement formed between 67 and 68°C. Shorter-maturing beers, such as ales, generally receive conversion at lower temperatures.

runoff from the mash/lauter tun, as well as leading to subsequent problems in filtration and chill haze (32). Since beta-glucans tend to be soluble in hot wort but insoluble in cold beer, they may contribute to chill haze. Beta-glucan-related problems are not often encountered with well-modified two- and six-row North American malts; they are more likely when under-modified malt or high levels of unmalted barley are used (35). For under-modified malts, low temperature rests at 35°C (suitable for the temperature sensitive beta-glucanase enzyme) can be employed to degrade beta-glucans (19). Some malsters list beta-glucan content as part of their malt specifications.

Factors Affecting Mashing Conditions

Temperature

As mentioned, the temperature influences the amount of extract produced (yield) and the fermentability of the wort during mashing. In general, the higher the temperature, the greater the yield but the lower the fermentability of the wort. At lower temperatures less extract is produced, but fermentability is higher. Only at very high temperatures will extract begin to drop off. Consequently, high temperatures favor alpha-amylase enzymes producing less fermentable worts, ones that are high in maltotriose and dextrins. Amber and gold lagers rely on the heavier, richer dextrin complement. Lower temperatures, on the other hand, favor beta-amylase enzymes producing more fermentable worts that are rich in maltose. Mild ales require the extra maltose that can be supplied by mashing at lower temperatures.

The rate in the rise of the temperature has an influence too in controlling the amount of fermentable extract. In general, a low conversion temperature (65°C) and a slow rise to conversion (0.5° C/min) will produce a more fermentable extract than a high conversion temperature (71°C) and a rapid rise to conversion (1.5°C/min) (32). The strike temperature is essential, as a variation of two or three degrees for five to ten minutes can dramatically alter the maltose/dextrin ratio of the extract.

Time

Mash times are another factor influencing yield and the fermentability of the wort. In general, longer mash times increase the concentration of the extract, but the rate of increase becomes slower (19). In general, short mash times at high mash temperatures will produce more dextrinous worts, while longer mash times at high temperatures produce more fermentable worts. Likewise, short mash times at low temperatures will produce less fermentable worts, while longer mash times at low temperatures will produce more fermentable worts. The fermentability of worts and hence the alcohol content of the final beer may be regulated, therefore, by adjusting mashing temperature and mash times.

Mash pH

For most beers, the optimum pH range for mashing is from 5.1 to 5.6, although values toward the lower end (5.1–5.3) are usually considered optimum (19). The "normal" mash pH, however, depends on the type of malts employed, the pH of the water, and the method of mashing. The mash cycle should not be started until the proper initial mash acidity is approximated (within pH 0.2). A favorable pH will result in (1) gains in extract and permanently soluble nitrogen, (2) reduced wort color, (3) improved mash filtration and wort runoff, (4) better break during boiling, and (5) a clearer final beer.

If the pH is too high, the hydrolysis of starch and proteins is adversely affected, resulting in beers with increased color and a sharp palate. In addition, a high pH will tend to increase the dextrin/maltose ratio and result in less-fermentable worts. These harmful effects must be corrected, either by adjusting the pH of the mash or by water treatment. If the pH is too low, peptidases are inhibited and haze troubles can develop.

Acidification of Mash

To lower the pH of the mash it can be corrected by the addition of acid malts, salts, or acids, or by biological acidification.

Brewers can rely on the natural acidity of dark-roasted malts to bring about the desired reduction of pH. Generally, higher-kilned malts have a greater effect on wort pH; the darker the malt, the greater the effect on wort acidity.

Many ale brewers add calcium in the form of calcium sulfate or calcium chloride either to the mash water or in dry form to malt as it is milled to lower the pH (31). Generally, if the water contains adequate amounts of calcium (50–100 mg/l or ppm), acidification occurs at virtually any mash-in temperature by the reaction of the calcium ions with phosphate ions from the malt.

Acids commonly used in the acidification of mash are phosphoric and sulfuric acids, which give beers a drier palate, or hydrochloric and lactic acids, which give beers a fuller palate.

Biological acidification, a practice common with German brewers due to terms of the Reinheitsgebot, involves introducing either *Lactobacillus delbrueckii* or *L. amylolyticus* to the mash; it is most effective when the brewing water is relatively soft. Biological acidification is also done to enhance the sharpness of the flavor, a practice common with Bavarian-style weizen.

Malt Modification

The temperature used for mashing is a function of malt modification. This is because the extent of modification influences the rate of starch solution. Lower temperatures are needed for well-modified malts because most of the enzymes are destroyed in the kiln, particularly beta-amylase, which is necessary for starch conversion. If the temperature is too high, starch conversion may not go to completion, resulting in low extract and possible starch haze in the finished beer. Higher temperatures are needed for poorly modified malts to achieve adequate extract because the starch is not as modified and hence is less soluble. Lewis reports that well-modified malts require lower temperatures (62–65°C), whereas poorly modified malts require higher temperatures (65–68°C) to achieve adequate extract yield (25).

Mash Water

The nature of the mashing water has an important influence on mash reactions. The ions of major importance at mashing are those of calcium, carbonate, and bicarbonates (8). Calcium lowers the pH of the mash mainly by its interaction with phosphates and to a lesser degree with protein from the malt. This lowering of the pH is critical in that it provides an environment for alpha-amylase, beta-amylase, and proteolytic enzymes. Carbonate and bicarbonate ions operate in the reverse direction. Excessive amounts of temporary hardness due to the presence of calcium, magnesium, and sodium salts of carbonates and bicarbonates in water are harmful in mashing waters. This is because these substances act as weak bases and raise the pH of mashes to an undesirable level. Water used in the main mash should be essentially free of bicarbonate hardness (50 mg/l or less) (32).

Mash Thickness

Thin mashes (i.e., its ratio of grist to brewing water) favor the conversion of starch to sugars, while in thick mashes the rate of saccharification is retarded, probably because the accumulating sugars competitively inhibit the hydrolytic enzymes (11). Therefore, thick mashes produce more glucose and maltotriose, which contain dextrins, while dilute mashes favor the production of sucrose and maltose, and thus wort attenuation. Thin mashes should give greater fermentability. On the other hand, thick mashes provide greater temperature stability and thus protect the amylase enzymes from the effects of rapid thermal inactivation (25). A thick mash also favors proteolytic action and so favors FAN. Usually, ales require thicker mashes than do lagers, where the mash is thinner. Of the lagers, Pilsners require a dilute mash and bocks a more concentrated mash (32).

Mashing Systems

Historically, mashing systems have evolved largely as a result of the advantages or for that matter the disadvantages inherent in the available brewing ingredients: water, malt, and adjuncts. Mashing systems were conveniently grouped under three methods: infusion mashing, decoction mashing, and double-infusion mashing. Infusion mashing is used by British breweries in producing ales, decoction mashing is most commonly associated in the production of lagers in Germany, and double mash-infusion mashing is a system used by large North American brewers in the production of lagers. A fourth method, temperature-programmed ("Step") mashing is now very common too.

Over the years the traditional mashing systems have been progressively refined and there has been a convergence in practices, although still recognizable, of ale and lager breweries.

Infusion Mashing

Infusion mashing, although not as popular, is a classical British method of brewing ales and stouts, where the brewer produces and recovers the extract at a single mash temperature, called the conversion temperature. Unlike other mashing methods, it doesn't require a series of different temperatures and rests. The conversion temperature represents a compromise between the optimal temperatures for alpha- and beta-amylase. Any deviations from the intended temperature could affect the beer since there is only one temperature determining makeup.

Infusion mashing requires only a single, unstirred vessel referred to as the mash tun (see Figure 9.1), which is particularly attractive to craft breweries because it is simple, inexpensive, and easily scaled to their low output. Larger ale breweries in order to achieve higher production capacity commonly use mash mixers and lauter tuns.

Infusion mashing is best used with well-modified malts, two- or six-row, that require only a saccharification rest and no protein rest. Well-modified malts yield sufficient extract when mashed properly at a single temperature. Some brewers prefer to use two-row malt because it is lower in protein than six-row malt and generally causes fewer problems with colloidal (haze) stability. If under-modified malts are infusion-mashed, the finished beer may have numerous filtration, clarity, flavor, and foam problems. This is because a single temperature infusion mash does not favor degradation and solubilization of the protein or the beta-glucans. This mashing method is not suitable for adjuncts that require gelatinization.

Advantages of this system are simplicity of operation and lower capital, labor, and energy costs (20–50% less) compared to other mashing systems. Also,

Figure 9.1
Mash tuns
Courtesy of Plzensky Prazdroj Brewery

there is minimal risk of air uptake during the mashing and wort separation operation since these two operations occur in the same vessel, the mash tun. One disadvantage with infusion mashing is lower extract recovery due to its passive treatment of mash. In practice, infusion mashes give 3 to 10% lower extract than decoction mashes. Another disadvantage is that infusion mashing requires greater precision in temperature control than does a multiple-temperature mashing systems.

Mash-In

The process itself begins by preheating the mash tun with steam or hot water to minimize heat loss. All taps are closed, and some hot water is admitted into the base the mash tun to drive out air from underneath the false bottom. The plates are covered with a layer of water 2–5 cm deep to cushion the falling mash and to prevent the mash from blocking the filtration slots (3). The next step is to "mash-in" the brewing water and the grist.

> ### Note | Quantity of Water for Mashing
>
> The total quantity of water required for mashing varies between 2 and 5.5 hl/100 kg of malt, depending upon the mashing method adopted and the required gravity of the wort. The total quantity of water is usually lower for infusion mashing with a water/grist ratio of 1.7–2.5 hl/100 kg of malt. The grist-to-water ratio is higher for decoction mashing ranging from 3 to 5.0 hl/100 kg of malt (26). In decoction mashing, the proportion of water to malt is higher for pale beers (e.g., Pilsner) and lower for dark beers (e.g., Munich). Generally, the ratio is 4 to 5.0 hl/100 kg of malt for pale beers and 3 to 4.0 hl/100 kg of malt for dark beers.

Whatever type of premasher is used—foremasher, Steel masher or spray ring—it is very important that the incoming grain is evenly wetted. If the grain is not fully wetted, this can result in "dough balls," which are nearly impossible to break up.

In a mash tun, unlike in a lauter tun, the wort is usually not circulated to assist in the establishment of the filter bed and clarification.

Strike-Temperature – The strike temperature is the temperature of mash water to be infused with grist to hit the desired mash temperature during mash-in. Keep in mind the mash temperature once established is difficult to correct because of the thick mash and the fact that the vessel is not stirred. However, minor changes in mash temperature can be accomplished either by underletting (pumping water in below the plates of the mash tun) or using sparge water to raise the temperature of the mash during run-off.

Stand Time

The mash is allowed to stand, without mixing, or addition of heat, for a given time to allow starch conversion to take place. The conversion temperature is usually in the range of 60 to 65°C for British malt and 63 to 70°C for American two-row malt (31). The higher mash-in temperatures for American malts is due to higher levels of alpha- and beta-amylase in the malt. The temperatures are above the starch gelatinization temperature for malt, and are designed to optimize the effectiveness of both alpha- and beta-amylase enzymes. However, at these mash temperatures proteolysis is limited, and most of the protein break down will have occurred during malting. Similarly, the reduction of beta-glucans will also be quite limited at these temperatures. The stand time can be as short as 15 minutes for American two-row malts or as long as 60 to 90 minutes if not longer for British malts that are mashed at lower temperatures.

Mash-Off

Following mashing, the temperature of the mash is raised between 75 and 78°C, the "mash-off" temperature. At this temperature, enzyme activity is deactivated and bacterial action is precluded.

Temperature-Programmed ("Step") Mash

Sometimes referred to as "step-infusion," this mashing procedure involves a series of rests at various temperatures in a mash mixer. It involves a protein rest and in some cases even an acid rest. The modified-infusion method mimics the traditional decoction-mash sequence, but with less satisfactory results (primarily because no part of the mash is ever boiled). However, temperature-programmed mashing is much simpler to employ and is less costly than decoction mashing in terms of number of vessels and energy costs. Temperature-programmed mashing can be used in lieu of infusion mashing too. One advantage of temperature-programmed mashing over single-temperature infusion mashing is that both specific enzyme activity and fermentability of the wort can be promoted by controlling the temperature and duration of stands at selected points. Temperature-programmed mashing also allows a higher extract yield, a 4 to 5% increase in extract over that obtained from a single-temperature infusion mash (38). Some brewers believe that step mashing is more effective than a single-infusion mash in dealing with under-modified malts, such as German and Belgian malts.

A typical step-infusion for under-modified malt starts with mashing-in at 35°C and holding for 30 minutes for an acid rest. The temperature is then raised to 45 to 50°C by 1 degree centigrade per minute, and maintained. For highly modified malts, the protein rest is for 10 to 15 minutes and for under-modified malts up to 30 to 45 minutes. The temperature is then raised 1 degree centigrade per minute until the conversion temperature is reached (65–70°C), followed by a 50 to 60 minute saccharification rest. Following the rest the temperature is again raised to the mash-off temperature (76–78°C). Higher temperatures are employed at the end to arrest enzymatic activity, facilitate solubilization of materials, and reduce viscosity. Alternatively, the brewer can mash-in at 45 to 50°C rather than at 35°C.

Another step-infusion program starts with mashing-in at 35°C and holding for 30 minutes before raising the temperature to 40 to 45°C for a protein rest. The temperature is then either raised slowly to 55 to 65°C, followed by a rise in temperature to 70°C for starch conversion, or raised quickly to a temperature of 70°C in one step. The long, slow rise with the intermittent step between 55 and 65°C favors wort with a high maltose to maltotriose ratio or highly attenuated worts. A very quick rise in temperature raises the maltotriose ratio and lowers fermentability. The temperature is than raised to a mash-off temperature between 75 and 78°C.

Some brewers recommend mashing-in at 45°C, followed by a 15 minute rest, which is optimal for peptidase. The temperature is then raised to 60°C for another 15 minute rest, which is optimal for glucanase and protease. The temperature is then raised to 65°C for starch conversion for a 5 to 10 minute rest, followed by a brief rest of 5 minutes at 75°C for mash-out.

Wort separation is done by using a lauter tun or mash filter, which is discussed in the following chapter.

Decoction Mashing

Decoction mashing is the traditional method used in the production of wort for bottom-fermentation beers and is most commonly associated with the production of lagers in Germany. It was developed largely as a result of the use of under-modified or enzymatically weak malts. This system involves removing a portion of the mash (the decoction) for heating to conversion temperature in the mash kettle. After the decoction has undergone starch conversion, it is brought to a boil to inactivate the enzymes and returned to the main mash (i.e., mash mixer) to increase the overall temperature.

Decoction mashing involves one-, two-, or three-decoction steps, which are referred to as single- double or triple-decoction mashes. The choice of the number of decoctions and the duration of the boil provides control over protein coagulation and malt sugar composition.

Today, most German malts are well-modified, and triple decoction mashing has largely given way to a simpler double decoction mashing or more commonly to single decoction mashing and even temperature-controlled mashing. It is even possible to brew a majority of lager styles successfully using an infusion mash. In fact, a growing number of German breweries are using single-infusion mashing to produce their pale lagers. However, there are some styles for which double- and triple-decoction is preferable because of the impact on flavor. Examples include German wheat beers and dark German lager styles, including bocks, dunkels, and märzen beers. These beers have higher levels of melanoidin, which is desirable in many German lagers, especially bocks.

Unlike infusion mashing, decoction mashing requires three separate vessels: 1) a mash mixer, where water and grist are mixed, 2) a mash kettle for heating the decoction, and 3) a lauter tun which is used for wort separation.

The chief advantage of decoction mashing is that it yields more extract with under-modified malts than with infusion-type mashes. In decoction mashing, part of the mash is boiled which reduces the size and complexity of the malt starch. Infusion mashing does not involve boiling. Furthermore, during decoction mashing the decoction passes through the diastatic-enzyme temperature range a number of times, depending on the number of decoctions. Finally, an important advantage of decoction mashing is that it allows enzymes to break down excessive high-order nitrogen malt into simpler substances. This reduces the haze potential.

Two disadvantages of this system are the energy requirements and the long processing time—up to six hours for a triple-decoction mash. In addition, the repeated boiling extracts more tannin from the grain husks than the gentler, shorter, infusion mash, which is especially a problem with six-row malts. There is also an increased risk of exposing the wort to air when transferring between the mash and lauter tun.

Three-Step Mash Decoction System

The three-step decoction involves acid, protein, and saccharification rests. The overall processing time is about 6 hours from mashing-in to lautering. There are many minor variations on the classical triple-decoction mash, especially with regard to the time in heating the decoction, the rate at which the decoction part of the mash is mixed with the main mash, and the length of the stands.

Mashing-In – Mashing begins by mixing-in the malt and cold water in the mash mixer. Mashing-in with cold water prevents starch balling by allowing the water to evenly permeate the mash. Generally, the malt is mashed-in with the brewing water at a temperature between 14 and 21°C, with the mash allowed to stand for approximately 15 minutes (28). Longer stands are recommended, up to 30 minutes, for malts that are dark, weakly enzymatic, hard tipped, or poorly modified.

Acid Rest – After the mash-in stand, the temperature of the mash is raised to 35 to 40°C to acidulate the mash (acid rest), which is done by the infusion of hot water or by direct heating. Once the pH has dropped below 5.8 (preferably 5.2–5.5), proceed with the first decoction (28). If sufficient acidification has not occurred, an addition of lactic acid mash may be necessary.

First Decoction – Once the pH has dropped, about one-third of the mash (the decoction) is withdrawn, transferred to the mash kettle or otherwise known as the decoction vessel, and quickly heated within 10 minutes (or as rapidly as possible) to 65 to 70°C. The decoction should always be thick, with just enough water to prevent scorching. The thickest part of the mash contains the heaviest and least accessible concentration of starch and protein, whereas the thinnest part of the mash is enriched with enzymes already passed into solution. The temperature is maintained for 10 to 15 minutes to allow starch conversion to take place (to reduce malt starch to sugars) before boiling the decoction.

Following starch conversion, the mash is brought to a boil and held there for 10 to 30 minutes for pale beers and 20 to 45 minutes for dark beers. During the boil, enzymes are inactivated, but residual starch is fully gelatinized and protein gum dissolved to albuminous fractions. At lower temperatures, protein gum is unaffected by enzyme activity and passes through mashing largely unconverted. Boiling also deoxygenates the mash, reducing hot-side aeration and allowing it to settle in well-defined layers in the lauter tun.

After the first decoction has been pulled, the rest mash (cold settlement) in the mash mixer is occasionally mixed to disperse temperature and enzyme activity. This is done while maintaining the temperature between 35 and 40°C.

As mentioned, about one-third of a thick decoction mash is sufficient to raise the entire mash to the next rest, but this is not necessarily the case. There are many variables that come into play, including the insulating qualities of the mash tun, the surface area-to-volume ratio of the mash in the mash tun, and the thickness of the mash. The volume of thick mash to be boiled, relative to the volume of the whole mash, is dependent upon mash thickness. In a thick mash, the favorable heat-to-volume ratio of the malt is such that the heaviest one-third can raise the temperature of the whole mash to the next rest. In a thin mash, however, the heat value of the malt is not enough to raise the temperature to the next rest. Consequently, thinner decoction mashes require that proportionally more mash be boiled. Some brewers believe that with a thin mash, the decoction should be about 40% of the total mash volume.

Protein Rest – After the first decoction has been boiled in the mash kettle, it is then pumped back into the mash mixer, where it is continually stirred with the remaining thin rest mash. In order to avoid harmful aeration of the mash, the boiled mash is pumped back in from below, rather than from above. The addition of heat energy brought about by boiling the decoction should raise the temperature of the rest mash to 50 to 55°C for the protein rest. The protein rest should be accomplished in less than 2 hours, ending with a rest period between 5 and 25 minutes before drawing off for the second decoction.

Second Decoction – Following the protein rest, about one-third of the mash (the thickest part) is withdrawn for the second decoction and transferred to the mash kettle. The temperature of the second decoction is then raised to between 62 and 65°C within 10 minutes, and then increased to 75°C for 10 to 15 minutes before boiling (28). The mash is typically boiled for 5 minutes, though very steely malts may require up to 25 minutes (28). During the second decoction, alpha- and beta-amylase reduce starch to simpler fractions, yielding dextrins and fermentable sugars. As the mash saccharifies, it becomes thicker, brighter, and browner. The mash can be thinned to speed up saccharification (beta-amylase is more effective in the looser mash); however, beta-amylase is more prone to destruction in thin mashes.

As with the rest mash in the mash mixer, it is stirred regularly to break up any pockets of unmodified starch and to ensure uniform conversion.

Saccharification Rest – After the second decoction has been boiled in the mash kettle, it is then pumped back into the mash mixer, where it is continually stirred with the remaining thin rest mash. The addition of heat energy brought by the boiling decoct raises the temperature of the rest mash to 65 to 68°C (called the "conversion temperature") within 5 minutes for the saccharification rest. Some brewers slowly pump the decoction over a period of 15 to 30 minutes to favor maltose production. The saccharification temperature can be maintained by infusing small amounts of boiling water into the mash. These temperature-maintenance infusions serve to improve filtering. Thinning the mash assures that its density does not interfere with filter bed settlement in the lauter tun. Usually, no more than 3.1 hl of water per 100 kg of malt is used for these infusions (28). After 15 minutes at the rest temperature, testing for saccharification with iodine should begin. Continue mashing until there is no color change.

Third Decoction – When the starch endpoint has been verified, a very thin portion of the mash is removed for boiling in the mash kettle. The third decoction is usually 40% of the volume, though for thinner mashes it can be up to 50% (28). Because there are fewer starch and albuminous particles in the thinner portion, there is less risk of these being decomposed during the boiling of the runoff. The boil usually takes from 5 to 10 minutes, but it can last up to 15 to 45 minutes. According to De Clerck, the third decoction is usually boiled for a slightly longer time than the first two decoctions (10).

Mash-Off – Following boiling, the third decoction is pumped from the mash kettle back to the mash mixer where the temperature reaches 75 to 77°C—the "mash-off" temperature. Temperature adjustments may be made by the infusion of either cold or boiling water, as required. The mash-off temperature inactivates the enzymes and provides a little extra heat to counter that lost during transfer to the lauter tun. If the mash-off temperature is too low, it will not terminate enzyme activity; if too high, the starch granules burst, and insufficiently modified carbohydrate and albuminous matter becomes dissolved and unfilterable. High temperatures also induce the extraction of tannins from the husk. After being allowed to stand for a few minutes, the mash is then pumped to the lauter tun for filtering.

Two-Step Mash Decoction System

The two-step decoction involves only protein and saccharification rests. The entire process usually takes 2 to 3 hours. As with the three-step decoctions system, there are many minor variations of the two-mash system, especially with regard to speeds of heating, the rate at which the decoction part of the mash is mixed with the main mash, and the lengths of the stands.

Kunze recommends, in *Technology Brewing and Malting*, mashing-in the malt and water at 50°C, followed by a 20 minute rest (19). Then the thick mash is removed and held for a short rest of 15 minutes in the mash kettle at 70°C before boiling. After boiling for 20 minutes, the mash is pumped back into the mash mixer, where it is mixed with the remaining thin rest mash. The addition of the first decoction raises the temperature of the rest mash to about 64°C for maltose formation. After a 10 minute rest, the second decoction mash is removed and, with a short rest period for 20 minutes at 71°C, heated to boiling. The second boiled mash is usually heated for a somewhat shorter time (approximately 15 minutes). It is then returned to raise the total mash to about 75°C for a 5 minute rest, followed by mashing-off at 78°C. This process takes about 3.5 hours.

Single-Step Decoction System

The single-step decoction involves only a saccharification rest. The entire process usually takes fewer than 2 hours. As with the other decoction systems, there are many minor variations of the single-step mash system.

A typical single mash system mashes-in at 50°C, followed by a 5 minute rest. After the rest, about 33 to 45% of the mash, the thickest part, is withdrawn and transferred to the mash kettle, where it is quickly heated to 66°C in 10 minutes, and then to boiling for 5 minutes. The boiling mash is then pumped back into the mash mixer, where it is mixed with the remaining thin rest mash. The decoct mash raises the temperature of the rest mash to 67 to 68°C within 5 minutes for the saccharification rest. After 15 minutes, the mash is tested for starch conversion. Once the iodine starch test is negative, the temperature of the main mash is raised to a mash-off temperature of 76.5°C before being transferred to the lauter tun.

Double-Mash Infusion System

The double-mash infusion system is the most widely used system in North America. This system prepares two separate mashes. It utilizes an adjunct cooker (i.e., cereal cooker) for boiling adjuncts and a mash tun for well-modified, highly diastatic malts. Most adjuncts must be boiled separately to gelatinize the starch. If pre-gelatinized adjuncts (e.g., flakes) are used, no boiling is required in the adjunct cooker. In this case, the flakes can be added directly to the mash tun during or at the end of the protein rest (11).

The double-mash infusion system was developed to utilize adjuncts as a source of cheaper extract and as a means of producing lighter, less-satiating beers that have traditionally found favor with North American consumers. Rice and corn are mostly used as adjuncts, though barley, wheat, and sorghum are used to a limited extent. Sugar and syrup are used as well as a source of extract, but they are added directly to the kettle, not the cereal cooker.

Adjunct Cooker

The first step, before adding adjuncts, is to mash-in a small quantity of barley in the hot water of the adjunct cooker. The alpha-amylase of the malt will hydrolyze the gelatinized starch, rendering the cooker mash fluid enough for pumping. The quantity of malt used depends upon the enzymatic power of the malt, the nature of the adjuncts, and the heating cycle of the cooker. A typical malt charge, according to Dougherty, is usually 10 to 20% of the total amount of cereal used (11). More specifically, if corn grits are used, the charge is about 10% and for rice about 20% (32). The high gelatinization temperature of rice destroys the malt amylase activity; consequently, more malt is required. As a result, brewers using rice may have higher specifications for alpha-amylase in the malt than brewers using corn. It is better to keep the cereal cooker malt charge as low as practical to avoid possible flavor impairment due to boiling of husk material.

The mash-in temperature is between 38 and 50°C for the malt (11). The agitator is operated continuously while mashing in the malt, and as low a speed as possible is maintained to minimize oxygen uptake. Following the mash-in, the cooker malt is allowed to rest or is mixed gently for a "peptonizing rest" period of 15 to 30 minutes (32). Some recommend a short stand at 70° C for well-modified malts that have sufficiently high enzymatic power. The principal reason for this rest is to liberate starches and proteins and activate the enzymes that were developed during malting. Also, this rest thins the mash.

At the end of the "peptonizing rest," the adjuncts are now added to the cereal cooker, while additional hot water is added in order to maintain fluidity. The mash is mixed as the adjuncts are transferred to assure immediate and thorough wetting of the adjuncts. At the same time, the temperature is gradually raised (with a possible rest) to boiling. During this rise from peptonizing to boiling, the adjunct starch granules gradually absorb water, thereby thickening and assuming a near gel consistency. Eventually the water-swollen granules rupture, which is termed "liquefaction" of the starch. The length of the boiling period depends upon the specific adjuncts. A relatively short boil of 5 minutes is considered adequate when the entire adjunct is refined corn starch, whereas a 30 to 45 minute boil is typically used for a rice mash (11).

Mash Tun

About the time the adjuncts are being added to the cereal cooker, the malt and water are added to the mash tun to begin a peptonizing rest within the same range, 38 to 50°C (11). After 15 to 30 minutes of peptonizing, the temperature is raised to 50°C.

The mixing of the two mashes—the adjunct cooker and mash tun mashes—is scheduled so that the completion of the required boiling time of the cereal mash coincides precisely with the completion of the protein rest of the mash tun mash. Following the transfer of the cereal cooker mash to the mash tun, there is a rise (approximately 20 minutes) in the temperature from 50°C to the saccharification-rest temperature of approximately 65 and 72°C. This phase of mashing is frequently called "the ramp." Often, the thermal contribution of the adjunct mash is not sufficient to reach conversion temperature and, so heat must be applied. After the combined mash has reached saccharification temperature, it is held there for 30 to 40 minutes until conversion is complete. Upon completion of saccharification, the temperature is quickly raised (1–2°C/min) to a "mashing off" temperature between 75 and 77°C (11). The mash is then ready for lautering–separating the wort from the spent grains.

Mashing Equipment

There are a number of configurations of mashing systems employed by brewers. Mashing is performed in a mash tun where the brewer produces and recovers the extract at a single mash temperature or in the special case of a pre-gelatinized or liquid adjunct being employed. If pre-gelatinized adjuncts are not used, a double-mash system that employs a mash cooker and mash tun is required. Although mash tuns can be used for lautering (wort separation), some breweries use lauter tuns for greater flexibility. Temperature-controlled and decoction mashing require separate conversion (mash mixers and mash kettles) and a lauter tun.

Holding Hopper

A holding hopper (often called a scale hopper) is used for grist and is mounted above the cereal cooker or the mash tun. Hoppers are usually equipped with vibrators to aid in the emptying process. They have automatic load cells for exact weight measurement (to an accuracy of +/- 1%). A dust collection system or a breather filter is installed on the holding hopper in order to minimize the dust.

Premasher

Premasher is used to mix the brewing water and grist in order to break up the clumps of grist before moving to the mash conversion vessel (i.e., mash mixer or mash tun). The three main types of premashers are: foremasher, Steele masher, or mashing ring. A foremasher is nothing more than a tube where the malt flows down and inside this tube there is a smaller tube where the water flows out and mixes with the grist. Another option is a Steele masher which uses an auger and blades to mix the grist and brewing water together. Steele mashers are preferable when milling/mashing rates are greater than 50 pounds per minute. For slower mashing rates, a simple mashing ring can be used. This sprays hot water on the crushed malt as it falls through a mill/auger delivery tube. Spraying devices are commonly used in small brew pub brewhouses. A spray ring works well when the water-to-grist ratio is in the 3:1 range and is less effective for thicker mashes (31).

Mash Mixer

Mash mixers or mash conversion vessels are usually stainless steel vertical cylinders equipped with a variable-speed propeller at the base of the vessel used in decoction mashing. A water station controls the temperature at mash-in as well as the proper grist-to-water ratio. Heating is done by steam jackets on the side-wall and bottom of the vessel, internal steam heated coils or by direct steam injection. The vessels are equipped with copper domes fitted with flues for carrying away steam. These vessels can also be used for cooking adjuncts.

A mash mixer requires usage of a lauter tun, in order to separate the sweet wort from the spent grains. It does not have a false bottom, which is different from that of a mash tun, in which both enzymatic degradation and wort separation take place.

Mash Kettle

A mash kettle or mash cooker is used for boiling part of the mash for decoction mashing. Mash kettles are similar in construction to mash mixers, except that they often have agitators that are more powerful and more steam heated areas, since their contents must be heated to boiling (3). Some mash kettles may have internal heaters to supplement the heating supplied by a steam jacket.

Adjunct Cooker

Adjunct cookers or commonly known as cereal cookers are used to cook adjuncts such a corn or rice at high temperatures to solubilize the starch. They are usually stainless steel vertical cylinders equipped with a variable-speed agitator. A water station controls the temperature at mash-in as well as the proper grist-to-water ratio. Heating is done by steam jackets on the side-wall and bottom of the vessel or by the less sophisticated method of direct steam injection. The side and bottom jackets can operate independently for optimum temperature control. Some adjunct cookers are equipped for pressure boiling.

The cooker is usually insulated and equipped with temperature probes accurate to +/– 0.25°C. Modern adjunct cookers have permanent cleaning-in-place (CIP) systems. A sparge ring that enables contents to be rinsed after the transfer is located at the top of the vessel.

Mash Tun

British brewers use the same vessel—commonly referred to as the "mash tun"—for both mashing and mash filtration. Mash tuns can only be used for infusion mashing (single temperature) and as a result under-modified malts or malts requiring a protein or glucanase stand cannot be handled. Among craft brewers, the mash tun is the preferred equipment for both mashing and wort filtration. It is the most cost-effective system in terms of capital investment and the easiest to operate.

Mash tuns are made of stainless steel and are equipped with a sparge arm for spraying high-temperature water (sparging) over the mash bed to displace the extract at the end of the mashing cycle. This dilutes the sugar solution and washes out the extract. The sparge arm consists of concentric pipes suspended just below the ceiling of the tun. The pipes are equipped with spray nozzles directed downward to deliver the sparge water in a uniform pattern over the entire surface area of the mash bed. The mash tun is usually equipped with a variable-speed, agitator paddles to stir the mash, as shown in Figure 9.2.

Heating is done either by steam jackets on the sidewall and bottom of the vessel or by direct steam injection. The side and bottom jackets can operate independently for optimum temperature control. Temperature probes are critical for the mashing process to maintain uniformity. Mash tuns have permanent cleaning-in-place (CIP) systems.

> **Note | Mash Tun Size**
>
> The vessel size of a mash tun for small breweries may have a diameter 1.2 to 1.6 meters with a grain bed depth of up to 1.2 meters, whereas larger breweries may have a vessel diameter 1.6 to 2.4 meters and a grain bed depth of up to 9.1 meters.

The wort collection system is fitted into the bottom of the mash tun. It is equipped with a slotted false bottom that is supported a few inches above the true bottom, with the total slot area typically totaling about 11% of the filter-plate area. It also consists of runoff pipes arranged in concentric rings.

Wort flows, by gravity, through the slots of the false bottom to the true bottom and then through the runoff pipes, which connect to a grant. After the grant, the wort flows to the underback or directly to the brewkettle, though in the

*Figure 9.2
Mash Tun Paddles
Courtesy of AAA Metal Fabrication*

early stages it might be re-circulated back to the mash tun until it runs clear. If the brewkettle is not available, the wort is run into the underback.

Underback

An underback is a small open topped vessel that allows the brewer to visually check the color and clarity of the wort. Once the brew kettle is available, the wort in the underback is pumped into the brewkettle, and the underback is filled with wort from the mash tun.

Brewhouse Systems Used by Craft Brewers

Craft brewers use two types of systems in brewing. The first is the traditional "American" style mash/lauter and kettle combination and the other is the "European" style mash/kettle, lauter tun, and whirlpool combination.

Combination Mash/Lauter Vessel and Kettle

Small breweries throughout the United States and Canada commonly use combination mash/lauter vessel and kettle systems (see Figure 9.3). The advantages of such a system are that it is simple to operate and costs less than other systems. Rice or corn cannot be used with this system; only sugar or some kind of pre-gelatinized extract can be used as an adjunct. Small brewers can increase production capacity using a combination mash/lauter vessel in conjunction with a combination kettle/whirlpool. Production capacity can further be increased by the addition of a separate whirlpool.

Figure 9.3
From Left to Right Mash/Lauter Vessel and Kettle
Courtesy of AAA Metal Fabrication

Combination Mash/Kettle, Lauter Tun, and Whirlpool

A combination mash/kettle, lauter tun, and whirlpool system allows for a more complicated brewing operation. The kettle becomes a multi-purpose vessel that is first used as a mash mixer and mash cooker. After mash-off, the mash is pumped to the lauter tun for sparging, and the kettle is quickly rinsed in preparation for the boil. Once the wort runs clear, it is diverted back into what is now the wort kettle for boiling. Following boiling, the wort is diverted to the whirlpool for hot trub separation. This system allows a single temperature infusion if desired, or a decoction or a step-infusion.

Calculations

Raw Material Requirements

The amount of malt (and adjuncts) required to produce a barrel of wort at a desired specific gravity (°Plato) must be calculated for each ingredient. The following formula can be used for calculating the pounds of extract per barrel for each ingredient:

Extract per barrel (in pounds) = [(259 + °Plato) x °Plato]/100

This calculation uses the figure 259 pounds as the weight of one barrel of water at approximately 4.2°C.

For example, if the brewer wants to produce 1,000 barrels of wort at 12°P the total amount of extract must first be calculated for the brew.

Extract per barrel (in pounds) = [(259 + 12°Plato) x 12°Plato]/100

Extract per barrel (in pounds) = 32.5

To produce 1,000 barrels of wort at 12°P 32,500 pounds of extract (32.5 x 1,000 barrels) is needed for the brew.

Now that the brewer knows the total amount of extract required to produce 1,000 barrels, the amount of extract from each ingredient can be calculated. This can be determined by multiplying the total extract by the percent of each raw material used in the brew.

For example, if the material bill consists of 90% barley malt and 10% wheat malt the brewer can calculate the amount of extract required from each material to produce 1,000 barrels at 12°Plato.

32,500 pounds of extract/brew x 0.9 = 29,250 pounds of extract/brew from barley malt

32,500 pounds of extract/brew x 0.1 = 3,250 pounds of extract/brew from wheat malt

The next step is to estimate the actual amount of material needed to achieve the desired volume and extract content taking into consideration the extract yield of each material and the brewhouse efficiency of dissolving the malt extract in water to form wort.

All raw materials used in brewing vary in extract availability based on the material itself and numerous other factors. Typically, extract availability is listed on the specification sheet provided by the vendor. Typically, brewhouse efficiencies range from 90 to 99% with larger brewers experiencing higher efficiency rates. Small breweries may achieve efficiencies in the low 90s while home brewers can expect rates as low as 75%.

If the average extract yield from barley malt is 78%, the number of pounds of extract to be derived from the malt is divided by 0.78 to determine the amount of malt required to produce 1,000 barrels of 12°P wort.

29,500 pounds of extract/0.78 = 37,820 pounds of barley malt/brew

If the average extract yield from wheat malt is 92%, the number of pounds of extract to be derived from the wheat malt is divided by 0.92 to determine the amount of malt required to produce 1,000 barrels of 12°P wort.

3,250 pounds of extract/0.92 = 3,533 pounds of wheat malt/brew

For instance, if the brewhouse efficiency is 95% the amount of barley malt and wheat malt can be determined using the following formula to achieve the goal of 1,000 barrels of 12°P wort.

37,820 pounds of barley malt/0.95 = 39,810 pounds of barley malt

3,533 pounds of wheat malt/0.95 = 3,719 pounds of wheat malt

Strike Temperature

To calculate the strike temperature several factors have to be taken into account, which are:

1. Temperature of the malt
2. Liters of water
3. Mash temperature

The equation used to calculate the strike temperature is:

$T_{strike\ temperature}$ = T_{mash} x (liters of water + (0.4 x kg malt)) − (0.4 x kg malt x T_{malt})/liters of water

where:

$T_{strike\ temperature}$ = strike temperature

T_{mash} = desired temperature of mash

T_{malt} = temperature of malt

0.4 = a pound of malt requires 40% of the energy required by a pound of water to increase the temperature by 1°

Example:

The recipe calls for 5,000 kg of malt with a temperature of 18°C using 13,000 liters of water to make up the mash, which should have a strike temperature of 67°C.

$T_{strike\ temperature}$ = [[67 x (13000 + (0.4 x 5000))] − (0.4 x 5000 x 18)]/13000

$T_{strike\ temperature}$ = [67 x (13,000 + 2,000) 36,000]/13

$T_{strike\ temperature}$ = 67 x 15,000 − 36,000]/13

$T_{strike\ temperature}$ = [1,005,000 − 36,000]/13

Tstrike temperature = 969,000/13,000

Tstrike temperature = 74.5°C

References

1. Alexander, Steve. 1997. "Fear of Phenols–A Guide to Coping with Brewing's Most Contrary Chemicals." *Brewing Techniques* 5 (6).
2. Barnes, Zane C. 2006. "Brewing Process Control." *Handbook of Brewing*, edited by Fergus G. Priest and Graham G. Stewart. Boca Raton, Florida: CRC Press, Taylor & Francis Group.
3. Briggs, Dennis E., Boulton, Chris A., Brookes, Peter A., and Stevens, Roger. 2004. *Brewing: Science and Practice*. Cambridge, England: Woodhead Publishing Limited.
4. Busch, Jim. 1996. "Stepping up to Lager Brewing – Part I: An Overview of the Brewing Process." *Brewing Techniques* 4 (3).
5. Busch, Jim. 1997. "Home Brewery Advancement: Step Mash for Customized Worts, Part I - The Enzymes that Breakdown Glucans and Proteins." *Brewing Techniques* 5 (3).
6. Busch, Jim. 1997. "Home Brewery Advancement: Step Mash for Customized Worts, Part II - The Starch Busting Amylases." *Brewing Techniques* 5 (4).
7. Carey, Daniel J. 1991. "Microbrewery Design and Performance." *MBAA Technical Quarterly* 28 (1).
8. Comrie, A.A.D. 1967. "Brewing Water." *Brewer's Digest* 42 (7).
9. Coors, Jeffery. 1976. "Practical Experience with Different Adjuncts." *MBAA Technical Quarterly* 13 (2).
10. De Clerck, Jean. 1957. *A Textbook of Brewing*. Volume 1, translated by K. Barton-Wright. London: Chapman-Hall Ltd.
11. Dougherty, Joseph J. 1977. "Wort Production." *The Practical Brewer*, edited by H. M. Broderick. Madison, Wisconsin: Master Brewers Association of the Americas.
12. Fahy, Ann and Jim Spencer. 1999. "Wort Production." *The Practical Brewer*, edited by John T. McCabe. Wauwatosa Wisconsin: Master Brewers Association of the Americas.
13. Fix, George J. 1989. *Principles of Brewing Science*. Boulder, Colorado: Brewers Publications.
14. Fix, George J. 1995. "The Role of pH in Brewing." *The New Brewer* 12 (6).
15. Heissner, Russ. 1999. Personal Communication. JVNW, Inc., Hingham, Massachusetts.

16. Hough, J.S. et al. 1982. *Malting and Brewing Science*. Volume 2. London, England: Chapman-Hall, Ltd.

17. Hough, J.S. 1991. *The Biotechnology of Brewing*. London, England: Cambridge University Press.

18. Johnstone, J. 1992. "Liquid Solid Separation in the Brewhouse." *MBAA Technical Quarterly* 29 (1).

19. Kunze, Wolfgang. 1996. *Technology Brewing and Malting*, translated by Dr. Trevor Wainwright. Berlin, Germany: VLB Berlin.

20. Kuplent, Florian. 1999. Personal Communication.

21. Lewis, Ashton. 1993. "Mashing Methods and Malt Compared." *Zymurgy – Special Issue: Traditional German, British and American Brewing Methods* 16 (4).

22. Lewis, David. 1998. "Biological Mash and Wort Acidification." *The New Brewer* 15 (1).

23. Lewis, Michael. 1987. "Three Types of Mashing Systems." *The New Brewer* 4 (1).

24. Lewis, Michael. 1993. "Getting the Most Out of the Mash." *The New Brewer* 10 (4).

25. Lewis, Michael J. and Tom W. Young. 1995. *Brewing*. London, England: Chapman & Hall.

26. Lewis, M. J. and Young, T. W. 2002. *Brewing*. New York, New York: Kluwer Academic/Plenum Publishers.

27. Noble, Stuart. 1997. "Practical Aspects of Fermentation Management." *The Brewer* 83 (991).

28. Noonan, Gregory J. 1996. *New Brewing Lager Beer*. Boulder, Colorado: Brewers Publications.

29. Ormrod, H. L. 1986. "Modern Brewhouse Design and its Impact on Wort Production." *Journal of Institute of Brewing* 92 (March-April).

30. O'Rourke, Timothy. 1999. "Milling and Mashing." *Brewers' Guardian* 128 (6).

31. Parkes, Steve and Kiesbye, Axel. 2006. "Brewhouse Operations: Ale and Lager Brewing." *Raw Materials and Brewhouse Operations*. Volume 1, edited by Karl Ockert. St. Paul, Minnesota: Master Brewers Association of the Americas.

32. Rehberger, Arthur J. and Gary E. Luther. 1995. "Brewing." *Handbook of Brewing*, edited by William A. Hardwick. New York, New York: Marcel Dekker, Inc.

33. Roberston, J. 1990. "Wort Production." *The Brewer* 76 (905).

34. Ryder, David S. and Joseph Power. 1995. "Miscellaneous Ingredients in Aid of the Process." *Handbook of Brewing*, edited by William A. Hardwick. New York, New York: Marcel Dekker, Inc.

35. Schwarz, Paul and Richard Horsley. 1996. "A Comparison of North American Two-Row and Six-Row Malting Barley." *Brewing Techniques* 4 (6).

36. Shahidi, F. and M. Naczk. 1995. *Food Phenolics: Sources, Chemistry, Effects, Applications*. Lancaster, Pennsylvania: Technomic Publishing Co., Inc.

37. Taylor, D.G. 1989. "Influence of Brewhouse Practice on Wort Composition." *The Brewers' Guardian* 118 (2).

38. Thompson, Donald, Karl Ockert, and Vince Cottone. 1999. "Craft Brewing." *The Practical Brewer*, edited by John T. McCabe. Wauwatosa Wisconsin: Master Brewers Association of the Americas.

39. Wainwright, T. 1971. "Biochemistry of Brewing." *Modern Brewing Technology*, edited by W.P.K. Findlay. Cleveland, Ohio: The Macmillan Press.

40. Wilkonson, R. 2003. "Mashing/conversion Technology." *The Brewers' Guardian* 132 (3).

Chapter Ten

Wort Separation

After mashing, when the starch has been broken down, the next step is to separate the liquid extract (the wort) from the residual undissolved solid materials found in the mash. Wort separation is important because the solids contain large amounts of protein, poorly modified starch, fatty material, silicates, and polyphenols (tannins). The objectives of wort separation (lautering) include the following:

- Produce clear wort
- Obtain good extract recovery
- Operate within the acceptable cycle time

The method of separating the wort from the mash solids and the equipment used is mainly a matter of choice on the part of the individual brewer, and sometimes of tradition. Wort separation may be carried out by any of a number of different methods: 1) the mash tun, 2) the lauter tun, 3) the mash filter, or 4) Strainmaster. The lauter tun is internationally accepted and continues to be the predominant wort separation device.

Mash Tun

The mash tun, traditionally used in Britain, is a vessel that serves the dual role of being used both for mashing and wort separation. They are the cheapest in terms of capital expenditures and the simplest to operate, with little or no automation. The combination of these factors makes the mash tun particularly attractive to craft brewers. The disadvantages of mash tuns are their requirement for coarsely ground malt, the need for well-modified malts, their limitations in working with adjuncts, and their inflexibility in regard to maintaining mash temperature (4).

The entire cycle from mashing, through wort runoff, sparging, and spent grain removal usually lasts from 3 to 4 hours—1 hour for mash conversion, followed by 3 hours of runoff, sparging, and grain removal. Excluding the time taken for the mash conversion, the mash tun is the slowest wort separation system.

Collection of Wort

After mash-off, the wort is withdrawn through the false bottom of the mash tun. The false bottom acts as a support for the grain bed and does not act as a filter bed. Initially, the flow rate of the dense wort is slow, to avoid pulling down or collapsing the suspended bed. Drawing the wort off too fast in the early stages of mash separation causes a suction effect that draws the spent grains closer together thus losing its particle stratification. This, in turn, can eventually collapse the suspended bed leading to filtration problems, which could result is long runoff times or even complete stoppage. Such a "set mash" may be cleared by re-floating the bed by pumping water from below into the false bottom of the mash tun, known as underletting.

Typically, brewers re-circulate the "first wort" in order to clarify the wort. This is accomplished by re-circulating the first runnings back over top of the grain bed as gently as possible so as not to disturb the grain structure. Once the wort is perfectly clear it is diverted to a holding vessel, an underback, where it is maintained at 71–82°C to minimize microbiological contamination or transferred directly to the kettle (4).

In single-temperature mash tuns, wort is always runoff continuously to recover the extract, while in lauter tuns the grain bed is re-suspended and resettled several times during runoff.

Sparging

Sparging extracts the fermentable liquid, known as wort, from the mash. Sparging begins just after the first worts have been collected and after the mash has settled but before the surface of the mash has become too dry. The sparge water should be applied at a rate that matches the runoff rate. The sparge water (around 75–78°C) is applied via the sparge arms that are centrally located (15). The temperature of the sparge water is higher than that of the mash in order to reduce viscosity of the wort thus facilitating run off and subsequent leaching of the extract from the grist particles. Sparge water should never exceed 80°C because of the possibility of extracting unwanted materials such as tannins, proteins, and unconverted starch (18). Regardless of the water temperature, the rate of flow must be maintained throughout sparging to dissolve and rinse the extract trapped in the hulls or adhering to malt particles.

A water pH greater than 7.0 is unacceptable for sparging; preferably, the pH should be around 6.0, which leads to better coagulation of proteins, better drainage of the grains, and a higher extract yield (21). Alkaline waters can result in an increase in the extraction of coloring material, undesirable compounds such as harsh-tasting polyphenols, haze-forming starch, lipids, and insoluble proteins from spent grains. At higher pHs, proteins form small masses that can clog the grain bed and retard wort filtration.

Specific Gravity of the Wort

While sparging is in progress, the gravity of the wort should be periodically checked. Sparging continues for about 90 to 120 minutes until the gravity has been reached and/or the required volume of wort has been recovered, these quantities being fixed by the original gravity of the beer and the capacity of the kettle. Briggs *et al.* report extract recoveries of 98% have been claimed, and 96–97% is usual but in smaller breweries the recovery rate can be as low as 85% (6). It is quite common to recover somewhat more wort than required because of evaporation during wort boiling, but trying to recover all of the wort increases amounts of undesirable material (polyphenols and bitter substances from the husks, silic acids, etc.).

Grains-Out

After wort separation is complete, the spent grains are either thrown out by hand, but more typically the grains are swept out of the discharge ports by horizontal arms that rotate over the false bottom of the tun. The spent grains are then collected and transferred by screw conveyor or compressed air to the collection silo.

Lauter Tun

The lauter tun is the most widely employed wort separation vessel system in North America and Europe. Lauter tuns are, in general, designed much like infusion mash tuns, but they are wider and shallower, as shown in Figure 10.1. Like the mash tun, filtering is through slots in a false bottom that supports the grain bed. However, there are some big difference between mash tuns and lauter tuns. Lauter tuns are suited for use of under-modified malts and high adjunct rates. However, if the recipe has less then 50% malt, there will be insufficient husk material to form an adequate filter bed (15). The grists used in a lauter tun are finer, the mashes are more dilute, and the bed depth shallower all of which helps in extract performance.

Like the mash tun, the lauter tun is also equipped with a sparging system to wash the extract from the mash. The top of the tun is usually spherical or conical and fitted with a vent for relieving the vapors of the hot mash to the atmosphere. The true bottom of the tun may be flat or sloped, or it may be constructed with several concentric valleys with intervening ridges. Suspended above the true bottom of the tun is a false bottom of milled, slotted, or welded wedge wire steel plates that act as the filtering system.

Figure 10.1
*Two-vessel Brewhouse Showing Lauter Tun in Foregound and Kettle
Courtesy of Drinktec U.S.A.*

Free surface area (the area through which the wort can flow) varies from 8 to 15% for milled bottoms and up to 25% on welded screen bottoms. The false bottom is not a filter plate but acts as a support for the grain bed.

Typically, false bottoms in craft brewery lauter tuns do not consist of the pie-shaped sections, but rather rectangular wedge sections of "wedge"- or "V"-wire screen. Their cost is considerably cheaper than machined bottoms, and the screens perform reasonably well.

The lauter tun, unlike the mash tun, is equipped with rakes (see Figure 10.2) to assist mash transfer and for leveling the bed and facilitating filtration of the liquid from the mash. European lauter tuns are fitted with zig-zag-shaped rakes with "shoes" at the lower ends while North American designs have straight rakes with lifting properties (15). Rakes are more important when the mash is stirred and mixed, such as with temperature-controlled infusion or decoction mashing.

Mash Transfer

In preparation for receiving a mash, the lauter tun is thoroughly rinsed, and underlet with hot water (often 75–78°C) to drive out the air underneath the false bottom (4). The plates of the false bottom are covered with hot water (1.2–2.5 cm) to help spread the mash (4). The foundation water prevents the intrusion of mash particles into the area under the false bottom, an intrusion

Figure 10.2
Lauter Tun Rakes
Courtesy of Sound Brewing Company

that can inhibit the flow of wort during the lautering operation. Unlike single-temperature infusion mashing, the mash loses its entrained air and sinks onto the false bottom in a dense bed.

The mash can be pumped either from above or from the bottom to the lauter tun. Filling from the bottom is usually faster than filling from the top. Pumping the mash from the bottom of the mash tun minimizes oxygen pickup and is considered advantageous for mash bed uniformity (29). Oxidation can result in off-flavors, thus reducing the shelf life of the beer. When filling the lauter tun from the top, separation of the spent grains into heavier and lighter fractions may result if the pumping is carried out too rapidly. An uneven grain bed often results in low yield and high particulate matter in the runoff wort. Best results are achieved using a low flow velocity at the beginning and then, as the lauter tun fills up, increase the rate of pumping. Also, dropping the mash into the lauter tun from too great a height can cause separation of spent grains and result in a consequent loss of extract. The lauter tun should not be overloaded, as this will bend the false bottom and lead to an uneven runoff and decreased yield.

Bed depth in the lauter tun is generally between 30 and 45 cm, though some can have depths up to 55 cm or be as shallow as 20 cm. Bed depth is a critical assumption in lauter tun design, as is the maximum loading capacity of the tun. Bed depth is inversely proportional to tun diameter, and loading capacity is directly proportional. Both factors should be taken into consideration when fabricating a well-designed tun for a specific brewing operation. In general, maximum runoff rates occur with shallow bed depths, those between 30 and 45 cm. Rapidly diminishing returns occur with bed depths greater than 55 cm.

Establishing the Grain Filter Bed

Two methods are generally employed to establish the grain filter bed uniformly over the entire floor of the lauter tun. The first method is to operate the lautering machine in a raised position at moderate speed with the blades in "grains out" position until all the mash is transferred. The second method is to operate the machine at a slow speed, with rakes rising slowly in a lautering position as the height of the grain bed increases.

Mash Rest and Underletting

Classical lautering operation provides for a rest period of from 15 to 30 minutes in the lauter tun after the bed is leveled (7). This rest allows the mash to settle prior to recirculation, which is necessary to avoid cloudy wort. Instead of providing a rest period, though, some brewers will circulate the wort almost immediately after the transfer of the mash into the lauter tun. Most procedures are intermediate between a rest and immediate circulation.

Many brewers perform an underlet at the completion of mash transfer, which re-suspends any settled mash particles into the liquid phase. The purpose is to help clear the area under the false bottom of the mash particles, and it also has the effect of preparing the entire filter bed for wort runoff. Some craft brewery lauter tuns are not set up for underletting.

Recirculation of Wort

Before the first wort is run to the brew kettle, it must be re-circulated (known as vorlauf) through the grain bed because it carries finely divided materials from beneath the false bottom. This assists in the establishment of the filter bed and wort clarification. During circulation in the lauter tun, particles form a filter bed composed of three distinct layers according to particle size and density. The largest and heaviest particles settle first to form a thin layer on the false bottom. The second layer is a mixture of mixed husks and other coarse particles. Finally, a thin layer is formed that is composed of fine, light particles (gelatinized starch and protein trub) of a grayish color. The upper layer is prone to collapsing and clogging if wort is drawn off too rapidly.

Turbid wort is transferred from beneath the false bottom and transported to the top of the mash bed via the vorlauf pump. The pumping clears out the area under the false bottom and establishes the filter bed for lautering. If the pumping is too fast, the runoff will cause turbulence between the false and real bottoms.

The vorlauf process lasts from 5 to 20 minutes, depending on the amount of mash, the depth of the spent grains, the type of lauter tun used, the filtration method, etc. A long recirculation will minimize carryover of malt solids and precipitated proteins into the kettle, but tends to increase the level of wort lipids. Lipids have a deleterious effect on foam but are important to the cell wall creation of yeast. On the other hand, shorter recirculation periods tend to carry over more solids. Some brewers prefer more solids, believing that they provide nucleation sites for the agglomeration of proteins during the boil and thus produce clearer worts. However, the haze that may be brought into the boil along with the solids carries higher-weight proteins, causing chill haze in the final product.

Collection of First Wort

After the filter bed is established and the wort has reached satisfactory clarity, circulation is stopped and the "first wort" is diverted to the kettle or underback (7). In some cases, the brewer may direct the wort to the kettle with some fine solids still present, knowing it will reach satisfactory clarity within a few minutes once transfer has started. DeClerck reports that the presence of a few particles in suspension is immaterial since they will be held back by the spent hops after boiling, or deposited along with coagulated protein in the wort cooler (6). Mashing with wheat is considerably more troublesome than mashing with barley malt, with sticky mashes and slow runoffs being common. Again if the wort is drawn off too fast it can cause a suction that draws the grain bed closer and can ultimately collapse the main bed. Most lauter tuns are fully automated in controlling the wort runoff rate as well as measuring and controlling the differential pressure above and below the false bottom of the lauter tun.

Sparging and Second Wort Collection

As the first wort is drawn off, sparging is initiated shortly before the wort level reaches the top of the grain bed. The top surface of the grain bed should not be allowed to dry out while sparging. This will avoid compaction and uptake of oxygen into the wort. Rehberger recommends a minimum of 2.5 cm of water on top of the grain bed (24), but water too deep can lead to stuck mashes. A temperature of 77°C should be carefully maintained within the mash during sparging (26).

Sparge water is proportioned in cycles or is continuous. Some brewers recommend proportioning in cycles of three distinct water additions, the first with 25% of the total, the second with 45%, and the third with 30%. This program limits hydrostatic pressure while maximizing the flow of extract. If continuous, the sparge water is metered and the flow is equal to the wort runoff rate. If sparging is done continuously, it takes less time; if done proportionately, the yield is higher because the sparge water has more time to extract the spent grain contents. Both procedures are commonly used.

During sparging, the rakes are operated to impart a lifting as well as a slicing action. This prevents compaction of the bed and ensures that no channeling occurs. In the initial stages, the lautering machine is usually operated at low speed, and the knives are positioned so that they never quite touch the plates or disturb the bottom layer. As turbidity increases, the knives are raised.

> ### Note | Wort Collection
>
> The runoff is transferred either to the kettle or to a grant vessel, i.e., a wort collection vessel. Bottom entry into the kettle allows for a gentler handling of the wort with less chance of oxidation. It is important that the heavy first runnings are not allowed to lie at the bottom of the kettle at high temperatures, which could result in caramelization. Some brewers recommend holding the wort at about 77°C in the interest of preventing enzyme activity. Worts must not be allowed to cool below 55°C because of the possibility of infection by bacteria, e.g., *Lactobacillus delbrueckii*.

If the bed becomes compacted, as noted in changes in the differential pressure and runoff, the brewer can temporarily shut off the wort flow, just as with the mash tun. This applies a gentle lifting action to the grain bed, which helps to restore porosity and permeability as well as serving to restore hydraulic equilibrium above and below the false bottom. The wort may need to be recirculated to restore clarity before running to the kettle. In more severe cases where compaction of the grain bed occurs, it may be necessary to shut off the wort flow and underlet (7). This procedure should lift the bed and push the grains away from the strainer grate. In this case, recirculation of the wort prior to running to the kettle is essential to minimize solid content. If no water still flows, try raking the grain bed with shallow cuts into the top of the bed, but with progressively deeper cuts as time passes. Try not to disturb the very bottom filter bed.

Compaction can also be the result of high percentages (>70%) of wheat or other glutinous adjuncts, over-crushing of malt, poor mashing techniques, and low mash temperatures. Compaction is particularly a problem after much of the dense wort has been displaced, making the grain bed less buoyant. Signs of a stuck mash include bubbles coming up from air pockets and dry clumps of malt in the mash bed.

Grains-Out

After the last wort and drainings have left the grain bed, spent grains are removed from the lauter tun and discarded. The grain-out doors are opened, the rakes are raised out of the grain bed, and the knife angles are changed to the grains-out position. The rakes at this angle will now push the grain from the center of the vessel toward the outside, where the grain-out doors have been opened. The rakes are gradually lowered from the top position to their lowest position on the bottom to clear out the grains. Many small breweries prefer rakes for their feature of removing spent grains more than for their ability to establish the grain bed.

Mash Filters

Although lauter tuns are widely employed for wort separation, some large-volume brewers prefer mash filters. Mash filters are very much like plate and frame filters consisting of a series of grid-type plates alternating with hollow frame plates that are suspended on side rails. Each grid plate of the filter is covered on both sides with a monofilament polypropylene cloth. Accordingly, the grist serves no purpose as a filter medium and their particle sizes are of no consequence. Mash filters are more likely to be used in larger breweries where throughput and floor space are priorities.

To initiate the wort separation cycle, the mash filter is flushed, then preheated with hot water. The mash is then pumped from the mash mixer into the filter through the top channel, completely filling the filter frames. The entire transfer to the mash filter is completed in 20 to 30 minutes (33). If the filter is under-filled, the efficiency of the extraction process suffers significantly because the sparge water will flow through the empty portion of the chamber (33). Overfilling, on the other hand, results in excessive density, adversely affecting filtration efficiency (33). When the filter is full, the wort collection system is opened, and the wort is drawn horizontally through the filter cloths. The first wort running time is around 12 to 15 minutes. To achieve satisfactory clarity, the wort is re-circulated through the filter. Filtration of the main wort takes 20 to 50 minutes. Sparging is initiated after the first wort is partially drained but before the filter cake becomes dry. Sparge water between 75 and 78°C is pumped into the filter from the bottom and lasts from 90 to 140 minutes (20). Just as with the lauter tun, the mash filter pressure differential is critical. After the last wort, the filter is opened automatically, plate-by-plate, and spent grains fall into a trough with a screw conveyor.

The mash filter has several advantages. These include its shorter filtering cycle, which results in a larger quantity of brews per day, greater extraction efficiencies (because it can process finer milled grists compared to lauter tuns), and a quicker turn-around time then either the mash tun or lauter tun. It is also easier to clean, requiring only one high-pressure cleaning per week.

Other advantages include possible filtration of high density worts, savings on brewhouse plant sizing, ease of automation, and low demands for effluent (1).

The major disadvantage is the clarity of the wort, although it is possible to improve this feature by inducing a pressure drop over the bed and by recirculating the worts (12). Another disadvantage is that it works better on well-modified malt and does not perform particularly well with gummy adjuncts or wheat malt (12). It is also considered labor-intensive when cleaning, requires a predictable mash volume, and discharges wetter grain (19).

In the competition for yield and brew frequency between lauter tuns and mash filters, a new generation of high-pressure filters (e.g., Meura 2001) was introduced to the market. While wort quality remains the same, the high-pressure filters have the advantage of significantly reducing cycle times in wort separation compared to conventional filters. As a consequence, high extract recovery is possible with high-pressure filters when using a two-stage milling system to prepare the grist. A hammer mill is used for a secondary grind of a separated fraction.

Strainmaster

The Strainmaster, which is not as widely used as the lauter tun or the mash filter, was patented by Anheuser Busch and developed by Nooter Corporation. The Strainmaster consists of a rectangular hopper-bottomed tank that is fitted at the base of the hopper and that has downward opening doors for discharge of spent grains. Within the vessel are header pipes running longitudinally down its length, and attached to these are perforated straining tubes. The header pipes are connected by wort draw-off mains to wort pumps. Wort withdrawal and circulation is initiated when the top row of tubes is covered. The wort is then recycled through the straining tubes to the top of the Strainmaster. The action of drawing wort through the tubes creates a filter bed around the tubes in much the same manner that occurs relative to the lauter tun. Recirculation continues until the desired wort brightness is achieved, than the wort is run to the kettle. As the level of the remaining first wort approaches the mash level, but while the mash is still covered with wort, the sparge is initiated over the top of the bed. Upon completion of the sparge, the grain bed is allowed to drain to relative dryness. The doors are then opened, and the grains fall into the grain receiving tank.

Advantages of the Strainmaster include the rapid runoff rate, greater flexibility in dealing with varying grist volumes, and low maintenance requirements. On the other hand, effluent demand of the Strainmaster is high; and worts are cloudier, due to higher solids and lipid contents, than with a lauter tun (12). Also, high lipid levels can lead to changes in fermentation performance in yeast. Finally, the extract efficiency is lower than for a lauter tun because of the uneven flow of wort through the grain bed and the inconsistent flow pattern of the Strainmaster.

Calculations

Assessing the Wort

Extraction efficiency is measured as "extract yield," which is the ratio of the mass of extract to the mass of malt or malt and adjunct. Rehberger reports that the extract yield normally runs from 70 to 75% (29). Wort assessment simply consists of taking a hydrometer reading of the wort and figuring the actual rate of extract. Once the specific gravity of the wort has been determined and corrected for temperature, the degrees of extract can be estimated using the following formula:

Degrees of extract = [specific gravity of wort (liters)/weight of grain (kg)]

Of course, this formula assumes that all of the extract (sugar) in the wort is derived from a single source.

References

1. Andrews, John 2004. "A Review in Progress of Mash Separation Technology." *MBAA Technical Quarterly* 41 (1).

2. Braekeleirs, R. 1991. Filters – Flexibility and Performance." *The Brewer* 77 (920).

3. Carey, Daniel J. 1991. "Microbrewery Design and Performance." *MBAA Technical Quarterly* 28 (1).

4. Briggs, Dennis E., Boulton, Chris A., Brookes, Peter A., and Stevens, Roger. 2004. *Brewing: Science and Practice*. Cambridge, England: Woodhead Publishing Limited.

5. Christiansen, Kent, Steve Borsh, and Gary Luther. 1993. "Lautering – Back to Basics." *MBAA Technical Quarterly* 30 (3).

6. De Clerck, Jean. 1957. *A Textbook of Brewing*. Volume 1, translated by K. Barton-Wright. London: Chapman-Hall Ltd.

7. Dougherty, Joseph J. 1977. "Wort Production." *The Practical Brewer*, edited by H. M. Broderick. Madison, Wisconsin: Master Brewers Association of the Americas.

8. Fix, George. 1989. *Principles of Brewing Science*. Boulder, Colorado: Brewers Publications.

9. Heissner, Russ. 1999. Personal Communication. JVNW, Inc., Hingham, Massachusetts.

10. Herrmann, H., B. Kantelberg, and B. Lenz. 1990. "New Lauter Tuns: Engineering Practice – Technology – Field Experience." *Brauwelt International* 1.

11. Hough, J.S., et al. 1982. *Malting and Brewing Science*. Volume 2. London, England: Chapman-Hall, Ltd.

12. Irvine, J.A. 1985. "Mash Separation Systems." *The Brewer* 71 (844).

13. Johnstone, J. 1992. "Liquid Solid Separation in the Brewhouse." *MBAA Technical Quarterly* 29 (1).

14. Kunze, Wolfgang. 1996. *Technology Brewing and Malting*, translated by Dr. Trevor Wainwright. Berlin, Germany: VLB Berlin.

15. Leiper, Kenneth A. and Michaela Miedl. 2006. "Brewhouse Technology." *Handbook of Brewing*, edited by Fergus G. Priest and Graham G. Stewart. Boca Raton, Florida: CRC Press, Taylor & Francis Group.

16. Lewis, Michael. 1987. "Three Types of Mashing Systems." *The New Brewer* 4 (1).

17. Lewis, Michael. 1993. "Getting the Most Out of the Mash." *The New Brewer* 10 (4).

18. Lewis, Michael J. and Tom W. Young. 1995. *Brewing*. London, England: Chapman & Hall.

19. Lewis, M. J. and Young, T. W. 2002. *Brewing*. New York, New York: Kluwer Academic/Plenum Publishers.

20. Moll, Manfred, ed. 1995. *Beers and Coolers*. New York, New York: Marcel Dekker, Inc.

21. Noonan, Gregory J. 1996. *New Brewing Lager Beer*. Boulder, Colorado: Brewers Publications.

22. Parkes, Steve and Kiesbye, Axel. 2006. "Brewhouse Operations: Ale and Lager Brewing." *Raw Materials and Brewhouse Operations*. Volume 1, edited by Karl Ockert. St. Paul, Minnesota: Master Brewers Association of the Americas.

23. Ormrod, H.L. 1986. "Modern Brewhouse Design and its Impact on Wort Production." *Journal of Institute of Brewing* 92 (March-April).

24. O'Rourke, Timothy. 1999. "Mash Separation: a review." *Brewers' Guardian* 128 (2).

25. O'Rourke, Timothy. 1999. "Mash Separation." *Brewers' Guardian* 128 (7).

26. Outterson, Donald R. 1999. Personal Communication. Outterson LLC., Cincinnati, Ohio.

27. Parks, D.R. 1989. "Comparing the Strainmaster to the Lauter Tun." *The Brewer* 75 (898).

28. Pfisterer, Egbert and Robert Gentry. 1999. "Improving Brewing Material Yield and Efficiency." *The New Brewer* 16 (1).

29. Rehberger, Arthur J. and Gary E. Luther. 1995. "Brewing." *Handbook of Brewing*, edited by William A. Hardwick. New York, New York: Marcel Dekker, Inc.

30. Reichert, Wolfgang. 1989. "Equipment Examined." *The New Brewer* 6 (6).

31. Roberston, J. 1990. "Wort Production." *The Brewer* 76 (905).

32. Samlesbury, D.R. 1989. "Comparing the Strainmaster to the Lauter Tun." *The Brewer* 75 (898).

33. Stewart, G.G. and I. Russell. 1985. "Modern Brewing Biotechnology." *Food and Beverage Products*, edited by Murray Moo-Young. Oxford, England: Pergamon.

34. Taylor, D.G. 1989. "Influence of Brewhouse Practice on Wort Composition." *The Brewers' Guardian* 118 (2).

35. Thompson, Donald, Karl Ockert, and Vince Cottone. 1999. "Craft Brewing." *The Practical Brewer*, edited by John T. McCabe. Wauwatosa Wisconsin: Master Brewers Association of the Americas.

36. Vernon, P.S. 1985. "Wort Clarification." *Brewers' Guardian* 114 (3).

37. West, C.J. 1986. "Mash Filters at Romford Brewery 1977-1986." *The Brewer* 72 (856).

Chapter Eleven

Wort Boiling

Following wort separation and extraction of the carbohydrates, proteins, and yeast nutrients from the mash, the clear wort must be conditioned by boiling in the kettle. This chapter covers the biochemical changes that occur during wort boiling, the types of kettle additives, hop and trub removal as well as the types of wort boiling systems used in brewing beer.

Biochemical Changes

The purpose of wort boiling is to stabilize the wort and extract the desirable components from the hops. The principal biochemical changes that occur during wort boiling are as follows:

- Sterilization
- Enzymes inactivation
- Protein precipitation
- Color development
- Isomerization
- Dissipation of volatile constituents
- Concentration of wort
- Reduction in wort pH

Sterilization

Although the wort separation and sparging processes are normally conducted at elevated temperatures (76–80°C), wort entering the kettle contains numerous microorganisms—yeast, molds, and bacteria—which can result in off-flavors and numerous other problems. Thus, it is important to eliminate these microorganisms by boiling the wort above 100°C for at least 45 minutes. In spite of the apparent destruction of bacteria in the kettle, heat resistant or thermophilic organisms (e.g., *Lactobacillus delbrueckii*) can frequently be found in trub deposits that may occur in hot wort lines, hop separators, and wort tanks (18).

Enzyme Inactivation

Boiling fixes the carbohydrate composition of the wort by inactivating residual enzymes that are responsible for carbohydrate and protein degradation and that may have survived mash-off or sparging. If left unchecked, these enzymes will continue to alter the carbohydrate composition in the wort. As a result, the dextrin/maltose ratio continues to change upsetting the balance obtained during the mashing stage (49).

Protein Precipitation

During the boiling process, it is necessary to decrease the level of high molecular weight nitrogen found in the malt. Protein precipitation occurs as the wort loses its turbidity during boiling, and material breaks out of suspension and precipitates as proteins coagulate. The wort needs to be sufficiently agitated to cause the denatured proteins to coagulate to form flocks. If the proteins are allowed to persist it could effect the pH, colloidal stability (chill haze and permanent haze), fining and clarifying properties, fermentation, and taste of beer (36). Removal of protein can be achieved by the addition of copper finings to the boiling wort, which is discussed later in the chapter.

> ### Note | Kettle Times
> Traditionally, kettle times lasted between 90 and 120 minutes, with a minimum of 10% evaporation per hour. However, today kettle times for an all-grain beer last from 60 to 90 minutes, with a 4 to 8% evaporation rate. Boil times of 2 hours or longer are usually reserved for special beers, such as strong Scottish ales.

Color Development

During boiling, the color of the wort becomes darker. This is brought about by the formation of pigments (melanoidins), the oxidation of polyphenols, and the caramelization of sugars.

Production of Melanoidins

The production of melanoidins—the Maillard reaction—occurs when reducing sugars from carbohydrates react with amino acids that are derived from the proteins during mashing. Melanoidin production is most active in the malting process, but it continues to some extent during mashing and to a somewhat lesser extent in boiling. Melanoidins are formed in malt during the kilning operation and are what give high-kilned malts their deep color and rich malty flavor. About one-third or less of the melanoidins is formed during the kettle boil (39). Toffee, nutty, malty, and biscuity flavors are associated with melanoidins.

Oxidation of Polyphenols

The oxidation of polyphenols is another source of color formation. Polyphenols are sometimes referred to as "tannins" and may be derived from malt husks and hops. High-carbonate water can also hasten color formation by increasing the extraction of polyphenols from husks and hops during mashing and sparging and by increasing caramelization. To reduce the presence of polyphenols, brewers can select two-row malts with thinner, lighter husks. Using hop pellets as well as high-alpha-acid hops will reduce the level of polyphenols during the boil. Other alternatives are to reduce the breakage of the husk while grinding the grain or to wet-condition the malt before grinding. Polyphenol extraction can further be minimized by reducing mash times, not over-sparging, and avoiding the use of sparge water that is too high in temperature.

Carmelization of Sugars

Caramelization is a chemical process that affects sugars subjected to temperatures of 200°C or greater. Direct gas fire heating and kettle design may increase the degree of caramelization, as may long boils and high-gravity worts. Caramelization is a desirable characteristic in Scotch ales.

Isomerization

The major flavor contribution of hops in beer is bitterness from iso-alpha acids. During the boil, the insoluble alpha acid extracted from hops is converted to a more soluble iso-alpha acid. One factor that greatly influences isomerization is a long, vigorous boil.

On average, only about one-third of the alpha-acid added is recovered in the boiled wort in the form of iso-compounds (23). There are several variables that affect isomerization of the alpha-acids in the kettle, which are:

1. Hopping Rate – A high hopping rate reduces utilization.
2. Wort Strength – Hop utilization is generally better in low gravity worts than in high gravity worts.
3. pH – A high pH results in a greater utilization of humulone to isohumulone, a process facilitated by magnesium.
4. Type of Hops – Generally, the more the hops are processed the greater the utilization.
5. Length of Boil – Most of the alpha-acids are isomerized at the start of boiling, and the isomerization rate progressively decrease as boiling continues. Approximately 90% of the wort bitterness occurs

within the first 30 minutes of boil (34). Prolonged boiling may be detrimental because isomerized beta acids form (46).
6. Losses in Trub – Iso-alpha acids combine with proteins that precipitate during the boil forming hot trub. Hence all-malt worts will decrease the overall hop utilization due to their higher protein content compared to adjunct worts. Likewise, some of the iso-alpha acids will be adsorbed onto the yeast cells during fermentation.

Dissipation of Volatile Constituents

Boiling of wort drives off volatile wort compounds, including dimethyl sulfide, aldehydes, and hydrocarbon components of the hop oils. The principal malt derived volatile lost during boiling is dimethyl sulfide or DMS which is rapidly lost through evaporation. However, the breakdown of S-methyl methionine, its precursor, can continue during the period between the end of boiling and wort cooling and can survive to the final beer. DMS gives beer a strong "sweet corn" or "lagery" flavor. The level of sulfur compounds (DMS, in particular) should be fewer than 120 µg/L at the end of the boil (46). DMS levels can be controlled in a number of ways: (37)

1. Using malt with low S-methyl methionine levels
2. Long wort boiling to decompose precursor and vaporize DMS
3. Short whirlpool stand time to reduce decomposition of precursor
4. Rapid wort cooling to minimize the time the wort is hot

All kettles have a condensate return in the throat of the stack to prevent condensed volatiles returning to the boiling wort. The condensate return should typically be about 2 feet above the hip of the kettle or about 2 feet above the first section of the straight wall before any 90 degree bends. One difficulty with pressure boiling is that undesirable volatiles are not steam-volatilized sufficiently, especially when whole hops, powder, or pellets are added to the kettle (17). Consequently, some breweries assist the removal of undesirable volatiles by keeping manholes open during boiling.

Concentration of Wort

The wort must be concentrated by evaporation since the water used in mashing and sparging has produced wort lower in specific gravity than the target gravity. The amount of water removed during the boil is directly proportional to the rate of evaporation. Evaporation rates vary between 4 and 10% and are a function of vessel design, vessel material, energy system, and air pressure. In many small brew pub systems, a 5% evaporation rate is common (46).

Some brewers make up the kettle to a set volume and evaporate down to a second set volume, then strike the kettle at that level (18). A more common method requires taking a gravity reading of the kettle after the last runnings and calculating the required evaporation to reach the desired strikeout gravity.

> **Note | Heating the Kettle**
>
> In order to save time, most brewers begin applying heat as soon as the wort covers the bottom of the kettle to minimize charring (or scorching), and to prevent damage to the kettle. Some systems may require that the kettle be more than halfway full before applying heat. Care must be taken when using direct gas fire since the first runnings are easily caramelized. If steam jackets are used, heating may be started as soon as several inches of wort are in the kettle by shutting off the side jacket.

Reduction in Wort pH

As previously mentioned, the pH of the wort starts to decrease during mashing and continues to fall during wort boiling to reach a final pH of between 5.2 and 5.3 (25). The drop in pH is due to the reaction of Ca^{2+} compounds with phosphates and polypeptides to form insoluble compounds that release hydrogen ions. A decrease in the wort pH is important for the following reasons (37):

1. Improves protein coagulation
2. Improves beer flavor in particular VDK (diacetyl) reduction
3. Encourages yeast growth
4. Inhibits the growth of many beer spoilage organisms
5. Reduces formation of excess color

Calcium sulfate, calcium chloride, phosphoric acid, and sulfuric acid can be added to the kettle to aid in pH reduction. Alternatively, lactic acid can be used if German purity laws must be adhered to.

Formation of Hot Break

The interactions of denatured proteins and simpler nitrogenous constituents with carbohydrates and/or polyphenolic constituents (tannins, flavanols, and anthocyanogens) are what form the trub or "hot break." When boiling is completed, the wort should be brilliantly clear, with mostly large particles of trub and spent-hop material. If the break is too fine, this can be the result of using less-modified malt, of insufficient boiling action, or of brewing water that is high in carbonates.

Factors Affecting Quantity of Hot Break

The amount of trub will be on the order of 200 to 400 g/hl wet weight (approximately 80–85% water). The quantity of trub depends on type and amount of malt, use of adjuncts, mashing program, duration and vigor of the boil, wort pH, and type of polyphenols.

All-malt worts will have correspondingly more trub but malts kilned at higher temperatures will have fewer high-molecular-weight proteins in the wort and therefore less trub. Worts brewed with adjuncts produce proportionately less trub, whereas higher gravity worts produce more trub.

It has also been shown that quicker, less intensive mashing routines yield worts containing more trub. Conversely, decoction mashing produces worts with considerably less trub, which is due to the mash boiling process, extended protein rest, and the removal of trub before the kettle boil.

The length of the boil can affect the amount of trub with longer wort boils forming more trub, albeit at an ever-decreasing rate. Long boiling times encourage polyphenols to form complexes with larger proteins. Boiling times of 1 to 2 hours under normal atmospheric pressure are usually needed for sufficient hot break formation. However, boiling too long may result in the excessive precipitation of fine materials. This can cause a number of problems, such as poor sedimentation in the whirlpool, poor filtration, and excessive color formation.

Factors Affecting Quality of Hot Break

It is important, at this stage of the process, that most of the proteins and polyphenols have been precipitated. If not, this could complicate fermentation and filtration of beer, and persist into the finished product, causing haze. However, even under ideal conditions, precipitation of these materials is not complete.

Optimum protein flocculation occurs above pH 5.5, but a pH between 5.2 and 5.5 is more appropriate, as it satisfies the other pH requirements of wort boiling and fermentation. Lower pH values produce fewer, smaller flocks, and at a pH below 5.0 the protein does not coagulate. When the pH is less than optimal, agitation and movement within the kettle become increasingly important in promoting protein flocculation.

Qualitative Assessment of Hot Break

At the end of the boil, a sedimentation test is carried out on 1 liter of wort in an insulated Imhoff cone to assess how well the protein has been coagulated. After 5 minutes, the wort should have a brilliant background clarity and the trub should have compacted to a level below 100 ml/l (4).

Kettle Additives

Hops

Timing

The hops may be added all at once; but more commonly, they are metered out in portions throughout the boil either at the start of the boil or near the end of the boil. Bittering hops are usually added at the beginning of the boil to maximize isomerization of their alpha-acid content and to drive off undesirable flavor compounds. Aroma hops are added part way through or at the end of the boil for the delicate hop flavor components. Some bittering hops contribute to hop aroma, even when added early in the boil (11).

In the past, British ales were typically hopped in a single lot, usually at the beginning of the boil, but ales today are more often double-hopped. Lagers are often hopped in multiple additions, usually three or four times, with the number of "charges," the timing, the type of hops used, and the amount of each charge varying from brewer to brewer. Some brewers even add hops after the heat is turned off. Each hopping is done at a different time in order to extract a different chemical composition, which builds to a more complex flavor profile. For example, the first addition achieves the greatest isomerization of alpha acids with most of the hop oils being evaporated off by the end of the boil. The second addition of hops will result in sufficient isomerization with some of the hops oils present at the end of the boil. While a third hop addition before the end the boil will result in very little isomerization but more hop oils.

Some brewers add a small amount of bittering hops (less than 1/10 of the hop charge) just before boiling to prevent excessive foaming (18). This may be necessary if the boiler capacity is close to the batch size. However, if the boiler is at least 25 to 30% larger than the volume of wort, boil-overs can be avoided by simply regulating the heat under the kettle.

Hops are added early in the boil to aid in the precipitation of proteins, thus providing a cleaner fermentation. Moreover, if hops are added early in the boil, undesirable volatile components (e.g., myrcene) are driven off (23). On the other hand, some brewers prefer boiling the wort vigorously for 15 to 30 minutes before adding hops to minimize the precipitation of alpha-acids. Beers that are hopped early in the boil are usually more stable than those with lower hopping rates or hopped later in the boil. Beers hopped toward the end of the boil are less stable because delicate flavors and aromas fade within a short time in packaged beers.

Finish Hopping

The primary method used to get hop flavor and aroma in the beer is to add hops very late in the boil. This practice is called "late hopping," "late kettle hopping," "aromatic hopping," or "finish hopping." Finish hopping not only gives the beer a unique character (often described as floral and fragrant), it also enhances the body and mouth-feel of the beer. Pilsners are traditionally brewed with finishing hops; and light lagers will, in general, benefit from a late addition of hops, whereas dark and amber lagers will not.

Unfortunately, the results from late hopping can be somewhat inconsistent since the composition of essential oils changes over time for many hops, and the amount of oil varies from year to year within the varieties. In addition, hop compounds are affected by brewing conditions such as evaporation rates and fermentation temperatures.

Some brewers prefer using whole hops to achieve optimal flavor and aroma in finish hopping, while others prefer to use non-isomerized hop pellets or kettle extract because of the short contact time. When whole hops are used for finishing, it takes a certain amount of time for the oils to be extracted from the lupulin glands; with pellets, though, the oil is readily available. Aroma varieties are the preferred choice, but some bittering hops are used for finishing, such as Centennial and Northern Brewer. Most late hop additions are done 15 minutes or less before the end of the boil. With aromatic varieties, it is not uncommon for several late hop additions to be made at the end of the boil and another just as the heat is turned off, allowing the hops to steep as the wort cools. The longer the hops are in the boiling wort, the less hop flavor and aroma they will impart and the greater will be the chance isomerization of alpha acids will form, thereby increasing the beer's bitterness.

To achieve hop flavor and aroma, hops can even be added to the hop back or grant. In this practice, which is common with British brewers, the hot wort is passed through a bed of hops. This is an extreme example of late hopping.

Dosage Procedures

In smaller breweries, whole hops or pellets are manually added to the kettle. When hops are added to the kettle, the heat should be shut off or the hops added very slowly to prevent a boil-over. Breweries using whole hops must use a hopjack or a hop strainer when casting the wort to remove spent hops. Large breweries are more likely to add pellets by an automated system. From a buffer tank, the pellets are transported in dry condition pneumatically, via a scale, to the wort kettle. Alternatively, the pellets can be weighed directly into a tank, where they are stirred with water or slurry and pumped to the kettle as required.

The procedure for adding hop extract varies from simply piercing the extract cans and hanging them in the kettle (either by a chain or in a basket) until drained, to elaborate premixing in buffer tanks and use of automatic dosing systems. On a small scale, the most convenient way of adding hop extract is to place the punctured cans, without preheating, in a perforated metal basket suspended in the kettle during the boil. It is important to inspect the cans to ensure that the extract has been completely washed out. Premixing of the extract in a buffer tank is found to give faster dissolution of the resins into the wort and a consequent increase in utilization.

Rates

The quantity of hops is determined by several factors: alpha-acid content, desired hop flavor and aroma in the finished beer, condition of the hops, efficiency of the hop extraction, brewing process, and type of brewing water. The hopping rate for bitterness can be calculated, whereas the rates at which hops should be added for flavor and aroma are less accurate. Usually, decoction-mashed wort is hopped more lightly than infusion-mashed wort so as not to mask the malt character developed during the decoction-mash brew. A higher hop rate is used with a short boil for waters high in chlorides or sulfates or if the beer has to be conditioned for a long time (9). Hop rates are usually higher for high-density beers since the solubility of alpha acid decreases as density increases. Normal hop rates in the United States are from 80 to 130 g/hl of wort (15).

Copper Finings

One way of enhancing floc formation is through the addition of kettle finings. Copper finings (also referred to as Irish moss) is a form of seaweed consisting mostly of a complex starchy polymer called K-carrageenan. K-carrageenan, like polyphenols, has a negative charge and is effective in precipitating positively-charged proteins from the wort solution.

Use of copper finings is helpful in some cases—especially with ales that have not been given a protein rest—and is common in the United Kingdom. Not only do copper finings stimulate the coagulation of proteinaceous materials, they also contribute to reduced wort boiling times and to better trub separation; hence a quicker and cleaner runoff of wort with fewer losses. This all should result in improved fermentations, improved colloidal stability, and improved beer filtration (45).

Poor hot break formation, i.e., a fine break that is slow to settle, may interfere with copper finings performance due to the high level of suspended solids (poor wort clarity). Conversely, good hot break formation, i.e., large protein flocs that settle rapidly, will produce good hot wort clarity and lead to good copper fining performance. Low pH wort, below 5.0, will adversely affect copper fining performance too (51).

Dosage Rate

Mathews reports that the rate of use of copper finings varies widely from brewery to brewery but generally is found in the range of from 10 to 80 mg/l (or ppm) (28). The major reasons for the widely differing rates of use are variations in grist formulations, mashing systems, and specific gravities of the wort (28). Exceeding the optimum dosage rate will not improve clarification, but it can lead to increased acidity of the wort during fermentation because of the carryover of unused moss into the kettle. The brewer should follow the manufacture's dosage recommendation when adding finings.

The optimum dosing rate is determined by preparing a solution of copper finings at a given proportion and adding it to a known volume of wort that is taken from the kettle at the end of the boiling process. After the wort break has been allowed to settle, the supernatant wort is decanted and cooled rapidly. The wort is periodically agitated in a cold-water bath to stimulate the formation of cold break. Allowing the hot wort to cool naturally will give erroneous and misleading results. After cooling, the cold wort is allowed to settle for a period of time, then visually assessed for clarity and volume of bottoms produced.

As the copper fining dose rate increases during the trial, so does the clarity of the cold wort, until the optimum fining dosage rate is reached. This is the point at which there is no benefit obtained by increasing dosage rates. Following the optimization trials, scale-up can be carried out.

To verify the correct dosage rate, cold wort can be taken from the cold side of the wort chiller to confirm optimum performance on the plant. Thompson recommends that the sample be allowed to settle for at least 2 hours, and for up to 16 hours, prior to assessing it for floc formation, clarity, and volume of sediment formed, and that it be kept close to sampling temperature (maximum 15°C) (51). At this stage, it may be necessary to make fine adjustments to the dose rate, depending upon the clarity of the cold wort and the volume of bottoms formed.

Choice of Material

Copper fining products come in a range of physical forms and degrees of purification, including powders, tablets, granules, refined carrageenan, and alkali-washed seaweed. Refined copper finings generally produce brighter clarified worts than do alkali washed products, and have lower optimum dose rates.

Timing

Timing the addition of copper finings is essential because k-carrageenan denatures and becomes less active with prolonged treatment at high temperature. The addition must, therefore, be timed to ensure that the copper finings dissolve completely but do not denature. Copper finings are most frequently added to the kettle midway through the boiling period so that they all dissolve but late enough so that they are not significantly degraded. Reportedly, some brewers add finings to the kettle between 5 and 15 minutes before knockout. Generally, the more highly refined and finely divided the product, the more rapid the dissolution (51).

Dosing Procedures

Powdered products are usually thoroughly mixed in cold water to provide a suspension that can then be added directly to the kettle or the whirlpool. If powdered products are added directly to the kettle, there is likelihood that a significant amount of product will be sucked up the chimney, which results in under-dosing and loss of performance (52). In addition, direct dosing of powdered products presents a potential health hazard in the form of dust generation. Copper finings in tablets can be added directly to the kettle or the whirlpool without loss of fining activity. Copper fining granules are usually added directly to the kettle.

Acids

Tannic Acid – Some brewers add tannic acid to reduce hop-bittering utilization, to reduce color reactions, and to improve wort clarity through enhancement of hot break formation.

Lactic Acid – Lactic acid can be added to lower the pH of the wort. Breweries produce the lactic acid in propagation tanks. The pH value is in the tank needs to be monitored to dose the right amount into the wort kettle.

Calcium Sulfate or Calcium Chloride

Calcium sulfate (gypsum) or calcium chloride is often added to lower the pH of the wort during the boil by 0.1–0.2 units (6). The calcium ion reacts with phosphates and polypeptides to form insoluble compounds releasing hydrogen ions. British brewers often add calcium sulfate to lower pH and thereby minimize color development when producing pale ales. This is because Maillard reactions, which produce melanoidins, are discouraged at lower pHs. Phosphoric or sulfuric acids can be used too in lowering the wort pH.

Syrups and Sugars

Syrups and sugars can be added to the kettle in either dry or liquid forms. They are called "wort extenders" since they increase the extract without little or not added investment in brewhouse vessels. Care must be taken when adding sugars such that they are added evenly without caramelization.

Flavorings

Ale brewers, especially brewers in Belgium, sometimes add flavorings such as fruits, spices, and herbs to the brew kettle either during or after the boil (46).

Wort Boiling Systems

Boiling is the most energy intensive operation in the brewery. Various wort boiling systems are in use, including direct-fire kettles, kettles with external heating jackets, kettles with internal heating systems, and kettles with external wort boilers. To speedup the heating process wort can be preheated using a plate and frame heat exchanger.

Kettle Designs

Direct-Fired Kettles

Traditionally, wort was boiled in direct-fired kettles, often made of copper. Since the heat source is localized at the bottom of the kettle, these vessels are not efficient in transferring heat into the wort, can scorch the wort, and are restricted by the volume of wort that can be boiled at anyone time. High evaporation rates are required to produce sufficient turbulence and typical boils will take more than 90 minutes with an evaporation rate at approximately 10% per hour (36). The localized heating surface promotes caramelization and burning of the wort, requiring frequent cleanings. Generally, the maximum amount of wort that can be boiled at any one time is 200 barrels (36).

Kettles with External Heating Jackets

Kettles with heat jackets are very common and are usually the next step after direct fired kettles. The heating jackets are symmetrically arranged around the vessel, or the jackets may be placed asymmetrically in order to encourage a rolling boil. Kettles often have two separate jackets located on the side-walls and the floor of the kettle. However, heating jackets suffer problems similar to those of direct-fired kettles in achieving efficient heat transfer, especially in larger kettles. Consequently, most kettles require agitators for a satisfactory boil. These vessels are also prone to foam formation during boiling, which can be controlled by an extractor fan and with a cold air draft over the wort surface.

Kettles with Internal Heating Systems

The advent of internal heating systems (see Figure 11.1) allowed for more efficient heat transfer and larger kettles for boiling larger volumes of wort. In many designs, the heaters are located in the center of the kettle to give a turbulent boil. The wort is heated by flowing upwards through the heating unit into a constricting tube (sometimes called a venturi tube) and emerges above the level of the wort striking a deflector plate. The deflector plate directs the wort back as a spray onto the wort ensuring a good circulation of the wort.

Kettles with internal boilers offer a number of advantages. A circulation pump (and its additional energy consumption) is not necessary since natural convection of the wort takes place in the kettle. The degree of efficiency of an internal boiler is significantly higher than that of an external boiler even without considering the forced circulation of the latter. This is because the internal boiler is always completely surrounded by the wort. Wort is subject to less thermal stress too. The boiling temperature of wort using an internal boiler is approximately 101°C. Using an external boiler, outlet temperature of the wort is approximately 106 to 107°C. Such high wort temperatures in external boilers are tantamount to a pseudo-pressure cooking effect, which many brewmasters would rather avoid. The investment costs for an internal boiler are significantly below those for an external boiler, not to mention that an internal boiler will not require extra space nor any circulation pump and pipes. Operation costs are lower because of less energy consumption and lower energy losses. Furthermore, this process and technology lends itself ideally to modernization of existing brewhouses.

Figure 11.1
Kettles
Courtesy of JVNW Inc.

The disadvantage of internal heating systems is that they are difficult to clean, prone to corrosion (if manufactured from copper), and limit the circulation of wort.

Kettles with External Wort Boilers

Many breweries have an external boiler outside of the kettle, sometimes called a calandria, through which wort is pumped. The most common type of external wort boiling system is the tube and shell heat exchanger system. These tubes provide an enormous surface area on which vapor bubbles can nucleate, and thus provides for excellent volitization. Wort is heated by pumping it from the bottom of the kettle through the external wort boiler and back into the kettle through a vertical tube called a "fountain." The wort strikes the bottom of the spreader on top of the fountain, fanning out in a ring onto the wort surface around it, knocking down foam, and creating a turbulent boil and convection. Once boiling is achieved, the circulation pump can be by-passed. The wort will freely circulate due to the density change between incoming wort to the boiler at 98°C, and the outlet wort and vapor from the boiler at around 105°C (36). The total volume of wort is circulated seven to twelve times an hour through this external boiler, ensuring that the wort is evenly boiled by the end of the boil. While the contents of the kettle boil are under normal atmospheric pressure, in the external boiler a slight overpressure is produced so that the wort is heated between 102 and 106°C (23).

External boilers are available with in-line density measuring devices for controlling the rate of evaporation and length of the boiling period. External boilers were originally designed to improve performance of kettles which did not provide adequate boiling effect, but have since been adopted by the industry as a sole means of boiling wort.

Although the initial investment cost can be high as well as higher energy costs, external calandria offer several advantages. Heating of the worts can start once 15% of the total kettle contents have been collected, especially if the wort has been preheated. Vigor of boil can be controlled mechanically (by wort circulation), thereby controlling the rate of evaporation and boil time. Evaporation rates up to 25% have been realized with these systems (39). Because of the advantages of using low-pressure steam, boiling time can be reduced by 20 to 30% (23). The use of low-pressure steam also means there is less fouling, allowing more brews to be processed between cleaning. This decreases brew house downtime thus improving throughput. Cleaning is easily achieved by circulating solution in the same manner as the wort. One external calandria may serve two kettles. Fobbing is kept to a minimum, which avoids some of the potential product defects, such as loss of foam and haze stability. The higher temperatures produced in the external heater result in improved hop utilization, and the coagulable proteins are completely precipitated. Finally, the improved boiling reportedly produces lower pH, lighter color, cleaner taste, and improved flavor stability.

Some of the disadvantages include wort damage due to pumping and passing through valves, fouling of narrow tubes, and the high cost of operating the pump (25).

Methods of Heating

The original method of heating a kettle was by direct fire, using coal or wood, but over the years this has given way to heating by direct gas or steam.

Most craft brewers usually use direct-fired gas vessels because the costs are much lower compared to purchasing, installing, and maintaining a steam-jacketed vessel. Also, direct-fire vessels do not require an engineer or the maintenance requirements and do not fall under the same insurance parameters and zoning regulations as steam vessels. If a direct-fire kettle is employed, then it is recommended that a high-fire stage be initially used too quickly bring the wort to a boil. A low fire should then be switched to in order to maintain a good rolling boil without boil-over. This procedure should minimize the risk of scorching the wort.

> **Note | Low-Pressure Boiling**
>
> Kettles with both the internal and external heating systems can be operated with an increased pressure during the boil usually up to 1 bar (36). The low operating pressure elevates the boiling temperature between 106 and 110°C, which has the effect of accelerating the various wort reactions, and reducing the boil time. At the end of boil, the excess pressure is released allowing the escape of the volatile compounds.

If steam is used, the heating of the kettle is done by jackets, an internal percolator, or an external calandria. From a brewing standpoint, steam is better since it releases a great quantity of heat when it condenses on the metal walls of the vessel. Steam is also much easier to control than direct fire in the brewhouse, having nearly instantaneous on-off action. Steam gives more even heat, and it subjects the kettle to fewer stresses. Another advantage of steam heating is that the same boiler may also heat a hot water tank, generate hot CIP water, or even generate steam for sterilization. A steam-fired system should be set up with high heat for quick boil and low heat to maintain a good rolling boil without boilover. Low-pressure steam systems (<15 psi) are available and usually do not require boiler inspections or specially trained personnel, nor do they usually create insurance issues. High-pressure steam systems exceeding 15 psi will often require a licensed operator. Heating with steam becomes more feasible at about 20 bbl capacity since at that point direct-gas-fired kettles become less efficient.

Hot Wort Clarification

Following the boil, the next step is to separate the hop debris and the trubaceous matter (hot break) so that the wort is bright and clear before cooling. If whole hops are used, the amount of spent hops will be between 0.7 and 1.4 kg/hl wet weight and the trub will weigh in the region of 0.21–0.28 kg/hl (6). Hot break will contain 80–85% crude protein, 20–30% tannin, 15–20% resins, and 2–3% ash (6).

Whole Hop Separation

The methods of separating whole hops vary considerably. In smaller breweries hops are removed by straining the wort through a cloth bag or a sieve while in larger breweries a number of whole hop separation systems are employed—such as a hop back or hop separator. They are usually placed in-line after the kettle before the hot break separation vessels such as coolships, settling tanks, whirlpools, and centrifuges.

Hop Back

A more refined device for separating whole hops is the hop back that resembles a mash tun and is often employed by British brewers. Wort from the kettle is run into the hop back and strained by the slotted base. The wort is recycled from under the plates through the filter bed of hops until the wort runs clear. As the spent hop material accumulates, it progressively improves the straining action so that hot trub is retained. Briggs *et al.* report a bed depth of 30–60 cm for optimum performance (6). Wort flow through the hop back is controlled by adjusting the valve on the positive side of the wort pump, which applies back pressure. An in-line sight glass is very useful for checking clarity. The filter bed is later rinsed to recover the wort retained in the spent hops. After the wort flows through the hop back, it is transferred either to a coolship, a settling tank, a whirlpool, or a centrifuge and then pumped through a heat exchanger for cooling. Sometimes aroma hops are added to hop back, to impart a more "hoppy character" to the beer (6).

Hop backs are less in use nowadays partly due to lower hopping rates and the extensive use of pellets and extracts. Hop backs also suffer from high labor costs and effluent loading, and they require disposing of the spent hops. In general, hop backs are not suited for large breweries and are more likely to be used in small-scale operations. Because of its disadvantages, the hop back has been replaced in some instances by the hop separator.

Hop Separator (Strainer)

The hop separator or strainer is only employed where whole hops are used and by large-scale breweries. The hop separator is an apparatus—a primary screen or slotted plates—through which the bulk of the wort flows. The hops retained on the screen are then forced or tumbled onto another set of screens from which they are removed by means of a chain belt or a worm conveyor, with residual wort passing through the screen. In some cases, water is sprayed over the hops to wash out absorbed wort. All of the wort passing through the screen falls into a small holding tank. There it is continuously pumped to the settling tank or directly to the whirlpool or centrifuge before being transferred to the heat exchanger. The hop separator can also be used to enhance flavor and aroma by adding fresh, whole hops to the strainer basket.

Although the spent hops do retain reasonable quantities of wort, some brewers feel that the deleterious effect on flavor due to the leaching of hops by a water sparge outweighs the cost savings. Like the hop back, hop strainers are rapidly disappearing.

Hot Trub Separation

If pellets or extract are used, the wort may be clarified by sedimentation in a coolship, by a settling tank, a centrifuge or more commonly by a whirlpool or a combined kettle/whirlpool. It is recommended that clarification take place at the highest possible wort temperature in order to obtain optimum trub removal (52).

There are different schools of thought regarding the degree of hot trub separation, with some brewers arguing that it is not necessary, especially when brewing ales. However, in the case of lagers it is almost always done to ensure a crisp taste. Removal of hot trub before fermentation minimizes the adsorption of protein-tannin on yeast cell walls. If protein-tannin is in excess it may reduce the yeasts' fermenting power and cause head retention problems, poor flavor stability, dark color, and bitterness or an onion-like astringent taste in the palate of the beer. Other off-flavors in the finished beer associated with excessive trub carryover are "vegetable" or "stewed" flavors. Also, if the wort is not racked off the trub, which contains sterols and unsaturated fatty material, the yeast may use some of those materials to reproduce, even without dissolved oxygen. This may contribute to higher alcohol levels (fusel alcohols) in the beer. Suffice it to say, the standard in the industry today is to strive for as complete a separation as possible, and thus for a cleaner fermentation with less solid suspended material. However, it has been reported by some brewers removal of too much hot trub by wort filtration can result in a beer with a thin taste.

Coolship

In the traditional method, the wort is cooled in shallow open vessels or coolships. The boiling wort is cast to a depth of 15 to 35 cm, where the hot break settles out, followed by the precipitation of some cold break. The wort remains in the coolship from 1 to 3 hours (depending on conditions) and in some cases up to 12 hours, cooling to approximately 60 and 77°C. In Germany, brewers occasionally inject air from the bottom of the tank to aid the sedimentation process. The air produces super-saturated wort that releases air bubbles that become attached to trub particles and float to the surface where they are skimmed off. Coolships are still quite prevalent in small, traditional Bavarian and Belgian breweries, but are seldom used in modern breweries.

Once sufficient sedimentation has occurred the wort is then transferred to the heat exchanger for further cooling or, if it is cool enough, it is transferred directly to the fermenter. The sludge on the bottom of the cooler contains an appreciable amount of wort that can be recovered by filtration.

The risk of biological contamination is much higher with the use of coolships than with other methods. To prevent infection, coolships are sometimes housed in rooms with sterile air ventilation. Additionally, since the wort is in contact with air, it will darken the color and increase hop bitterness of the finished beer.

Settling Tanks

Successors to coolships are settling tanks that have flat or conical bases. Settling tanks take less space because they are deeper, and they are enclosed to reduce the risk of microbiological contamination. They are also often equipped with coil or jacket cooling. The entire process is usually faster with settling tanks than with coolships but since settling tanks are deeper, the break does not settle as well as it does in coolships. Floats are often installed in the tanks to shorten the required settling time prior to the initiation of cooling (39). Although the goal is to produce a brilliantly clear wort, settling times should be restricted due to the continuing formation of aldehydes and the continuing transformation of SMM precursor to DMS (4). One problem with settling tanks is that compact wort can be difficult to remove.

After sedimentation, the clear wort is drawn off and transferred to the heat exchanger for cooling. The transfer process can be optimized with the addition of an in-line turbidity meter that shuts off the pump if the solids content is too high, or by using a site-glass in a non-automated brewery. If the solids content is not monitored, this could negatively impact downstream activities such as fermentation and conditioning.

Centrifuge

Sometimes the wort is clarified by centrifugation, which spins the wort with high velocity, forcing the hop and trub debris to the side of the vessel. Figure 11.2 illustrates a typical centrifuge. If the wort is heavily hopped, it may be necessary to pre-treat it through a strainer to remove some of the abrasive hop solids. Qualitatively, a centrifuge does not guarantee better hot break removal than a whirlpool, but it allows for a more consistent turbidity level. One nice feature in using a centrifuge is that the brewer can immediately begin centrifuging after the boil without waiting for a concurrent temperature drop, thereby shortening the entire break removal and cooling process. Other advantages include reduced beer losses since centrifuges are ideally suited to recover entrained wort in the trub.

Drawbacks of using a centrifuge include its initial capital cost and the fact that it requires regular maintenance and can break down in the middle of a brew. For these reasons, some breweries have a parallel operation of several centrifuges with a backup in place. In extreme cases, the centrifuge has been known to modify protein molecules because of the extreme forces in play, thereby producing a slightly sulfuric-yeasty aroma (13). It can also be difficult to optimize these systems if the entire kettle volume is clarified by a centrifuge since there is a large range in particle mass from large hop debris to small protein flocs. Other disadvantages may include an increase in beer temperature, potential for oxygen pickup, and an increase in fine haze particles. Consequently, these systems are not as popular as whirlpool systems.

Whirlpool

The whirlpool, as shown in Figure 11.3, is the most widely used method for hot trub removal

Figure 11.2
Centrifuge
Courtesy of Alfa Laval Inc.

particularly in breweries that use hop pellets, powder, and extracts. Whirlpools cannot be used with whole hops unless a hop back or strainer is used initially. It represents the most efficient method for trub removal for small brewers.

Whirlpools come in various designs, and opinions vary as to which is the best geometry. Some designs call for a conical bottom, while others prescribe a flat or even inverted cone bottom. Flat bottom designs, particularly if they have trub retaining rings, allow virtually all of the wort to be drained from the trub. However, subsequent removal of the solids is very difficult, requiring excessive amounts of flushing water or using a mechanical device (43). Sumped bottom designs generally allow for easy trub removal but entrap significant volumes of wort. The flat bottom design is probably the best for small breweries. Whirlpools have a depth-to-diameter ratio of from 1:1 to 1:5, but more often a 1:3 ratio is preferred by most brewers (23).

To minimize shear force breakup of trub flocs during transfer, the wort velocity should be less than 5m/sec and the transfer time fewer than 10 minutes, which can be achieved with large diameter pipework (4). As the hot wort enters the tank tangentially, the trub moves out to the periphery of the tank, sinks down the sides, and is then propelled to the center of the tank, where it forms a hard conical cake. The consistency of the solids changes with time, and eventually the mass will set like concrete due to oxidative copolymerization of proteins and polyphenols. If the in-feed velocity is too low, the deposit of trub will not form in a central cone; or if it does so, the

Figure 11.3
Whirlpools with External Calandria
Courtesy of Briggs of Burton

compactness will be inadequate, which will prevent the trub from draining out of the vessel with the last runnings. If the speed is too high, the velocity will break up the particles and make separation difficult. The rotation time is usually between 10 and 40 minutes, with smaller tanks ranging from 10 to 15 minutes because the surface to center area—the surface area ratio—is much greater than in larger tanks (38).

Certain additions made to the kettle, such as polyvinylpyrrolidone, Irish moss, and particularly bentonite, may assist in clumping together hot break particles, thereby improving the degree of compaction of the trub pile. Anaerobic conditions are also said to increase the amount of material deposited, particularly if carbon dioxide is injected into the wort line while being pumped into the whirlpool tank (17). In general, the effectiveness of whirlpools decreases as the original gravity of the wort increases. This is because the relative difference in density between the trub particles and the wort decreases (5).

After settling for 25 to 30 minutes, the clear wort is drawn off and transferred to the heat exchanger for cooling (13). If the stand time is too short, separation will be incomplete. Chapman reports that stand time should not exceed rotation time; otherwise, the trub cone may spread (8). Long stand times may increase the risk of infection and may increase the breakdown of S-methylmethionine, raising dimethyl sulfide levels. In fact, a major source of dimethyl sulfide in breweries is extended holding of the wort in the whirlpool. Extended hold times can also result in oxidation of phenolics and fatty acids, producing a "cardboardy" flavor in the final product (13).

As in the case of settling tanks, an in-line turbidity meter is used to monitor wort solids as it is transferred to the heat exchanger for cooling. If the wort is drained from the vessel faster than the wort is drained within the trub cone, the hydrostatic pressure within the cone results in wort being pressed out. Consequently, the trub may dissolve, and there is also an increased likelihood of the cone collapsing.

Trub removal is carried out using high-pressure water jetting. In this procedure, a cleaning-head unit breaks up the core then powerful directional jets flush it across to the outlet. It is absolutely critical that the trub be removed within 5 minutes of the vessel emptying; otherwise, cooling causes the trub to become sticky and significantly more difficult to remove (42).

> **Note | Wort Recovery**
>
> Wort recovery from the trub recovered from the whirlpool can be achieved by using centrifuges and filters but the brewer runs the risk of microbiological contamination. Alternatively, the trub can be used in sparging if using a lauter tun. This way, the extract is recovered and the trub solids disposed of with the spent grain (4).

Brew Kettle/Whirlpool

The latest development in small brewhouses is the use of a combination whirlpool and brew kettle. The vessel, as its name implies, is used for wort boiling and as a whirlpool. The vessel can have an external wort boiler or calandria and no internal fountain or spreader. One external heater can supply two kettles; with one is acting as a whirlpool, the second is used as a kettle. These systems are particularly relevant in larger or multi-vessel installations. Some systems can be direct-fired or steam-jacketed. A pump is used to create the whirlpool action by drawing off the bottom of the kettle and injecting it into a tangentially inlet port. An advantage of using the whirlpool and brew kettle is that trub flocs are not damaged, as they might be in a separate vessel system where the mechanical action of transfer can break up the trub. Other advantages include a combined kettle and whirlpool saves space and enables the process of brewing to be shortened with resultant savings in both energy and brewing time. Also, there is the cost savings that result from buying one vessel instead of two.

Calculations

Hop Additions

The bitterness of a beer is expressed in IBUs (International Bitterness Units), which is a measure of iso-alpha acid in a beer, measured in mg/l. To calculate the amount of hops needed, the brewer must know several factors: bitterness units for the beer style (IBUs), batch size, and production loss, utilization yield of alpha-acids, and types and proportions of hop products required.

For example, suppose we want to know how much alpha-acid must be added to obtain a target of 34 IBUs if the batch size is 500 hl, there is an expected production loss of 10.8%, and the alpha-utilization rate is 31%. We can calculate the needed quantity as follows.

Amount of cast wort = 500 hl
Amount of wort yield = 500(100 − 10.8)/100 = 446 hl
Amount of alpha-acid = 34 mg/l(446 hl x 100 l/hl)
Amount of alpha-acid = 1,156,400 mg or 1,516.4 g

Utilization factor = 31%

100% bitterness = 1,516.4/(31/100) = 4,891.6 g or 4.89 kg

How much of the alpha-acid is supplied if the brewer is using Type 45 pellets containing 9% alpha-acid and whole leaf hops containing 7% alpha-acid at a ratio of 3:1, respectively.

Amount of Type 45 pellets = 4.89(0.75)/(9/100) = 40.75 kg
Amount of whole leaf hops = 4.89(0.25)/(7/100) = 17.46 kg

References

1. Archibald, Hugh. 1981. "The Workings of the Whirlpool." *The Brewers' Guardian* 110 (110).
2. Barchet, Ron. 1993. "Hot Trub: Formation and Removal." *Brewing Techniques* 1 (4).
3. Barchet, Ron. 1999. Personal Communication. Victory Brewing Co.
4. Barnes, Zane C. 2006. "Brewing Process Control." *Handbook of Brewing*, edited by Fergus G. Priest and Graham G. Stewart. Boca Raton, Florida: CRC Press, Taylor & Francis Group.
5. Carey, Daniel J. 1991. "Microbrewery Design and Performance." *MBAA Technical Quarterly* 28 (1).

6. Briggs, Dennis E., Boulton, Chris A., Brookes, Peter A., and Stevens, Roger. 2004. *Brewing: Science and Practice*. Cambridge, England: Woodhead Publishing Limited.

7. Busch, Jim. 1995. "Kettle Reactions – An Introduction to Better Beer." *Brewing Techniques* 3 (4).

8. Chapman, J., J. L. Woods, and J. R. O'Callaghan. 1997. "Whirlpool Separation - Parts I and II." *The Brewer* 63 (751 & 752).

9. De Clerck, Jean. 1957. *A Textbook of Brewing*. Volume 1, translated by K. Barton-Wright. London: Chapman-Hall Ltd.

10. European Brewery Convention. 1997. *Hops and Hop Products, Manual of Good Practice*. Nürenberg, Germany: Getränke-Fachverlag Hans Carl.

11. Fix, George and Laurie Fix. 1997. *An Analysis of Brewing Techniques*. Boulder, Colorado: Brewers Publications.

12. Garetz, Mark. 1994. *Using Hops - The Complete Guide to Hops for the Craft Brewer*. Danville, California: Hop Tech.

13. Gordon, Dan. 1991. "The Trouble with Trub." *The New Brewer* 8 (1).

14. Grace, Michael. 1994. "Making Thing Clearer." *The New Brewer* 11 (6).

15. Grant, Herbert L. 1977. "Hops." *The Practical Brewer*, edited by H. M. Broderick. Madison, Wisconsin: Master Brewers Association of the Americas.

16. Haase, Nicolas. 1999. Personal Communication. Drinktec USA, Ltd./ Jacob Carl GmbH.

17. Hough, J. S., et al. 1982. *Malting and Brewing Science*. Volume 2. London, England: Chapman-Hall, Ltd.

18. Hudston, H. Ranulph. 1977. "Wort Boiling." *The Practical Brewer*, edited by H. M. Broderick. Madison, Wisconsin: Master Brewers Association of the Americas.

19. Klein-Carl, Gunter H. 1998. "Requirement of Energy in Brewhouses." *Poster Session IBS*, Atlanta.

20. Klein-Carl, Gunter. 1999. Personal Communication. Jacob Carl.

21. Johnstone, J.T. and D.J. Whitmore. 1971. "Clarification of Beer." *Modern Brewing Technology*, edited by W.P.K. Findlay. Cleveland, Ohio: The Macmillian Press.

22. Kollnberger, Peter. 1984. "Wort Boiling Systems – New Developments." *MBAA Technical Quarterly* 21 (3).

23. Kunze, Wolfgang. 1996. *Technology Brewing and Malting*, translated by Dr. Trevor Wainwright. Berlin, Germany: VLB Berlin.

24. Leather, R.V. 19984. "The Theory and Practice of Beer Clarification – Part 1 – Theory." *The Brewer* 80 (960).

25. Leiper, Kenneth A. and Michaela Miedl. 2006. "Brewhouse Technology." *Handbook of Brewing*, edited by Fergus G. Priest and Graham G. Stewart. Boca Raton, Florida: CRC Press, Taylor & Francis Group.

26. Lewis, David. 1997. "Hot Break Separation – Whirling the Wort." *The New Brewer* 14 (6).

27. MacDonald, J., et al. 1984. "Current Approaches to Brewery Fermentations." *Progress in Industrial Microbiology*, edited by M. E. Bushell. Guildford, United Kingdom: Elsevier.

28. Mathews, J.D. 1986. "Copper Finings – A New Insight." *The Brewer* 72 (864).

29. Moll, Manfred, ed. 1995. *Beers and Coolers*. New York, New York: Marcel Dekker, Inc.

30. Noonan, Gregory J. 1996. *New Brewing Lager Beer*. Boulder, Colorado: Brewers Publications.

31. O'Brien, Mike. 1999. Personal Communication. Pico-Brewing Systems, Inc.

32. Ormrod, H.L. 1986. "Modern Brewhouse Design and its Impact on Wort Production." *Journal of Institute of Brewing* 92 (March-April).

33. O'Rourke, Timothy. 1984. "A Modern Approach to Wort Boiling." *The Brewer* 70 (831).

34. O'Rourke, Timothy. 1999. "Wort Boiling (part 1)." *Brewers' Guardian* 128 (8).

35. O'Rourke, Timothy. 1999. "Wort Boiling (part 2)." *Brewers' Guardian* 128 (9).

36. O'Rourke, Tim. 2002. "The Process of Wort Boiling." *The Brewer International*. Retrieved September 08, 2007, from http://www.ibg.org.uk.

37. O'Rourke, Tim. 2002. "The Function of Wort Boiling." *The Brewer International*. Retrieved September 08, 2007, from http://www.ibg.org.uk.

38. Outterson, Donald R. 1999. Personal Communication. Outterson LLC.

39. Rehberger, Arthur J. and Gary E. Luther. 1995. "Brewing." *Handbook of Brewing*, edited by William A. Hardwick. New York, New York: Marcel Dekker, Inc.

40. Rehberger, Arthur J. and Gary E. Luther. 1999. "Wort Boiling." *The Practical Brewer*, edited by John T. McCabe. Wauwatosa Wisconsin: Master Brewers Association of the Americas.

41. Reichert, Wolfgang. 1989. "Equipment Examined." *The New Brewer* 6 (6).

42. Roberston, J. 1990. "Wort Production." *The Brewer* 76 (905).

43. Robson, F.O. 1974. "Wort Boiling." *The Brewer* 60 (712).

44. Royston, M.G. 1971. "Wort Boiling and Cooling." *Modern Brewing Technology*, edited by W.P.K. Findlay. Cleveland, Ohio: The Macmillian Press.

45. Ryder, David S. and Joseph Power. 2006. "Miscellaneous Ingredients in Aid of the Process." *Handbook of Brewing*, edited by Fergus G. Priest and Graham G. Stewart. Boca Raton, Florida: CRC Press, Taylor & Francis Group.

46. Sidor, Larry. 2006. "Hops and Preparation of Hops." *Raw Materials and Brewhouse Operations*. Volume 1, edited by Karl Ockert. St. Paul, Minnesota: Master Brewers Association of the Americas, 2006.

47. Stewart, G.G. and I. Russell. 1985. "Modern Brewing Biotechnology." *Food and Beverage Products*, edited by Murray Moo-Young. Oxford, England: Pergamon.

48. Strauss, Karl M. 1977. "Wort Cooling." *The Practical Brewer*, edited by H. M. Broderick. Madison, Wisconsin: Master Brewers Association of the Americas.

49. Thomas, R.D. 1982. "The Economic Benefits of Copper Finings." *The Brewer* 68 (814).

50. Thompson, Donald, Karl Ockert, and Vince Cottone. 1999. "Craft Brewing." *The Practical Brewer*, edited by John T. McCabe. Wauwatosa Wisconsin: Master Brewers Association of the Americas.

51. Thompson, G.J. 1994. "The Theory and Practice of Beer Clarification – Part 2 – Practice." *The Brewer* 80 (961).

52. Vernon, P.S. 1985. "Wort Clarification." *Brewers' Guardian* 114 (3).

53. Wainwright, T. 1971. "Biochemistry of Brewing." *Modern Brewing Technology*, edited by W.P.K. Findlay. Cleveland, Ohio: The Macmillian Press.

54. Warner, Eric. 1991. "Overview for Designing a Brewery." *Brewer Planner – A Guide to Operating Your Own Small Brewery*, complied by The Institute of Brewing Studies. Boulder, Colorado: Brewers Publications.

55. Webster, B.J. 1994. "The Whirlpool in the 90's." *The Brewer* 80 (962).

Chapter Twelve

Wort Cooling and Aeration

After hot trub separation, the wort is preferably cooled to a temperature of 5 to 15°C for bottom-fermented beers and to 15 to 18°C for top-fermented beers (pitching temperature). The wort is then aerated in preparation for the addition of yeast and subsequent fermentation.

Wort Cooling Systems

From the physical point of view, cooling is an easy process. In practice, however, breweries can choose different ways of heat transfer using either a single-stage (chilled water only) or multiple-stage (ambient water, glycol) plate heat exchangers (see Figure 12.1). Wort enters the heat exchanger at between 96 and 99°C and exits cooled to pitching temperature.

Single-Stage Cooling

The cooling operation can be achieved with a single-stage system using a glycol-jacketed cold-water tank to cool the wort (8). In this case, a separate cold water well-insulated and jacketed tank is needed equipped with a glycol-cooled heat exchanger to pre-chill the water.

Two-Stage Cooling

In a two-stage system the first stage utilizes water to remove the bulk of the heat, cooling the incoming wort to within 3°C of fermentation temperature. In the second stage, the wort is cooled to the fermentation temperature by a secondary refrigerant (e.g., glycol, ammonia, etc). The energy requirement is less with two-stage cooling than with one-stage cooling (12). A two-stage system also reduces the demand on the secondary refrigerant system. However, if a secondary coolant is being used care must be taken to regularly check that the plates and seals are not leaking. Some craft brewers, in an attempt to

Figure 12.1
Wort Cooler
Courtesy of Steinecker

reduce capital expenditures, will use the same glycol system that provides the cooling for fermentation. Most small brewers would prefer two-stage cooling but use one-stage cooling requires less of capital investment.

In both systems, the heat from the wort that is transferred to the water can then be used for other purposes such as mashing and sparge water for the next brew.

Formation of Cold Break

As the clear hot wort is cooled, the previously invisible coagulum loses its solubility and precipitates. The precipitate is referred to as the cold break and begins forming at about 60°C (7). The cold break mostly consists of protein-polyphenol (tannin) complexes, whereas the hot break is mostly proteinaceous. The cold break also has a higher level of carbohydrates (primarily beta-glucans) than hot break (10). Highly modified malts yield a higher percentage of polyphenols in cold trub than do less-modified malts, while under-modified malts yield more protein and beta-glucans and relatively fewer polyphenols (1).

> **Note | Microbiological Stability**
>
> It is important at this stage to maintain complete microbiological sterility, since wort in its cooled, unpitched state is most vulnerable to competing microorganisms. The most dangerous period for bacterial development is when the wort is between 40 and 20°C. Later, during fermentation, bacterial development is inhibited as the yeast starts to actively grow. The disappearance of fermentable sugars and readily assimilated nitrogenous matter, along with the production of alcohol and acids, makes the medium much less susceptible to bacterial infection and development; and the final beer is also more resistant to contamination.

It is essential that the cold break is rapidly cooled in order to precipitate as much as the proteinaceous material as possible. A long, slow cooling does not give a good cold break because more protein is trapped in suspension; this gives rise to a finer trub, chill haze, and harsh, sulfur-like aftertastes in the beer. Coarse trub is essential for good separation and good beer stability. In addition, a rapid cooling will minimize the development of dimethyl sulfide (DMS), which is more likely, to form when using lager malt.

If the wort is reheated, cold trub will go back into solution, forming a chill haze. Cold break particles are extremely small and do not flocculate as well as hot break.

Factors Affecting Quantity of Cold Break

The total dry weight of cold break is between 15 and 30 g/hl, accounting for about 15 to 30% of the total trub material precipitated during cooling and boiling of the wort. The actual amount produced in a given wort depends on numerous factors, such as malt modification, mashing program, wort temperature, the presence of hops, and the rate of hot break removal. Wort produced using finely milled malt contains a large quantity of cold trub because the finely ground husks allow greater polyphenol extraction (10). On the other hand, more intensive mashing programs, such as double- and triple-decoction, degrade proteins more extensively and yield less cold trub than do infusion-mashing programs. Infusion mashing can yield higher molecular weight proteins that pass through to the chilled wort. Adding hops, which contain polyphenols, increases cold trub. Furthermore, adding hops late in the boil increases the amount of cold trub formed. It is generally true that the higher the hot break volume, the lower the cold break volume and vice versa (20).

Removal of Cold Break

After the wort is cooled, the cold break must be removed before fermentation, or else the beer will taste wort-like, bitter, and even harsh. Opinions vary as to whether cold break should be removed at all before transferring the wort to the fermenter.

Traditional lager brewers advocate the removal of cold break prior to fermentation, and some even filter cold worts prior to pitching (14). Lager brewers believe cold break removal aids in colloidal stability in the beer, circumvents the formation of sulfury flavors, and removes harsh bitter fractions derived from hops. De Clerck suggests that clearer worts lead to beers that clarify better (7). It is reported that if significant amounts of trub are carried over into the ferment, the yeast will produce an excess of fusel alcohols. It has also been reported that suspended solids can lead to adverse effects on foam and flavor stability.

On the other hand, traditional ale brewers tend to leave cold break in the wort because they believe that excessively bright worts can result in slow, or "sticking" fermentations due to the removal of essential yeast nutrients and fatty acids (probably lipids) [14]. Furthermore, the physical performance (especially flocculation characteristics) of some top-fermenting yeast can be impaired in very bright worts (14). However, for traditional cask beers, the wort should be reasonably bright going to the fermenter because anything carried over at this stage only adds to the finings load during casking. The trend is to run the wort through a whirlpool and then a heat exchanger before transferring it to the fermenter.

Today, as might be expected, opinions still vary as to whether or not it is necessary to remove cold break.

Some believe partial removal of cold trub has been shown to extend the shelf life of lager by removing oxidation catalysts and haze-forming constituents in the finished beer (1). Kunze recommends a residual amount of 40 to 60 mg cold break per liter of wort and reports that levels too low could result in an empty-tasting beer (12). Kunze also suggests that values as high as 150 mg/l are sometimes regarded as normal (12). Pale lagers such as Pilsners are more likely to show improvement with partial removal of cold trub than are robust, darker styles. Taking these extra steps reportedly will yield subtle improvements in mellowness of flavor and haze stability (6).

However, some brewers permit some trub carryover because the particles in trub provide nucleation sites for carbon dioxide bubble formation, thus improving fermentation vigor (18). It has also been shown that trub contains zinc and other minerals and nutrients beneficial to yeast performance. They also believe that removing cold trub might actually slow fermentation and harm the finished beer, giving it an onion-like flavor. These same brewers report having experienced slower fermentations, higher acetate ester levels, and lower yeast growth and viability after removing all cold trub.

Methods of Cold Break Removal

The brewer has several options in removing the cold break before transferring the wort to the fermenter.

Starter Tank

The simplest way to remove cold trub is through cold sedimentation, which requires only one vessel—a starter tank—which can be used with pitched or unpitched wort. Because wild yeast and bacteria are a problem, care must be taken to ensure that the starter tank is sanitary, particularly if the wort is unpitched. Shallow vessels rather than cylindroconical tanks make the best settling tanks. Deeper vessels require more time but can benefit from an addition between 10 and 20 g/hl coarse diatomaceous earth (1). Reportedly, as much as 70% of total cold trub precipitates when using diatomaceous earth.

Many traditional European lager brewers, especially German and Scandinavian, use starter tanks. Yeast is added to the cooled wort, and the contents are then allowed to stand for a period of up to 24 hours before being transferred to the fermenter (14). Reportedly, the beer is smoother, rounder, and pleasantly bitter.

When sedimentation is complete, the wort is transferred to the fermentation vessel. The sludge at the bottom of the vessel is subsequently discarded but this can create problems since 2–5% of the wort is trapped in the sludge. The wort can only be recovered by centrifuging the mixture of trub, spent hops, and wort employing a centrifuge. The wort recovered in this way can be added to the same batch of wort or to the next brew.

Removing Cold Break from Pitched Wort – For long lag times, Barchet recommends pitching the yeast and allowing the chilled, aerated wort to settle between 6 and 24 hours before racking to the fermenter and leaving the sediment behind (1). Racking should be done before the active phase of fermentation begins because natural circulation of carbon dioxide in fermentation will rouse the settled trub not allowing for further sedimentation (1). If highly flocculent yeast strains are used, over-pitching by the amount that settles out in the cold trub is recommended (1). Brewers have reported that as much as 30% of the total cold trub settles out of pitched wort after a 12 hour rest.

Removing Cold Break from Unpitched Wort – Some brewers believe that it is better to remove the cold trub before pitching if the lag time to fermentation is fewer than 6 hours. After the wort has stood for less than 6 hours, it is then moved to another vessel, aerated, and pitched with yeast. Brewers report that sedimentation of unpitched worts is more complete than that of pitched worts (approximately 50%).

Flotation

Another method of cold trub removal involves the flotation using vertical vessels with an even or slightly angled bottom. Cold break particles adhere to the surface of air bubbles that are forced continuously through cooled wort from the bottom of the tank by a Venturi tube and a mixing pump. A flow rate of 20 to 60 liters of air (not oxygen) per hectoliter is required (6). The cold break particles attach themselves to the air bubbles, and a layer of foam is formed on the surface of the wort. Flotation affects the surface active compounds, including alpha-acids, and this has the effect of reducing hop utilization in the finished beer (27). After a 2 to 6 hour rest, the clear, aerated wort underneath is subsequently pumped from the bottom of the tank and sent to a fermentation vessel. The compact foam-trub layer remains in the tank. It is important not to allow the cold trub cake to break and fall into the clean wort, which will occur if the wort stands too long.

According to Kunze, about 60 to 65% of the cold break is removed by flotation (12). This method is claimed to be superior to sedimentation in starter tanks and to filtration in removal of cold trub. Flotation also has the advantage of reducing the time the wort stays in the separation tank as compared to using the sedimentation method. It should be emphasized that great care must be taken to ensure that wort held is not infected.

For effective flotation, leave at least 30 to 50% headspace to accommodate foaming that occurs during flotation. Excessive foaming can be minimized by employing counter pressure, thereby saving valuable vessel capacity. Flotation is better done in shallow vessels than in cylindroconical tanks since the latter do not disperse air uniformly, and they cause the trub cake to break and enter the clear wort as it approaches the cone.

Flotation can be performed with or without yeast. Nowadays it is usually done with yeast because the flotation time can be employed as the initial fermentation phase. At the same time, the holding time with yeast minimizes the risk of infection, and oxygen can be used by the yeast. The presence of yeast does not compromise the capability of flotation in removing cold trub. In fact, only a small amount of yeast becomes trapped in the trub cover. The positive effects of flotation on fermentation outweigh any loss of yeast trapped during the process.

Diatomaceous Earth Filtration

Diatomaceous earth (DE) filtration is the most effective means of removing cold trub and is likely to be used by larger breweries. DE filtration is not effective in removing hot break since the filter aid dissolves in hot wort. Typically, coarse diatomaceous earth is used, and the best results occur at low temperatures, i.e., 0°C. Because of thermal influences, the filter must be sized to filter all of the wort within a relatively short time. On the average, diatomaceous earth filtration will remove 75 to 85% of the total cold trub, with higher rates (95%) occurring when fine diatomaceous earth is used.

The main side effect of DE filtration is that it "strips" some of the flavor out of the wort, leading to less full-bodied beer. Diatomaceous earth filtration can also result in poor yeast growth and fermentation. Other problems associated with DE filtering include the cost per brew, the difficulty in maintaining sterile conditions, and DE disposal, which can be very expensive for large breweries.

Centrifugation

The hot wort can be clarified by means of efficient, self-cleaning centrifuges; these, however, will only work reliably if the wort—trub mixture is added homogeneously. This can usually be achieved by means of an intermediate hot wort tank with a stirring mechanism. Centrifugation has the major advantage in that the times for sedimentation and transfer can be drastically shortened. Centrifugation, popular among some Continental Europe brewers, is used not only to remove cold break but sometimes to remove hot break too. Approximately 50 to 60% of total cold trub can be removed using centrifugation. As with DE filtration, care must be taken not to remove too much trub since the resulting beer may taste empty and lack character. Centrifugation can also result in poor yeast growth and fermentation.

Whirlpool

An alternative solution is the whirlpool, where the wort is pumped tangentially into a cylindrical vessel. This creates an even, rotating stream. The solid particles suspended in the rotating liquid will separate due to friction (tea cup effect), migrate to the bottom center, and coalesce to form a cake. It is also possible to achieve this effect in whirlpool kettles, where both boiling and trub separation take place, provided that the bottom of the kettle is appropriately shaped.

Aeration of Chilled Wort

After cooling the wort is aerated to increase yeast activity and to start the fermentation process. The amount of oxygen required depends on yeast strain, wort temperature, wort gravity, amount of trub in the wort, and a number of other factors. For example, worts at high temperatures and high specific gravities have greater oxygen requirements than worts at lower temperatures and specific gravities. Worts low in trub generally will have high oxygen requirements, while worts with high trub levels have lower oxygen requirements.

Oxygen Requirements

The oxygen requirements for individual brewing strains can range from 3 to 30 mg O_2/l but usually it is in the range of 7 to 18 mg O_2/l. Yeast strains with low oxygen requirements can be aerated using sterile air since it contains approximately 8 mg O_2/l, while strains with high oxygen requirements must be aerated with pure gaseous oxygen.

Oxygen levels that are too low can result in abnormal off-flavors. At low oxygen levels, respiring yeast produces significantly more esters—principally ethyl acetate, but also isoamyl acetate and ethyl caproate—and this irreversibly flavors the beer with fruity/solvency aromatics. Inadequate oxygenation also causes pyruvic acid, fatty acids, and amino acids to decarboxylate to aldehydes. Acetaldehyde, the aldehyde of pyruvic acid, usually predominates, giving the beer an odor similar to that of green apples. Elevated diacetyl levels can also occur in chilled wort low in oxygen. Poor wort aeration can result in longer fermentation times and high final gravities.

If the oxygen levels are too high, over-oxygenation can lead to a vigorous yeast growth that can result in off-flavors—increased development of esters and acetaldehyde (22). Kunze reports that over-oxygenation should not be of much concern since oxygen is rapidly consumed by yeast in the initial stages of fermentation (12).

> **Note | Brewing Practices Influence on Wort Aeration**
>
> With traditional brewing operations, aeration of the wort was not so much of an issue since traditional practices were more conducive to air-saturated worts. It is not surprising, with modern fermentation practices, aeration of the wort is needed.

Point of Injection

The point of injection of oxygen has been the subject of many debates, and the choice is largely a matter of tradition. Many brewers believe that gas injected into the hot wort prior to cooling improves cold break formation. However, when injected into wort at the hot end of a cooler the wort allows oxidation of wort components, resulting in darker-colored worts. In addition, according to Moll, oxygenation of hot wort contributes to off-flavors in the beer, e.g., garlic flavor (16). Some breweries compromise by adding oxygen at a point within the cooler where the wort is at an intermediate temperature (14). Moll reports that most breweries oxygenate the wort after cooling while it is on its way to the fermenter, since at that point its solubility is higher and there is less risk of oxidation (16).

Both air and oxygen must be sterile. If plant air is used, it must be filtered through a set of water and oil trap filters and then through a sterile grade (0.25 micron) filter (26). Although reputedly free from microorganisms, bottled gases (food or hospital grade) should still be micro-filtered. If the air is not sterilized, it can possibly introduce microorganisms to the yeast.

Methods of Oxygenation

Wort oxygenation is most commonly achieved by direct injection of either sterile air or oxygen at the time it exits the wort cooler. Devices used to aerate the wort are:

1. Ceramic or sintered metal elements
2. Venturi tubes
3. Two-component jets
4. Static mixers
5. Centrifugal mixers

Ceramic or Sintered Metal Elements

With ceramic or sintered metal elements, air is injected as very small bubbles through fine pores into the flowing wort. This is a very simple and effective method, but care should be taken that pores are kept clean. The oxygen should be injected against a head of wort, producing a backpressure of at least 15 psi (15).

Venturi Tubes

With Venturi tubes, there is a substantial increase in flow velocity, and the air is thoroughly mixed in the wort.

Two-Component Jets

Brewers employing two-component jets inject the air (or oxygen) as very small bubbles under high pressure into the wort stream, which is then forced through a small orifice and into an expansion chamber. On expansion, the pressure drops, then the gas bubbles expand and are rapidly dissolved in the wort.

Static Mixers

In static mixers, the initial mixing of the wort and air is achieved in a reaction section with built-in angle bands (12). The many angles force the wort to constantly change direction, thereby producing a turbulent flow.

Centrifugal Mixers

If a centrifuge mixer is employed, air can be introduced into the wort stream via the centripetal pump impeller, which results in very fine bubbles.

Although extensive aeration occurs, it is counterbalanced by relatively high energy costs.

Measuring Oxygen Levels

While aerating the wort, it is useful to have a pressure gauge integrated into the aeration device to monitor the amount of oxygen or air that is being taken up by the wort. Craft breweries often use a sight glass or listen to the formation of bubbles. Oxygenation control equipment commonly consists of either a gas volume or mass-flow regulator, or a dissolved oxygen meter. The oxygen meter should be located in the wort line downstream from the gas injection point located as near to the fermenter as possible to confirm the level of dissolved oxygen. Moll recommends measuring the oxygen content in line at the fermentation vessel to avoid erroneous values taken too near the air or oxygen injection point (16). In some cases, air or oxygen is added to wort in the fermenter itself either before or after pitching, or both.

References

1. Barchet, Ron. 1994. "Cold Trub: Implications for Finished Beer, and Methods of Removal." *Brewing Techniques* 2 (2).

2. Barchet, Ron. 1999. Personal Communication. Victory Brewing Co.

3. Barnes, Zane C. 2006. "Brewing Process Control." *Handbook of Brewing*, edited by Fergus G. Priest and Graham G. Stewart. Boca Raton, Florida: CRC Press, Taylor & Francis Group.

4. Briggs, Dennis E., Boulton, Chris A., Brookes, Peter A., and Stevens, Roger. 2004. "*Brewing: Science and Practice.* Cambridge, England: Woodhead Publishing Limited.

5. Boulton, Chris. 1991. "Yeast Management and the Control of Brewery Fermentations." *Brewers' Guardian* 120 (3).

6. Carey, Daniel and Grossman, Ken. 2006. "Fermentation and Cellar Operations." *Fermentation, Cellaring, and Packaging Operations.* Volume 2, edited by Karl Ockert. St. Paul, Minnesota: Master Brewers Association of the Americas, 2006.

7. De Clerck, Jean. 1957. *A Textbook of Brewing.* Volume 1, translated by K. Barton-Wright. London: Chapman-Hall Ltd.

8. European Brewery Convention (EBC). 1997. *Brewery Utilities, Manual of Good Practice*. Nürenberg, Germany: Getränke-Fachverlag Hans Carl.

9. Fix, George and Laurie Fix. 1997. *An Analysis of Brewing Techniques.* Boulder, Colorado: Brewers Publications.

10. Gordon, Dan. 1991. "The Trouble with Trub." *The New Brewer* 8 (1).

11. Knudsen, Finn B. 1985. "Fermentation Symposium – Part II – Fermentation Variables and Their Control." *MBAA Technical Quarterly* 22 (4).

12. Kunze, Wolfgang. 1996. *Technology Brewing and Malting*, translated by Dr. Trevor Wainwright. Berlin, Germany: VLB Berlin.

13. Leiper, Kenneth A. and Michaela Miedl. 2006. "Brewhouse Technology." *Handbook of Brewing*, edited by Fergus G. Priest and Graham G. Stewart. Boca Raton, Florida: CRC Press, Taylor & Francis Group.

14. MacDonald, J., et al. 1984. "Current Approaches to Brewery Fermentations." *Progress in Industrial Microbiology*, edited by M. E. Bushell. Guildford, United Kingdom: Elsevier.

15. Martin, E.C.B. 1972. "Lager Fermentation, Storage, Filtration." *Brewers' Guardian* 101 (4).

16. Moll, Manfred, ed. 1995. *Beers and Coolers*. New York, New York: Marcel Dekker, Inc.

17. Munroe, James H. 1995. "Fermentation." *Handbook of Brewing*, edited by William A. Hardwick. New York, New York: Marcel Dekker, Inc.

18. Munroe, James H. 2006. "Fermentation." *Handbook of Brewing*, edited by Fergus G. Priest and Graham G. Stewart. Boca Raton, Florida: CRC Press, Taylor & Francis Group.

19. Noble, Stuart. 1997. "Practical Aspects of Fermentation Management." *The Brewer* 83 (991).

20. O'Rourke, Timothy. 1999. "Wort Boiling (part 2)." B*rewers' Guardian* 128 (9).

21. O'Rourke, Timothy. 1999. "Wort Cooling." *Brewers' Guardian* 128 (10).

22. Rehberger, Arthur J. and Gary E. Luther. 1995. "Brewing." *Handbook of Brewing*, edited by William A. Hardwick. New York, New York: Marcel Dekker, Inc.

23. Royston, M.G. 1971. "Wort Boiling and Cooling." *Modern Brewing Technology*, edited by W.P.K. Findlay. Cleveland, Ohio: The Macmillian Press.

24. Strauss, Karl M. 1977. "Wort Cooling." *The Practical Brewer*, edited by H. M. Broderick. Madison, Wisconsin: Master Brewers Association of the Americas.

25. Strauss, Karl M. 1999. "Wort Cooling." *The Practical Brewer*, edited by John T. McCabe. Wauwatosa Wisconsin: Master Brewers Association of the Americas.

26. Thompson, Donald, Karl Ockert, and Vince Cottone. 1999. "Craft Brewing." *The Practical Brewer*, edited by John T. McCabe. Wauwatosa Wisconsin: Master Brewers Association of the Americas.

27. Vernon, P.S. 1985. "Wort Clarification." *Brewers' Guardian* 114 (3).

Chapter Thirteen

Fermentation

Fermentation is the process by which fermentable carbohydrates are converted by yeast into alcohol, carbon dioxide, and numerous other byproducts. It is these byproducts that have a considerable effect on the taste, aroma, and other properties that characterize the style of beer.

Pitching Yeast

Strain

The yeast strain itself is a major contributor to the flavor and character of the beer. Thus the choice of yeast strain depends on such things as the oxygen requirements, cropping methods, attenuation limits, fermentation rate, fermentation temperatures, flocculation characteristics, and the flavor profile (sulfur compounds, esters, fusel alcohols, etc.). For example, the flocculation characteristics of yeast strains are very important. If secondary fermentation is intended than a more, flocculating yeast strain is needed in order to assure a greater number of cells are in suspension at transfer from the fermenter to the conditioning tanks. On the other hand, a slowly, flocculating yeast might not be ideal if a fully attenuated wort is desired at the end of fermentation. A slowly flocculating strain would leave to much yeast in suspension at the end of fermentation making beer separation more difficult.

Pitching Rates

The brewer's ability to pitch the correct number of yeast cells to initiate fermentation is crucial to consistently producing a product of superior and constant quality. Pitching rates are governed by a number of factors, including yeast strain, fermentation capacity of the yeast, yeast viability, flocculation characteristics, previous history of the yeast, and desired beer flavor characteristics. Other considerations when choosing pitching rates include wort gravity, wort constituents, fermentation temperature, and the degree of wort aeration. For example, highly flocculent yeast strains may settle prematurely, requiring the brewer to either over-pitch or to mix and aerate the yeast by "rousing."

Low pitching rates can result in long lag and reproductive phases. This, in turn, can increase the production of byproducts such as higher alcohols, esters, and diacetyl. Munroe reports poor yeast growth can also lead to high sulfur dioxide levels (38). Furthermore, since many sugars are not metabolized at low pitching rates, the resulting beer tends to have a residual sweetness (lower attenuation). Insufficient yeast growth can eventually lead to slower and even stuck fermentations.

High pitching rates run the risk of developing off-flavors (from higher alcohols, esters, and diacetyl), as does under-pitching. Some brewers have found that if the pitching rates are too high, there is also an increased risk of developing so-called hot spots in the cone of the cylindroconical fermenters. Under these conditions, temperatures may increase 10°C or more over the intended fermentation temperature, creating an environment for the development of more off-flavors (23). Beers that are over-pitched are also hard to filter and, once bottled, display poor stability with respect to clarity. Finally, over-pitching multiple generations leads to an aging of the yeast population and produces fewer new cells. New cells are considered to have higher attenuating characteristics.

The pitching rates most often employed vary between 5 and 20 million cells per milliliter depending on the specific gravity of the wort (5). As a general rule, the pitching rate is 1 million viable cells per degree of Plato per ml of wort (24). For example, a 12°P, wort would require 12 million cells per milliliter. Others like to pitch slightly below this level to stimulate the formation of new cells. Generally, ales require lower pitching rates because their fermentations are conducted at higher temperatures than are lager fermentations. Increasing the pitching rate results in fewer doublings since the yeast cells under given conditions multiply only to a certain level of cells per unit volume, regardless of the original pitching rate (47).

Viability

The pitching rates should be adjusted to account for the number of viable cells rather than the total number of cells. Dead cells can be determined by staining with methylene blue and counting with a hemacytometer. Although this method suffers from subjectivity, the brewer should discard the yeast if methylene blue values are lower than 95% since there is the risk of adding an unacceptably high number of dead yeast cells to the wort. High dead cell counts may be due to a lengthy storage in the yeast brink, harsh treatment in the brink, or lengthy delay in removal of yeast after fermentation (40). High levels of dead yeast cells lead to unfilterable hazes, poor foams, high metal contents in the beer, altered organoleptic profiles, and poor flavor stability (43).

Methods for Determining Pitching Rates

Traditionally, brewers determined pitching rates on the basis of either the volume or the weight of yeast slurry to a known quantity of wort. Today most breweries determine pitching rates by a number of alternative methods including the use of a counting chamber (hemacytometer), centrifugation of the yeast slurry, electron cell counting, or of in-line biomass sensors.

Weight-Based

A weight-based pitching is preferable to volume-based since volume varies with temperature, carbonation, liquidity of the slurry, and degree of yeast flocculation (31). Recommended pitching rates by weight range from 0.14 to 0.40 kg/hl for ale yeast and from 0.22 to 1.2 kg/hl for lager yeast.

Although pitching rates based on weight are more accurate than those based on volume, they are still quite inaccurate given the presence of solids. Consequently, the correct pitching rate, by either weight or volume, requires a calculated dilution of the yeast slurry based on an estimate of solids in the slurry. Total solids can be estimated by an oven-drying method, centrifugation, volumetric or gravimetric wet solids, or turbidity methods.

Hemacytometer

A better method in estimating cell concentration is to use a hemacytometer, which is a specialized slide with a counting chamber of known volume. Using a hemacytometer requires placing a known volume of slurry, usually diluted, on a microscopic slide and counting the number of cells. This method is further enhanced by using methylene blue stain to identify only viable cells. Hemacytometers are popular with craft brewers because of their simplicity and low cost, and they do a reasonably good job of determining cell concentration. However, counting microscopically using a hemacytometer is time-consuming and subject to between-analyst readings. There are also problems associated with dispersing the yeast, getting accurate dilutions, and cells sticking to the glassware. To minimize the sampling errors many cells need to be counted to get an accurate value.

Centrifugation

Another simple method is to centrifuge a sample to determine the ratio of solids in the sample thus the correct amount of yeast to pitch.

Electron Cell Counting

Although electron cell counting is very accurate, it records budding cells, chains of cells, and non-yeast aggregates as single cells (64). Noble reports that the Tuchenhagen NIR turbidimetric system gives good results but that it is not as accurate in determining yeast count with dark beers, as it is with light beers (41).

In-Line Biomass Sensors

More recently, commercially available in-line biomass sensors have become available. These devices use the passive dielectrical properties of microbial cells and can discriminate between viable and nonviable cells and trub (47). According to Noble, the Kell Biomass probe gives good results, but careful calibration is needed for each strain, which can be difficult to achieve when using several strains (41).

Dosing

Yeast can be directly added to the fermenters (batch), mixed with the wort in a starter tank prior to transfer to the fermenter, or continuously metered into the cold, aerated wort stream on the way to the fermenter. Typically, large breweries have automatic dosing systems in which the amount of yeast is controlled by a turbidity measuring system. Continuously metering the yeast allows for its uniform distribution in the wort, which makes fermenting vessel occupation times more predictable. Compared with the usual batch method of adding yeast to fermenters, constant dosing of yeast eliminates many problems, including yeast stratification, slow fermentation starts, and uncontrolled yeast growth (46). Yeast must always be colder than the pitching wort so that growth is stimulated on pitching (25).

Microbiological Contamination

Uninoculated wort is liable to contamination by many types of mold; bacteria such as *Pediococcus,* spp. and *Lactobacillus* spp.; and wild yeasts such as *Hansenula, Dekkera, Brettanomyces, Candida,* and *Pichia*. Other *Saccharomyces* species may be present. However, once the yeast inoculum is added, the selective effect against microorganisms other than brewing yeast begins. This is due to falling pH, the antimicrobial effect of hop compounds, the developing anaerobic conditions in the wort, and the production of carbon dioxide and ethanol (6).

Lager Fermentations

Starter Tanks

Traditional European lager brewers, especially in Germany and Scandinavia, use starter tanks in which yeast is added to the cooled wort and the contents then allowed to stand for a period of up to 24 hours before racking (30).

Traditionally, starter tanks were open vessels, but nowadays they are generally designed like regular closed-in fermenters.

The purpose in using a starter tank is to allow some undesirable solids to be removed by settling or flotation (35). The starter tank also removes trub particles that were not collected during trub separation, as well as hop resins that precipitate after cooling (25). Trub contains polyphenols, ketones, and sulfur compounds that may be absorbed into the ferment and thereby affect the flavor of the beer. Knudsen reports that starter tanks are also used to more accurately measure wort volume, to mix yeast and wort, and to equalize temperature differences prior to transferring wort and yeast to the fermenter (24).

Topping Up (Darauflassen)

If the brewer doesn't have enough yeast or if the brewer has a smaller mash tun and kettle than the fermenter, the brewer can employ a technique called topping up or darauflassen. Topping up, a technique common among German lager brewers, is the infusion of wort into a tank with strongly fermenting young beer (high kraeusen). As a result of the addition of fresh wort, the amount of healthy yeast cells produced will increase. When topping up, it is very important that the wort being added has the same temperature as the fermenting wort. If there is a difference in temperature, the yeast may suffer a shock and the fermentation may be retarded or stop completely. Scheer recommends starting a fermenting brew at a ratio of 25% kraeusen to 75% wort (49).

Traditional Lager Fermentation

While vertical cylindroconical fermenters are used to make the majority of the world's lager production, open square fermenters are still commonly used (especially in Eastern Europe) in traditional lager fermentation. In recent years, some major brewers have started fermenting their lagers in open square fermenters for quality reasons.

Traditional brewing methods involve pitching the yeast at lower temperatures between 5 and 6°C and allowing the temperature to rise between 8 and 9°C (25). If the fermentation is started to high of a temperature, it could exhaust the extract needed for secondary fermentation leaving too many yeast cells in suspension. Traditionally, beers that use large amounts of dark-roasted malts are fermented at higher temperatures since the wort has lower levels of fermentable sugars and amino acids for the yeast (7).

The first visible signs of fermentation are the appearance of fine bubbles on the wort surface. This stage, known as "low kraeusen," is marked by formation of a foam wreath near the center of the wort surface. The reduction in extract is accompanied by a decline in pH.

Within 24 to 48 hours after pitching, rocky cauliflower heads called "kraeusen" appear on the surface of the wort. After about 72 hours, the kraeusen produced breaks into cream-colored, less dense "rocky heads." This is the start of the most active phase of fermentation, the so-called "high kraeusen." High kraeusen occurs at the time of maximum rate of carbon dioxide and heat generation, which corresponds to the maximum decline in specific gravity. The temperature of the wort is controlled by attemperation. Typically, lager fermentations are characterized by a slower rate of gravity loss than ale fermentations. The pH decreases between 4.0 and 4.5 (38).

As the availability of fermentable extract drops during the post-kraeusen stage, yeast activity slows as well as the evolution of carbon dioxide. The head begins to fall, and the beer begins to become bright. As the head collapses, a ring of precipitated foam adheres to the fermenter walls, and most of the yeast falls to the bottom of the fermenter.

Finally, the beer is cooled (when cell count is at its highest and the apparent gravity is roughly half the original gravity) to facilitate settling and aid the transfer process by reducing yeast carryover. It is important that the ambient temperature is reduced at a rate that will not induce thermal shock to the yeast. Generally, the brewer reduces the temperature by 1 to 1.5°C per day and transfers the green beer to the lager cellar when the temperature is between 4 and 5°C (25). The yeast count should fall between 5 and 15 million cells per milliliter with a fermentable extract between 0.8 and 0.12°P (7).

Abrupt reduction in temperature can shock the yeast and may arrest fermentation. Consequently, yeast viability may be affected and the production of off-flavors (e.g., sulfur) may increase due to autolysis. However, if the cooling is too slow, yeast can release foam. This can damage proteolytic enzymes and markedly affect foam quality (41).

The cooler fermentation temperatures are believed to produce a better quality beer that is fuller, mellower, complex flavors, with superior foam stability. Cold fermentation also limits the development of byproducts, with the exception of esters which may increase.

Modern Lager Fermentation

Today modern lager fermentation typically uses vertical cylindrodroconical fermenters achieving similar flavor profiles compared to traditional fermentation systems. In modern fermentation systems, the yeast is pitched at higher temperatures between 7 and 8°C, and after a couple of days the temperature is increased to 10 to 11°C (25). After 3 to 4 days, at peak fermentation, the fermentation temperature is allowed to ramp up in order to facilitate a rapid reduction in diacetyl. Some brewers use the same starting temperature for pitching but then increase the temperature between 14 and 15°C (24). Other breweries are known to pitch between 10 and 13°C and then increase the temperature to as high as 17°C, with a short diacetyl rest (28).

Over the last 20 years there has been a trend to move to higher fermentation temperatures. While this has the advantage of decreasing fermentation times (from 3 weeks to as little as 5 days), it has the disadvantage of causing an increase in higher alcohol production, giving rise to loss of beer flavor. As a result, some brewers have reduced their fermentation temperatures for lager beers.

Once fermentation is complete, the contents are cooled to about 4°C to settle the yeast in the cone of the fermenter. Unlike traditional methods of collecting yeast, yeast cropping with cylindroconical fermenters involves collecting the yeast from the cone of the fermenter. It will normally take 24 hours for the yeast to settle, although cooling may be interrupted if the beer is going to undergo a diacetyl rest, if necessary, before being chilled to yeast cropping temperatures. The settling time is dependent on the flocculation characteristics of the yeast. Typically, brewers will wait up to 24 hours to further settle the yeast before cropping.

The green beer is then transferred to the cellar for conditioning and either chilled to − 2 to 4°C through a heat exchanger or in situ in the tanks. This practice allows for more efficient cooling and the freeing of the cylindroconical fermenters for fermentation.

Ale Fermentations

Starter Tanks

Like lager brewers, some ale brewers initiate fermentation in starter tanks in which the residence time may be as short as 3 hours or as long as 36 hours (22). Traditionally, the British have placed far less emphasis on starter tanks than Continental Europe brewers (30). Cold break removal is typically not necessary with the Yorkshire and Burton Union systems since in those systems the cold break is easily removed before the yeast is harvested.

Traditional Ale Fermentation

The brewing of traditional cask ale is fermented in shallow vessels that could be round, square or rectangular often referred by the type of fermentation system (e.g., Yorkshire square, the Burton Union, the ale top-skimming system, and the ale dropping system). The Burton Union system is now confined to a single brewery in Burton-on-Trent, and the Yorkshire square system is found in Yorkshire and Midland breweries.

Traditional ales are usually fermented at higher temperatures over a relatively shorter time. The higher temperatures and consequent shorter fermentation time (approximately 4 to 6 days) contribute to a product that has a different balance of various flavor compounds, then that of lagers, such as more fusel alcohols and esters. Typically, the yeast is pitched between 15 and 18°C, and the temperature is allowed to rise gently to 20 to 25°C, depending on the yeast strain (24). Some brewers will rouse the beer during fermentation to maintain the yeast in suspension and increase the level of dissolved oxygen and rate of attenuation.

As fermentation proceeds, a substantial crop of yeast collects on the surface of the beer. Removal of yeast at the end of fermentation (typically 3–4 days) is traditionally done by manual skimming or suction of the head of yeast that settled on the surface of the beer. Finally, with fermentation over, the ambient temperature of the beer is reduced to racking temperature at 12 to 15°C. It is important that the ambient temperature is reduced at a rate that will not induce thermal shock to the yeast.

The beer is rarely filtered and is primed with sugar or transferred with a small amount of residual extract directly into the casks. A secondary fermentation takes place in the cask to mature the beer.

Modern Ale Fermentation

Today modern ale fermentation typically uses vertical cylindrodroconical fermenters achieving similar flavor profiles compared to traditional fermentation systems. Typically, the yeast is pitched between 15 and 22°C, and the temperature is allowed to rise gently to 18 to 25°C, depending on the yeast strain (7). At the end of fermentation a diacetyl rest may be incorporated, although some popular ales (e.g., Irish stouts) have a perceptible diacetyl character that is are part of the flavor profile. Once fermentation is complete, the contents are cooled to about 4°C to settle the yeast in the cone of the fermenter. The settling time is dependent on the flocculation characteristics of the yeast. Typically, brewers will wait for an additional 24 hours for the yeast to further settle before cropping. Changes in yeast handling are required since collecting yeast off the surface is no longer possible.

The green beer is then transferred to the cellar for conditioning and either chilled to − 2 to 4°C through a heat exchanger or *in situ* in the tanks. This practice allows for more efficient cooling and the freeing of the cylindroconical fermenters for fermentation.

Diacetyl Rest

Employing high temperatures at the end of primary fermentation are for reducing diacetyl and 2,3-pentanedione (collectively referred to as vicinal diketones or VDKs). This procedure is known as a diacetyl rest. VDKs, which are assimilated by yeast toward the end of fermentation, are responsible for off-flavors in beer. The diacetyl rest reinvigorates the yeast culture so that it metabolizes those byproducts—such as diacetyl and 2,3-pentanedione—that are excreted early in the fermentation, thereby removing them from solution. Depending upon the yeast type, the medium, and the physical environment, this process is variable in time and temperature.

Employing a diacetyl rest is particularly important when producing beers with high adjunct fraction due to the high levels of diacetyl produced (11). It is also important in brewing lagers, as they do not have a heavy flavor impression. The technique reduces the time needed to achieve a product similar to beer brewed when using the more intensive cold-lagering process. With its use, lagering may require only 7 to 14 days to achieve the same clarity and flavor stability that would only be expected after 5 to 7 weeks with traditional lagering (42).

> **Note | Factors Influencing Diacetyl Reduction**
>
> During fermentation the yeast has a great capacity to remove diacetyl; however, toward the end of fermentation there may not be enough active yeast to remove diacetyl. This can be due to poor quality yeast (especially when reused), to an overly flocculent yeast, or to prematurely separating (racking) the yeast and beer before all the diacetyl is reduced at the end of fermentation. Premature temperature reduction at the end of fermentation can result in settling of the yeast, thereby reducing the yeast capacity to reduce diacetyl levels.
>
> Diacetyl not reduced during fermentation will probably be reduced during conditioning, providing there is a sufficient concentration of yeast. Campbell reports that increasing the yeast concentration during maturation will remove diacetyl and 2,3-pentanedione and at the same time shorten the storage period (6).

After the diacetyl rest, the beer is cooled to lagering temperature. Generally, cooling does not begin until the analysis indicates low levels (fewer than 0.1–0.2 mg/l or ppm) of VDKs and/or diacetyl for each fermenter. It is particularly important for accelerated fermentations that the concentration of VDKs is low enough before cooling to ensure that diacetyl levels do not rise above the taste threshold in subsequent processing, particularly during pasteurization (39).

Once the yeast has settled, it will begin to uptake diacetyl and 2,3-pentanedione that are excreted early in the fermentation. It is important in brewing lagers not to lower the fermentation temperature prematurely or rack the beer prematurely until the yeast sediment has reabsorbed the diacetyl, or else the beer will end up with a distracting "buttery" aroma and flavor.

High Gravity Fermentations

High-gravity fermentations involve worts of up to 18°P and even higher (38). Following fermentation and maturation, the beer is diluted with cool carbonated water to a prescribed original gravity or to a prescribed alcohol concentration. There are a number of advantages associated with high-gravity brewing. Munroe reports that it results in beers that are more consistent (% alcohol, original gravity, etc.) and more physically stable since the compounds responsible for haze are more easily precipitated at higher concentrations. Handling more concentrated wort results in increased utilization of equipment and lower energy costs (38). According to Munroe, the disadvantages are longer fermentation times, different flavor characteristics, and poorer hop utilization than normal gravity fermentations (38).

Yeast Collection

Aside from the need to remove most of the yeast from the beer prior to conditioning, yeast recovery for reuse in subsequent fermentations is an important process in the brewery. The total amount of yeast produced is dependent upon not only the yeast strain but also the level of aeration, the fermentation temperature, the specific gravity of the wort, and the pitching rate. Increases in any of these variables will lead to greater yeast crops. According to Knudsen, the yeast crop is usually about 2 to 4 times the amount pitched, and he recommends a yield of about 2.5 times, which is more than adequate for yeast collection (23). Generally, the total amount of yeast produced is greater with ale fermentations than with lager fermentations.

Fermentation Systems

Traditional Lager Fermentation System

The traditional method of collecting lager yeast is to decant the "green" (unconditioned) beer from the settled yeast on the bottom of the open fermenters. The yeast is collected manually from the middle layer of the sediment on the floor.

Traditional Ale Fermentation System

Traditionally, top-cropped ale yeast was harvested by skimming the head of yeast/foam that formed on top of the beer in the shallow, flat-bottomed fermentation vessel. Generally, the second crop that forms towards the end of fermentation is the one that is harvested since the yeast is pure, with very high viability. The first crop is often discarded because it contains most of the coagulated protein and dead cells. The yeast can either be harvested manually by scooping the yeast off the surface, in the case of small fermenters, or using an inverted funnel to collect the yeast in larger vessels.

Cylindroconical Fermenters

Today, with the advent of cylindroconical fermenters, the differentiation on the basis of bottom and top cropping has become less distinct. Cylindroconical tanks allow improved yeast separation and collection strategies for both lager and ale yeast. The angle at the bottom of the tank allows the yeast to settle into the base of the vessel at the completion of primary fermentation. This aids in yeast collection from the bottom of the cone without exposing the yeast to outside air, leaving the beer comparatively free of yeast. Temperature reduction at the end of primary fermentation further aids sedimentation of yeast into the bottom of the cone for easy beer transfer and yeast collection.

The removal of the yeast from the cone should be carried out as soon as possible to maintain viability levels. The yeast is collected from the conical bottom as a plug before the beer is removed from the fermenter. When the yeast is harvested, it is always cropped from the center (in terms of both space and time). Harvesting too early, or from the bottom of the cone, will likely result in early flocculating yeast that has not completely attenuated the beer. Cropping too late will result in collecting a non-flocculent yeast that stays in suspension too long and clouds the beer, thereby making clarification difficult. These negative characteristics will inadvertently be accentuated by re-pitching the yeast that has not been harvested from the center.

Care should be taken in controlling the rate in which the yeast is harvested. If the yeast is harvested too fast the solid content of the crop could be too low, which can also cause channeling within the fermenter. If too slow, the process takes too long and the yeast may compact. There are several ways to regulate the cropping rates, including use of an orifice plate, a Saunders or Ball valve, or the use of a variable speed cropping pump. Some brewers prefer a stop/go cropping policy; this involves harvesting the yeast plug for a short time and then stopping to resettle the yeast before cropping is resumed. Once the yeast is removed the tank can then be emptied of its beer.

Yeast Storage

Upon harvesting, the yeast can be transferred directly to another fermenter, transferred to a yeast brink and held for the next fermentation, or stored with its own beer at 0°C. The yeast will remain healthy while held at refrigerated temperatures, especially if the cells are kept suspended by gentle agitation so that "hot" spots do not accelerate cell death and autolysis. Small craft breweries may use nothing more than a white food-grade tub with a snap lid for storing the yeast.

One advantage in transferring yeast directly to the fermenter is that the viability of the yeast is normally excellent. There is also less chance of general infection pickup from the plant. Some of the disadvantages include pitching rates are difficult to estimate and the possibility of bacterial spoilage contamination, particularly by Pediococci (17). Advantages in using a yeast brink are many and the preferable method since it is equipped with cooling jackets for temperature control, agitators for uniform yeast concentration, load cells for measuring contents, and a built-in CIP system. Some yeast brinks may also have the ability to regulate the pressure as well as the ability to aerate the yeast with oxygen or air before pitching.

It is recommended that the yeast be kept in the holding tank for no longer than 24 hours at a temperature near but not below 0°C (23). Kunze recommends washing and sieving the yeast if it cannot be used immediately (25).

In order to keep track of the age and condition of pitching yeast, it is essential to keep accurate records of the history of cropped yeast. A commonly adopted method of doing this is to give each pure culture an identifying code and to give each subsequent yeast crop of that culture a sequential number.

Fermentation Systems

There are many different fermentation systems that are used worldwide that have evolved based on available technology, brewing materials, and perceived product quality. The following is a brief description of some of the more common fermentation systems in use today.

Cylindroconical System

Cylindroconical fermenters are the most commonly used fermentation systems used today to produce both lagers and ales. As the name implies, the enclosed vessels are vertical cylinders with a conical base and, normally, a dished top, as shown in Figure 13.1. This design allows for easy yeast collection and CIP cleaning. They range in size between 100 and 7,000 hl, have from a 1:5 to a 3:1 ratio of height to diameter, and work under pressures of from 1 to 1.3 bars above atmospheric pressure. In fermentation vessels with a ratio greater than 3:1, there is a tendency for increased production of higher alcohols at the expense of esters (5). Vessel geometry plays an important role in fermentation. As the height-to-diameter ratio increases, so does the mixing of yeast and wort, as well as the fermentation rate.

These fermenters (FVs) were originally intended to be dual-purpose, used for both fermentation and conditioning, but in practice they are mainly used for fermentation (64). These vessels are fitted with vacuum and pressure relief valves to compensate for rapid pressure changes when filling, emptying, or cycling cleaning solutions. Pressure and vacuum release mechanisms are needed to avoid explosion or implosion. For example, emptying such vessels without balancing pressure will lead to implosion, the most common cause of damage to tanks in craft breweries.

Figure 13.1
Cylindroconical Fermenters
Courtesy of Beraplan Härter GmbH

In cylindroconical tanks, the concentration gradient encourages the evolution of carbon dioxide from the base of the cone up through the center of the vessel. The return of the beer is mainly near the walls and is greatly aided by cooling jackets placed high up the wall of the fermenter. Circulatory patterns vary according to the rate of fermentation, the position of the cooling jackets, and the speed of cooling. Hydrostatic pressure will tend to lead to a greater content of dissolved carbon dioxide near the base; but during active fermentation and/or cooling, the concentration gradient in the tank tends to disappear.

An important characteristic of most of these vessels is the steep-angled cone at the base. It has been discovered that an included angle of 70 degrees is the best for yeast runoff into the base of the vessel, leaving beer comparatively free from yeast in the vessel. However, in some cases the included angle is greater than 70 degrees. Very large fermenters tend to have much shallower slopes, with angles of up to 105 degrees (64). Some breweries also specify a high polish for the cone surface to achieve good yeast plug slippage (53).

Yeast Strains

Many top-cropping yeast strains used in open fermenters will behave as bottom-cropping yeast strains in cylindroconical fermenters. This has allowed many British breweries to keep using their own "in-house" yeast strains with conical FV's. In spite of their popularity, some traditional ale brewers still do not use cylindroconical fermenters to produce beer (64). This is because the carbon dioxide produced is in excess of that required in traditional ales. Nevertheless, traditional lager brewers have experienced few problems with their yeast strains in using cylindroconical fermenters.

Foam Production

Foam production during fermentation tends to be excessive and is one of the reasons why some large breweries use antifoam agents. However, many breweries, instead of using antifoams, leave 15 to 25% of the volume as headspace to accommodate the foam during fermentation. Kunze recommends a headspace of 40% in fermentation vessels for wheat beers (25). Some control can also be exercised by using lower levels of dissolved oxygen in higher gravity wort (25). The use of pressure is another method of controlling foam height. On the other hand, some brewers prefer fermenters with little headspace so the foam at high kraeusen can stick to the tank roof. They claim that this removes coarse bitterness from beer.

Temperature Control

Cylindroconical fermenters lend themselves readily to temperature programming. The fermentation temperature can be automatically controlled on a preset program by means of a temperature recorder-controller operating the vessel jackets on the walls and the cone. Temperature probes are located at different depths to detect variations in temperature due to circulatory currents. A temperature probe should be located low on the tank so that it will be immersed when the tank is only partly full. If the vessel is used for conditioning in which the beer is cooled to near the freezing point, an additional probe higher up on the tank is advisable. This is to prevent the beer from freezing due to temperature inversion when the colder beer rises to the top of the tank.

Coolants

For temperature control of cylindroconical tanks the brewer can either use direct or indirect cooling:

1. Temperature control with ammonia gas is accomplished with direct expansion and evaporation of the liquefied gas in the cooling jackets integrated in the tank wall. Ammonia in the form of liquid is pumped through the cooling jackets where it evaporates causing cooling. The mixture of gas and liquid is then returned to the receiver where it is re-compressed and condensed to a liquid to repeat another cycle. Freon is often used too.
Direct-expansion refrigerants can give rise to higher electrical efficiencies but there are also operating and safety concerns to consider. These systems are generally more difficult to operate, more

costly to install, and if using ammonia gas it is a very corrosive, dangerous and pungent.
2. In indirect cooling a secondary coolant is used such as brine, ethanol, or propylene glycol in the cooling jackets integrated into the tank wall. The coolant is warmed as it passes through the cooling jackets and is returned to the refrigeration plant where it is cooled by a primary coolant. Craft brewers are more likely to use refrigerated glycol whereas large breweries often use ammonia.

Cooling Jackets

As mentioned, the coolant is circulated through cooling jackets and then returned to the plant. Cooling jackets are of several designs:

1. Pressed corrugated – widely used for secondary refrigerant systems
2. Dimple – very popular in beer and wine production
3. Inflated (quilted) – suited more to low volume primary refrigerant systems acting as an evaporator
4. Limpet – continuous coils welded to the vessel offer reduced heating/cooling times and a greater degree of accuracy in controlling product temperatures

Most tanks will have two cooling jackets on the sides and an additional jacket on the cone.

Traditional Ale Top-Skimming System

This is the traditional ale skimming system used in the United Kingdom. It utilizes top-fermenting yeast cropped for re-pitching by skimming from the surface. The vessels are normally shallow and flat-bottomed and may be round, square, or rectangular. Traditionally, these vessels are open-topped to facilitate skimming and are located in a well-ventilated room to disperse the carbon dioxide evolved during fermentation. Some British brewers maintain that there is a smoother, rounded flavor produced with open fermentation tanks that is lost in tall cylindroconical tanks. Today many of the vessels are enclosed, allowing carbon dioxide collection, and the skimming is carried out by suction. The fermentation is controlled by attemperation using cooling panels on the walls of the vessel.

Traditional Ale Dropping System

This is a variation of the ale top-skimming system, whereby the wort is pitched, collected, and partially fermented in one vessel (the settling tanks) and then dropped or transferred to a second vessel after 24 to 36 hours to complete the fermentation (53). Trub and cold break are left behind in the first vessel, and the transfer also rouses the fermentation.

Yorkshire Square System

This system, as shown in Figure 13.2, is used in the north of England to produce ales with a clean, round palate. It is a variation of the ale fermentation system adapted to enable the use of a flocculent yeast strain or a two-strain system with one of the yeasts being flocculent. The vessels are normally square or rectangular in shape, and constructed of stainless steel. The Yorkshire square system consists of a lower compartment with a gently sloping upper deck located just above the fill level of the vessel. This deck has one or more manholes with removable covers and small diameter pipes—"organ pipes," which lead from the surface of the deck down into the lower vessel. On the surface is an outlet with a "plug," which is used for skimming from the deck. The lower compartment is filled with wort and, usually, strongly flocculent yeast. The fermentation is controlled by attemperation coils in the vessel or by cooling panels on the wall. During the fermentation, the yeast head issues through the manhole and remains while the entrained beer drains back via the organ pipes. The yeast is commonly roused back into the beer by pumping wort from the fermenting chamber and spraying it over the yeast to assist attenuation. When the required attenuation has been achieved, the yeast is cropped by removing the skimming plug and scraping off the deck.

Figure 13.2
Yorkshire Squares

Burton Union System

This system was commonly used in Burton-on-Trent and produces pale ales with a fruity character by using "powdery" non-flocculent yeast. The wort is collected and pitched in a separate vessel and then transferred to the union set after 24 to 48 hours. The union set consists typically of 2 rows of wooden casks, 24 to 50 in number and each of 7 hl capacity, positioned under a cooled trough. Each cask is fitted with a swan neck that overhangs a slightly inclined trough mounted above the casks. The casks have individual attemperation coils used to control the progress of the fermentation. The carbon dioxide gas produced during the fermentation carries over the yeast froth up through the swan neck into the trough, where cooling induces the yeast to settle. The beer that collects in the trough runs to the lower end and returns to the casks. After fermentation, the beer is discharged to other vessels, the "racking backs." The yeast is recovered from the trough for re-pitching. Adjusting the level of an outlet cock in each cask controls the final yeast count in the beer at transfer.

Open Square Fermenters

Traditional lager systems in Continental Europe and still common in Eastern Europe utilize open square fermenters (see Figure 13.3) similar to those used in the traditional ale top system. Enclosed fermenters are more common. The bottom-fermenting yeast does not require surface skimming because the yeast is deposited on the floor of the vessel at the end of the fermentation. The flat bottoms of these various vessels make them less than ideal for yeast cropping as this involves scrapping the yeast off the floor, which is particularly difficult in large vessels. The fermentation is controlled by attemperation using cooling panels on the walls of the vessel.

Dual Purpose Vessel System

The vessel characteristics that make cylindroconicals so suitable for fermenting vessels also make them ideal for conditioning tanks. This has led to the installation of dual-purpose vessels where primary fermentation and conditioning are carried out in the same vessel. Dual-purpose vessels have cooling jackets located high in the vessel for fermentation and low in the vessel for conditioning. Other differences are that conditioning tanks are normally required to be top-pressured to maintain carbonation levels, which is not required in fermenters.

The obvious advantages of dual purpose-vessels include flexibility in operation (they can be used either for fermentation or conditioning), reduced labor requirements, lower detergent costs, and reduced beer losses due to elimination of transfer (53). However, some of the disadvantages of dual-purpose systems include the production of beer with a different flavor

Figure 13.3
Open Square Fermenters

profile from that of two-vessel systems. The two-vessel systems, with separate fermenting and conditioning vessels, results in better vessel utilization (33). Also, two-vessel systems are better suited in removing yeast and, diminish the potential for off-flavors since fewer solids is present in the beer allowing for shorter filtration runs.

Calculations

Wort Fermentability

The proportion of wort that can be fermented is called the percentage of fermentability of the wort as expressed by the following equation:

Fermentability (%) = [(Original gravity − final gravity)/Original gravity] x 100

The original gravity is expressed as °P (Plato), which is a measure of the density of beer wort in terms of percentage of extract by weight. The final gravity or apparent attenuation limit means the gravity of the wort when it is fully fermented. Therefore if wort with an original gravity of 12 °P is fermented to a gravity of 4 °P the percentage fermentability is:

1. Fermentability (%) = [(12 − 4)/12] x 100 = 67%

References

1. Allen, Fal. 1994. "The Microbrewery Laboratory Manual – A Practical Guide to Laboratory Techniques and Quality Control Procedures for Small-Scale Brewers, Part I: Yeast Management." *Brewing Techniques* 2 (4).
2. Barnett, J.A. 1979. *A Guide to Identifying and Classifying Yeasts*. New York, New York: Cambridge University Press.
3. Boughton, Richard. 1987. "Practically Managing Yeast." *MBAA Technical Quarterly* 24 (4).
4. Brenner, M.W. 1970. "A Practical Brewer's View of Diacetyl." *MBAA Technical Quarterly* 7 (43).
5. Briggs, Dennis E., Boulton, Chris A., Brookes, Peter A., and Stevens, Roger. 2004. *"Brewing: Science and Practice*. Cambridge, England: Woodhead Publishing Limited.
6. Campbell, I. 1987. "Microbiology of Brewing: Beer and Lager." *Essays in Agricultural and Food Microbiology*, edited by J.R. Norris and G.L. Pettipher. Chichester [West Sussex], England: John Wiley & Sons Ltd.
7. Carey, Daniel and Grossman, Ken. 2006. "Fermentation and Cellar Operations." *Fermentation, Cellaring, and Packaging Operations*. Volume 2, edited by Karl Ockert. St. Paul, Minnesota: Master Brewers Association of the Americas.
8. Coors, Jeffery H. 1977. "Cellar Operations." *The Practical Brewer*, edited by H. M. Broderick. Madison, Wisconsin: Master Brewers Association of the Americas.
9. De Clerck, Jean. 1957. *A Textbook of Brewing*. Volume 1, translated by K. Barton-Wright. London: Chapman-Hall Ltd.
10. Dowhanick, Terrance M. 1999. "Yeast – Strains and Handling Techniques." *The Practical Brewer*, edited by John T. McCabe. Wauwatosa Wisconsin: Master Brewers Association of the Americas.
11. Fix, George J. 1989. *Principles of Brewing Science*. Boulder, Colorado: Brewers Publications.
12. Fix, George J. 1993. "Diacetyl: Formation, Reduction, and Control." *Brewing Techniques* 1 (2).
13. Fix, George J. 1995. "The Role of pH in Brewing." *The New Brewer* 12 (6).
14. Fix, George and Laurie Fix. 1997. *An Analysis of Brewing Techniques*. Boulder, Colorado: Brewers Publications.
15. Gilliland, R.B. 1981. "Brewing Yeast." *Brewing Science*. Volume 2, edited by J.R.A. Pollock. London, England: Academic Press.
16. Hammond, John. 1986. "The Contribution of Yeast to Beer Flavor." *Brewers' Guardian* 115 (9).

17. Henson, M.G. and D. M. Reid. 1987. "Practical Management of Lager Yeast." *The Brewer* 74 (879).

18. Hewson, Bob. 1998. "Managing Traditional Ale Fermentations." *Brewers' Guardian* 127 (8).

19. Hoggan, John. 1980. "Influence of Conditioning and Fermentation." *Brewers' Guardian* 109 (11).

20. Hough, J.S., et al. 1982. *Malting and Brewing Science*. Volume 2. London, England: Chapman-Hall, Ltd.

21. Kirsop, B. W. 1975. "Wort Aeration," *Brewers Digest* 34 (Dec.).

22. Knudsen, Finn B. 1977. "Fermentation Principles and Practice." *The Practical Brewer*, edited by H. M. Broderick. Madison, Wisconsin: Master Brewers Association of the Americas.

23. Knudsen, Finn B. 1985. "Fermentation Symposium – Part II – Fermentation Variables and Their Control." *MBAA Technical Quarterly* 22 (4).

24. Knudsen, Finn B. 1999. "Fermentation, Principles and Practices." *The Practical Brewer*, edited by John T. McCabe. Wauwatosa Wisconsin: Master Brewers Association of the Americas.

25. Kunze, Wolfgang. 1996. *Technology Brewing and Malting*, translated by Dr. Trevor Wainwright. Berlin, Germany: VLB Berlin.

26. Lampinen and Young. 1973. "Experiences with Conical Fermenters," *MBAA Technical Quarterly* 10 (172).

27. Lewis, Michael J. and Tom W. Young. 1995. *Brewing*. London, England: Chapman & Hall.

28. Lewis, M. J. and Young, T. W. 2002. *Brewing*. New York, New York: Kluwer Academic/Plenum Publishers.

29. Linley, P. A. 1978. "Lager Fermentation and Maturation." *Brewers' Guardian* 107 (1).

30. MacDonald, J., et al. 1984. "Current Approaches to Brewery Fermentations." *Progress in Industrial Microbiology*, edited by M. E. Bushell. Guildford, United Kingdom: Elsevier.

31. Mallet, John. 1992. "Pitching Control." *New Brewer*. 9 (4).

32. Martin, E. C. B. 1972. "Lager Fermentation, Storage, Filtration." *Brewers' Guardian* 101 (4).

33. Maule, D. R. 1977. "Reflections on Cylindro-Conical Fermenters." *The Brewer* 63 (752).

34. Maule, D. R. 1980. "Propagation and Handling of Pitching Yeast." *Brewers Digest* 55 (2).

35. Miedaner, H. 1978. "Optimization of Fermentation and Conditioning during the Production of Lager." *The Brewer* 64 (2).

36. Meilgaard, M. C. 1974. "Flavor Chemistry of Beer, Part I," and "Flavor Interaction Between Principal Volatiles," *MBAA Technical Quarterly* 11 (250).

37. Moll, Manfred, ed. 1995. *Beers and Coolers*. New York, New York: Marcel Dekker, Inc.

38. Munroe, James H. 1995. "Fermentation." *Handbook of Brewing*, edited by William A. Hardwick. New York, New York: Marcel Dekker, Inc.

39. Munroe, James H. 1995. "Aging and Finishing." *Handbook of Brewing*, edited by William A. Hardwick. New York, New York: Marcel Dekker, Inc.

40. Munroe, James H. 2006. "Fermentation." *Handbook of Brewing*, edited by Fergus G. Priest and Graham G. Stewart. Boca Raton, Florida: CRC Press, Taylor & Francis Group.

41. Noble, Stuart. 1997. "Practical Aspects of Fermentation Management." *The Brewer* 83 (991).

42. Noonan, Gregory J. 1996. *New Brewing Lager Beer*. Boulder, Colorado: Brewers Publications.

43. O'Connor-Cox, Erin. 1998. "Improving Yeast Handling in the Brewery." *Brewers' Guardian* 127 (3).

44. Pinkerton, John M., Jr. 1998. "Centrifugal Separation for the Regional Craft Brewery." *The New Brewer* 15 (1).

45. Pryor, Jim. 1999. Personal Communication. Vafac Inc.

46. Rises, S. 1986. "Automatic Control of the Addition of Pitching Yeast." *MBAA Technical Quarterly* 23 (1).

47. Russell, Inge. 1995. "Yeast." *Handbook of Brewing*, edited by William A. Hardwick. New York, New York: Marcel Dekker, Inc.

48. Ryder, David S. and Joseph Power. 1995. "Miscellaneous Ingredients in Aid of the Process." *Handbook of Brewing*, edited by William A. Hardwick. New York, New York: Marcel Dekker, Inc.

49. Scheer, Fred. 1989. "Methods for Carbonating Beer." *The New Brewer* 6 (2).

50. Schwarz, Paul and Richard Horsley. 1996. "A Comparison of North American Two-Row and Six-Row Malting Barley." *Brewing Techniques* 4 (6).

51. Seddon, A. W. 1975. "Continuous Tower Fermentation – Experiences in Establishing Large Scale Commercial Production." *MBAA Technical Quarterly* 12 (130).

52. Simpson, Bill. 1988. "Managing Lager Fermentations." *Brewers' Guardian* 127 (11).

53. Smith, Ian B. 1991. "Yeast Fermentation Systems – A Review." *Brewers' Guardian* 120 (3).

54. Sohigian, David. 1999. "Count on Your Hemacytometer for Consistent Pitching Rates." *Brewing Techniques* 7 (1).

55. Stewart, G.G. and I. Russell. 1985. "Modern Brewing Biotechnology." *Food and Beverage Products*, edited by Murray Moo-Young. Oxford, England: Pergamon.

56. Stewart, G.G. and I. Russell. 1993. "Fermentation – The 'Black Box' of the Brewing Process." *MBAA Technical Quarterly* 30 (159).

57. Tenney, Robert I. 1985. "Fermentation Symposium – Part I – Rationale of the Brewery Fermentation," *MBAA Technical Quarterly* 22 (4).

58. Thompson, Donald, Karl Ockert, and Vince Cottone. 1999. "Craft Brewing." *The Practical Brewer*, edited by John T. McCabe. Wauwatosa Wisconsin: Master Brewers Association of the Americas.

59. Van Der Aar, P.C. 1995. "Flocculation of Brewers' Lager Yeast." *MBAA Technical Quarterly* 32 (222).

60. VanGheluwe, Chen and Valyi. 1975. "Factors Affecting the Formation of Fusel Alcohols During Fermentation," *MBAA Technical Quarterly* 12 (169).

61. Voakes, Murray. "Yeast Handling." 1988. *Brewery Operations*, Volume 5, edited by Virginia Thomas. Transcripts from the 1988 Microbrewers Conference. Boulder, Colorado: Brewers Publications.

62. Wainwright, T. 1971. "Biochemistry of Brewing." *Modern Brewing Technology*, edited by W.P.K. Findlay. Cleveland, Ohio: The Macmillian Press.

63. Whitear, A.L. 1974. "Beer Stabilization – Past and Present." *Brewers' Guardian* 103 (3).

64. Young, T.W. 1987. "The Biochemistry and Physiology of Yeast Growth." *Brewing Microbiology*, edited by F.G. Priest and I. Campbell. Essex, United Kingdom: Elsevier.

Chapter Fourteen

Conditioning

Following primary fermentation, the "green" or immature beer is far from finished because it contains suspended particles, lacks sufficient carbonation, lacks taste and aroma, and it is physically and microbiologically unstable. Conditioning reduces the levels of these undesirable compounds to produce a more finished product. The component processes of conditioning are:

- Maturation
- Clarification
- Stabilization

Maturation

Maturation techniques vary from brewery to brewery but generally they can be divided into two general schemes for finishing beer after primary fermentation called secondary fermentation and cold storage.

Secondary Fermentation

Traditionally, maturation involves secondary fermentation of the remaining fermentable extract at a reduced rate controlled by low temperatures and a low yeast count in the green beer. During secondary fermentation, the remaining yeast becomes re-suspended utilizing the fermentable carbohydrates in the beer. The carbohydrates can come from the residual gravity in the green beer or by addition of priming sugar or by kraeusening. Yeast activity achieves carbonation, purges undesirable volatiles, removes of all residual oxygen, and chemically reduces many compounds, thus leading to improved flavor and aroma.

Secondary fermentation processes include lagering, kraeusening, and cask- and bottle-conditioned beers.

Lagering

Lagering was developed in Germany for bottom-fermented lagers, and it involves a long, cold storage at low temperatures. Although lagering refers to bottom-fermented beers, some top-fermented beers such as Kölsch and Alt beers also require periods of lagering.

For classical lagering, the temperature is held at 4 to 5°C until the VDK level is below 0.15 mg/l (or ppm) and then slowly cooled (5). It is during these warmer temperatures that much of the secondary fermentation takes place as the fermentable extract is slowly depleted. The temperature is then reduced one degree per day until the beer reaches freezing, and is held at freezing for the length of the maturation period. Very little fermentation takes place at this stage but the beer flavor improves due to continued yeast activity. Some brewers reduce the temperature to as low as – 1 to 2°C when lagering for long periods. As might be expected, lagering at higher temperatures will produce a shorter, vigorous secondary fermentation in a shorter period of time.

Secondary lager fermentation lasts any where from several weeks up to three months. Longer lagering is usually reserved for strong beers. Excessive lagering, especially at elevated temperatures, can lead to yeast autolysis and a subsequent loss in beer quality. Once the beer has undergone sufficient yeast settlement and precipitation of most of the haze-loading materials (e.g., protein and polyphenols), it will likely undergo chillproofing before filtration.

Today the trend is to minimize storage times between 2 and 3 weeks at higher temperatures given the advances in technology—refrigeration, stabilization agents, filtration, and carbonation.

Temperature – It is important that the temperature is brought down gradually (1–3°C/day) until the lagering temperature has been reached. Dropping the temperature too fast will inactivate the yeast, which could reduce the metabolizing of sugars and cause subsequent difficulties with diacetyl off-flavor. Gradually lowering the temperature encourages the yeast to come out of suspension and reduces yeast carryover.

Yeast – Selection of yeast with the proper flocculation characteristics is obviously important for a long aging process. Powdery yeast prolongs maturation times because it does not settle sufficiently at the completion of fermentation. On the other hand, very flocculent strains settle too soon, prematurely terminating fermentation. The quantity of yeast can also affect lagering. Most lager brewers work on a figure between 1 and 4 million cells/ml when employing the traditional lagering method (21). Some have even been reported working with 5 to 15 million cells/ml (28).

Sludge – Some of the short-comings in using this method is the sludge at the bottom of the tank can warm up, allowing yeast autolysis and producing off-flavors in the beer. The sludge also contains a considerable amount of beer, which could otherwise be recovered with filtration and/or centrifugation processes. In addition, the sludge must be removed from the tank manually, which increases operating costs and prevents utilizing the tank in a closed system.

Kraeusening

"Kraeusen" is the German word used to describe the infusion of a strongly fermenting young beer into a larger volume of beer that has undergone primary fermentation. Traditionally, the wort used for kraeusening is obtained from the high 'kraeusen' stage of primary fermentation and added in small portions (5–20% by volume) to the green beer to start a secondary fermentation (8). MacDonald suggests adding a volume of kraeusen equal to 10 to 12% of the "green" beer, containing approximately 2% (w/w) residual extract with a cell count of between 10 and 15 million (29). Usually, higher gravity beers require a larger proportion of kraeusen. Kraeusen may also be made from wort and a yeast culture, or from a sugar solution together with yeast.

> ### Note | Transferring Beer Between Vessels
>
> Carbon dioxide or nitrogen is often used to minimize uptake of oxygen when transferring beer from one vessel to another (e.g., fermenter to conditioning tank). This is accomplished by applying a top pressure of gas in the tank that contains the beer after flushing the mains and receiving tank with the gas. As the beer is transferred to the receiving tank, gas is injected into the beer providing a small blanket of gas above the beer in the receiving tank. As the vessel fills, air in the receiving tank is displaced upwards thus not coming in contact with the beer.

Kraeusening virtually eliminates the lag phase and reduces the risk of oxidation and contamination at racking. Kraeusening tends to give a more complete fermentation, which leads to lower terminal gravities and a dryer finish to the beer. This technique results in a fuller beer with mellower flavors and different physical stability characteristics than non-kraeusen beer. It produces more acetaldehyde and hydrogen sulfide than the traditional lagering process. Kraeusening is effective in reducing diacetyl, particularly if the yeast is of high viability.

Kraeusen is usually added at a temperature between 6 and 9°C and eventually reduced to 1°C by the end of fermentation (8). Of course, the temperature will depend on the yeast strain, with cold-tolerant strains preferring a colder secondary fermentation temperature (5–7°C). Some brewers recommend removing the yeast before transferring to the cellar and then lagering the beer at about 4°C for a week or two before kraeusening. Cooling to − 1°C may occur gradually during secondary fermentation; or it may occur rapidly at the end to promote yeast settling, especially in the case of powdery yeast.

Kraeusening usually requires a longer maturation period than the traditional lagering process in order to reduce undesirable flavor notes to acceptable levels. Some kraeusen beers are lagered for six to eight months before being bottled. Today only a few brewers practice kraeusening, but it continues to be popular with traditional Pilsner brewing. British brewers use a similar system, though with sugar syrup instead of beer wort, to start secondary fermentation in the cask.

Cask-Conditioned Beers

Casking has its origins in the British Isles and is most widely used to make pale ales (bitters), porters, and stouts. Beer is racked either directly from fermenting vessels into casks when fermentation is judged sufficiently complete (a residual extract of 0.75 to 2°P) or when the correct charge of yeast is present (0.25–4.00 million cells/ml) (30/21). If too little yeast is present in the beer, secondary fermentation is too slow and insufficient carbon dioxide is dissolved in the beer. However, if too much yeast is suspended in the beer, secondary fermentation may to violent. Although traditionally beer was racked directly to the cask, some brewers pass the beer through a rough filtration to improve clarity.

> **Note | Influence of pH on Beer Quality**
>
> The final beer pH (generally in the range between 4.0 and 5.0) not only increases beer stability and hastens the rate of conditioning during aging but also helps beer resist bacterial contamination. A pH too low is thought to favor the formation of compounds such as *trans*—2 nonenal or "cardboard" aromas while a pH that is too high causes the formation of caramel or sherry-type aromas.

Once racked onto the cask, various additions are made. These include 1) priming sugars (usually hydrolyzed sucrose, called "invert" sugar) of approximately S.G. 1.150 at 0.35 to 1.75 l/hl, 2) hops, 3) isinglass finings to promote sedimentation of yeast and clarification of beer, and 4) potassium metabisulfite, which is a bacteriostat at pH values below 4.2 (21). Thus all the objectives of maturation can be achieved in one simple and relatively short process.

Hops are added to give beer a fresh, herbal, hop aroma without adding additional bitterness. They are usually in the form of whole hops, Type 100 plugs, Type 90 pellets, or enriched hop powder. Some brewers recommend 50 to 100 g/hl of whole hops for dry hopping. In other instances, direct addition of hop oil or emulsion of essential oils of hops provides the aroma. Contact time for oil is usually for 2 to 3 weeks, whereas contact times greater than 3 weeks may develop cooked-vegetable and cheesy flavors. As an alternative, some brewers prefer to use enriched hop extracts. Hops and hop products are added through the shieve hole of each cask by means of a funnel.

The isinglass finings are added in the amount between 0.36 and 1.08 l/hl either at racking, in the cask at the same time as the primings, or at any stage up to putting the cask on stillage (21). Settling of the finings requires between 2 and 20 hours, depending on the yeast strain.

The beer is stored at the brewery for up to 7 days, preferably at a temperature between 13 and 16°C, to promote conditioning and fining (21). One brewer of cask-conditioned ales stores the beer for 54 hours at 5°C (30). The maturation period should result in the product being almost or fully fermented and should also allow time for the fining materials to precipitate the yeast and proteins. The casks are usually vented during this period, particularly if storage temperatures are high, either by removing the spile for rapid release of gas or by using a porous spile for a slower escape of gas (21). However, many brewers do not consider venting to be desirable (1). Nevertheless, breweries recommend venting of casks to release excess pressure due to beer warming.

Bottle-Conditioned Beers

The practice of using priming sugars for bottle-conditioning has been refined by British brewers and is still followed by some craft brewers as well as a few larger British brewers. Belgian brewers are also known for using this method to add unique flavors. Bottle-conditioning usually involves a short time in the conditioning tanks to improve overall stability and flavor before adding priming sugars. Some brewers allow some yeast to pass through for secondary fermentation, while others prefer to completely remove the primary fermentation yeast and re-pitch with ale or lager yeast. Some brewers use lager yeast because it generally has a smaller cell mass, is less likely to leave an autolyzed flavor, and flocculates and settles better than ale yeast. Thompson reports pitching rates between 0.75 and 3.0 million cells/ml for bottle-conditioned beers (51). After a brief conditioning period, the beer may be filtered before being sent to tanks with stirring agitators, where the priming sugar (e.g., glucose, dextrose, or invert sugar) is added. The beer is then bottled for secondary fermentation, which takes between 10 and 14 days. After secondary fermentation, the bottled beer is moved to cold storage (5–15°C) to protect the flavor and to expedite yeast sedimentation. The yeast typically settles at the bottom of the bottle.

Cold Storage

Today with the use of modern equipment for refrigeration, carbonation, and filtration, obviates the need for secondary fermentation and a long cold storage.

The green beer under going cold storage is fully attenuated and virtually free from yeast, which is achieved because of higher fermentation temperatures and a diacetyl rest. Cold storage comprises relatively short-term storage at temperatures – 2 to 4°C for several weeks or less compared to secondary fermentation and subsequent cold storage that took several months.

Although fermentable carbohydrates are exhausted in primary fermentation, the beer produced still has a green or immature flavor; and proper conditioning is still required. Thus, cold storage is principally for enhancing physical stability and providing time for flavor maturation. In addition, some stabilization (e.g., chillproofing) may be needed to encourage the precipitation of protein/polyphenol complexes (see below).

Clarification

Although much of the suspended yeast will settle to the bottom of the storage tank by gravity sedimentation, it can be very time consuming in preparing beers for filtration. Consequently, the brewer can add fining agents at the onset or during storage to speedup the sedimentation process. Alternatively, the brewer can use centrifugation to remove yeast and other solids after fermentation. Each of these processes is described in the following sections.

Fining Agents

Although good clarity can be obtained from simple sedimentation, better results can be obtained in less time by using fining agents—isinglass and gelatin. The use of finings is not universal. They find their widest employment in the United Kingdom with some ale brewers but there has been renewed interest in North America.

Isinglass

Isinglass is a traditional "real ale" clarifier used in the United Kingdom, where the style of beer benefits from a 48 hour clarification before or after casking. It is also used for fining chilled and filtered beers. Isinglass is a gelatinous substance derived from the internal membranes of fish bladders and comes in many different forms. The currently accepted mechanism involves a direct interaction of positively charged isinglass with negatively charged yeast to form flocs, which precipitate. Its effectiveness in settling ale yeast varies with the strain of yeast, and it is generally not recommended for precipitating lager yeast.

Isinglass has an advantage over gelatin because its fining action acts mainly by neutralization of the electrical charge of the yeast rather than by mechanical means as is the case with gelatin. As a result, less isinglass is needed than gelatin for the same degree of fining.

Some yeast strains tend to produce large flocs that settle rapidly to leave bright beer but that also tend to give relatively large volumes of sediment. Other strains result in smaller flocs settling less quickly, though they produce lower volumes of sediment. The isinglass types that form larger flocs are particularly useful to cask ales, where excellent clarity is required in a short period of time (25). Those types that form smaller flocs are suitable for fining where the object is to reduce the solids loading of the filter, rather than to achieve absolute clarity (25).

The performance of isinglass is affected by a number of factors, including pH, concentration of yeast, level of positively charged colloidal material, and hop rates. The optimum pH range for good fining performance is between 3.7 and 4.2. Poor fining action may result if the concentration of yeast is too high (e.g., more than 2 million cells/ml) [21]. An excess of positively charged colloidal material in the beer, such as positively charged calcium ions, may reduce the fining action of isinglass by reducing the charge on the yeast cell (47). High hop rates have been found to adversely affect fining performance too (47).

Storage, Preparation, and Dilution – Isinglass stored at 30°C denatures fairly rapidly, but it is relatively stable at 20°C, with no effect on either the collagen content or the fining performance (50). Preparation of isinglass is dependent on the physical form used as starting material. If the isinglass is in its raw form, it needs to be "cut" using acids. There is some evidence that organic acids provide a better product than inorganic, with tartaric acid being the most preferable acid (47). Freeze-dried powdered finings are also available, and are ready for use in less than 30 minutes. Alternatively, isinglass is available in concentrated paste and liquid forms. According to Thompson, the best method of dilution is by recirculation in a dilution tank, since in-tank mixers can lead to air pickup and under-dosing (50).

Dosage Rate – The optimum dosing rate is determined by adding isinglass at different rates to beer samples. As the isinglass dose rate increases, so does the clarity of the beer, until the optimum fining dosage rate is reached. This is the point that gives the brightest beer in the least time, with the minimum generation of bottoms. According to Ballard, if too much or not enough isinglass is added, loose sediment can form at the bottom of the tank after fining, a phenomenon referred to as "fluffy bottoms" (4). Following the optimization trials, scale-up can be carried out on a number of production brews. Ryder reports that a rate of 10 to 25 mg/l is typically applied for freeze-dried powdered finings (47). This rate has shown to be effective for a 3,000 hl vertical tank in clarifying beer within 3 days (48). Rates between 30 and 60 mg/l for isinglass in its raw form have been reported too (14).

Dosing – It is recommended that isinglass be added to the entire run of beer after fermentation and at the time of cooling or during transfer (50). The optimal method of adding finings is to proportionally inject it at a point of high turbulence, preferably against the direction of the beer flow. If finings are not dosed proportionally throughout transfer, the beer runs the risk of being over- or under-fined. Such beer may have poor clarity or a relatively large volume of sediment after fining. Alternatively, brewers can add isinglass to the storage vessel for good fining action prior to transferring the beer.

Auxiliary Finings – Auxiliary finings are often used in conjunction with isinglass finings for improved clarity (25). The negatively charged auxiliary finings come in two main types—acidified silicates and acidified polysaccharides. The negative charge enables the auxiliary finings to react with positively charged proteins and, subsequently, with isinglass finings. The auxiliary finings react quickly with proteins in the beer and more slowly with the isinglass; this gives rise to a denser, faster-settling coagulum that clears the beer and is more resistant to disturbance on the tank bottom (16). Silicate auxiliaries, unlike polysaccharide auxiliary finings, are capable of forming precipitate on their own, without the addition of isinglass finings.

Centrifugation

Centrifuging is a popular method of reducing the yeast content of beer and is often used where fining agents are not used or are used in conjunction with fining agents. Many brewers who practice accelerated cold conditioning use centrifuges, as they offer a greater degree of control over yeast count and eliminate the time needed for maturation and fining. The yeast count in beer ex-centrifuge can be controlled to a level of 0.05×10^6 cells/ml (29). The flocculation characteristics of the yeast are of less importance as long as the yeast remains in suspension for most of the fermentation. Optimal yeast separation is achieved at temperatures between 3 and 5°C (29). There are two principal ways to use centrifuges for yeast removal after fermentation.

Centrifuge During Beer Transfer

The first method is to use the centrifuge to separate the yeast crop as the beer is transferred from the fermenter to the conditioning tanks. This accomplishes beer transfer and yeast collection in one step. The one-step procedure is generally used when the physiological state of the yeast is not critical. There is concern among brewers that centrifugation causes mechanical shearing of yeast, resulting in the release of cellular contents and subsequent off-flavors. However, both research and testimony from brewers indicate that in most cases off-flavors from cellular damage is not a problem in the final product. If secondary conditioning or ruh storage is needed, the centrifuge can be adjusted to allow a sufficient amount of yeast to remain in the beer or, alternatively, to allow a sufficient amount of yeast in the beer to bypass the centrifuge.

Centrifuge Following Yeast Sedimentation

The second method is to use a centrifuge to clarify the beer after the yeast has been separated by sedimentation. Following sedimentation, some brewers will re-circulate the tank's contents, pulling from the bottom of the outlet and routing back into the tank before centrifuging. This ensures an even loading of the centrifuge. The two-step procedure is generally used when the collected yeast will be used for re-pitching (37).

Advantages and Disadvantages

Centrifuges have a number of advantages such as small space requirement, more consistent clarity, most are self cleaning, minimum oxygen pickup, and can be continuously operated for an indefinite period. In spite of their advantages, the big drawback in operating centrifuges is the capital investment required, which is why only large-scale commercial brewers use centrifuges. Other concerns include they are noisy, complex and difficult to maintain, and increase the temperature of the beer, which occurs during centrifugation (5).

Stabilization

In addition to clarification (i.e., removing yeast), beer must display physical stability with respect to haze. Colloidal instability in beer is caused mainly by interactions between polypeptides and polyphenols. Amino acids make up polypeptides which, in turn, make up proteins. Polypeptides and polyphenols combine to produce visible haze that reduces a product's physical shelf life. Reducing the levels of one or both of the precursors using suitable stabilizing treatments will extend physical stability.

Polypeptides responsible for haze formation originate mainly from barley. Polyphenols in beer originate from barley and hops. Polyphenols are mostly lost throughout the brewing process, particularly during mashing, boiling, wort cooling, and maturation.

Upon storage for one to three weeks, beer develops cloudiness which is only observable at about 0°C and which completely re-dissolves if the temperature of the beer rises. This type of haze is called "chill-haze" and is formed when positively charged polypeptides combine with polyphenols. Permanent haze forms by cyclic warming and cooling forming insoluble complexes, which will not dissolve at warmer temperatures.

To remove polypeptides or polyphenols and to improve its physical stability, a number of methods are employed for reducing chill haze. This procedure is often referred to as "chillproofing."

Chillproofing Agents

Over the years a number of chillproofing agents have been used to enhance beer haze stability (i.e., reduction of beer proteins and/or polyphenols). Each agent has its pros and cons, and many are used in combination, to suit the brewers own requirements and plant constraints.

Some of the processing aids used to reduce chill haze and in turn improve the beer's physical stability (removing positively charged proteins and polyphenols) are classified as follows:

- Polypeptide specific
  - proteolytic enzymes
  - tannic acid
  - hydrolysable tannins
  - silica gels
  - bentonite
- Polyphenol specific
  - Polyvinylpolypyrrolidone (PVPP)

Proteolytic Enzymes

Proteolytic enzymes are prepared from papain or pepsin and act on proteins by degrading them to smaller molecules. They do not increase the volume of solids, and thus do not create filtration problems and beer losses. Papain and pepsin are not specific to haze proteins, so mid-sized proteins (i.e., polypeptides), are often reduced, which affects beer foam. Some brewers add alginates as foam enhancers to counteract the reduction in foam (35).

Papain is sold mainly as a liquid preparation with application rates ranging from 10 to 50 mg/l (47). Normally it is added during the latter part of conditioning period, just before filtration and usually requires several days to chemically react with the proteins. Although there is some residual effect of papain in packaged beer, virtually all its chemical activity is destroyed during pasteurization. If the beer is not pasteurized, the enzymes reportedly can remove excessive quantities of proteins after packaging, thereby negatively affecting foam stability (13).

Tannic Acid

Tannic acid is traditionally used to remove haze material and is still employed in Germany. The advantage of this method is that tannic acid does not affect those proteinaceous compounds that contribute to foam. It also has a secondary effect as an oxidizing agent. It is usually purchased as a powder, mixed with de-aerated water, and added to the beer *en route* to the conditioning tanks. Tannic acid is usually applied at a rate of 20 to 30 mg/l and may have a negative effect if applied at rates above 50 mg/l (13). It is fast-acting, taking only about 24 hours to precipitate out the proteins (47).

Some of the drawbacks include the large volume of "tank bottoms," which requires careful movement of beer from above the sediment or the use of a centrifuge or filter to separate beer from the tank bottoms (5). For this reason, the practice in using tannic acid has fallen out of favor with many brewers. Moreover, there is concern by some brewers that tannic acid solutions are moderately dangerous to handle.

An alternative method of treatment involves proportioning tannic acid into beer as it is being transferred to the diatomaceous filter (48). The tannin-protein complexes are subsequently removed by filtration.

Hydrolysable Tannins

Hydrolysable tannins are also used to precipitate proteins. Beers treated with hydrolysable tannins show excellent colloidal stability as well as acceptable clarity values (47). Hydrolysable tannins are added to the beer during conditioning, or following conditioning as the beer is transferred to the diatomaceous filter (10). The efficiency of hydrolysable tannins is highly dependent upon the temperature and time. Delcour reports that contact times between 5 and 10 minutes at − 1°C are sufficient for highly purified hydrolysable tannins (10). Others report longer contact times between 15 and 20 minutes at a temperature below 0°C for precipitating proteins (47). The optimal dosage depends on the amount of nitrogen, polyphenol, and dissolved oxygen contents. It is recommended that the yeast count not exceed 5 million cells/ml for optimal efficiency of the fining action in beer (10).

Silica Gels

Silica gels are effective chillproofers in that they remove high-molecular-weight proteins responsible for haze formation without detriment to foam stability. The several types of silica gels are differentiated by surface area, porosity, and water content. Silica gels are broadly classified according to their water content as 'xerogels', 'semi-hydrogels' or 'hydrogels'. The different moisture contents of the gels means xerogels are dusty powders, while semi-hydrogels and hydrogels are generally free from dust. Xerogels contain more silica per given weight than hydrogels since they contain less water and thus can be used in lower dosages.

Silica gels are nontoxic and are completely removed from the beer by filtration. They reportedly have little effect on flavor or foam because of their selectivity (49). However, improper treatment may remove proteins important to foam, although this effect is of much less concern with silica gels than with proteolytic enzymes. Silica gels can be used along with polyvinylpolypyrrolidone (PVPP) for polyphenol precipitation that leads to a very effective chillproofing treatment. The chillproofing effect obtained is often better than that obtained with larger amounts of either material alone.

Silica gels can either be added to the beer during conditioning or dosed into the beer stream during filtration.

Added to Conditioning Tanks – The slurry of silica gel can be added to the beer during transfer from the fermenter to the conditioning tank. To ensure thorough contact, the slurry should be dosed proportionately into the beer. The adsorption of chill haze proteins by the silica gel is rapid, but not instantaneous. Contacts times will run anywhere from an hour up to 24 hours, particularly for beers that are difficult to stabilize. Longer contact times (even several days) will cause no harm. The gel particles settle rapidly forming a compact and stable sediment that can easily be removed with minimum loss of beer. Alternatively, silica gel can be combined with isinglass finings to improve the settling properties. Depending on the required stability and the type of beer dosage levels of silica gels are commonly 25–50 g/hl.

Incorporated into Body Feed – The most common way to add silica gel beer stabilizers is in-line, prior to filtration either as partial or total body feed replacement of diatomaceous earth (DE). This method will remove the adsorbent from the beer during the normal filtration process. Typically, silica gel is added as a partial feed along with the filter aid to achieve optimum filtration performance and initial clarity. Partial feeds have the effect of reducing the rate of growth of DE cake on the filter element, thereby extending the life of the filter run. To provide longer contact times, generally 10 minutes or longer, a special buffer tank is required. The buffer tank should be equipped with mixers for complete dispersion and thus prevent settling of the gel. The DE can eventually be replaced by silica gel. The beer should be slightly above 0°C so that haze-causing proteins can be better removed by the silica fining (51).

Application rates depend on the stability of the beer (e.g., original gravity, raw materials, etc.), the required shelf life, and the use of additional haze stabilizers (12). Generally the application rates are between 50 and 100 g/hl.

Bentonite

Bentonite, an insoluble alumino-silicate, can also be used as a protein adsorbent, but it is beset with many disadvantages. For one, it is nonspecific, adsorbing all groups of proteins and thereby impairing foam stability, though this effect is of much less concern with bentonite than with papain (35). Bentonite also requires lengthy cold storage for settling time. Moreover, the sediment produced in the tank is compact and difficult to remove, and it contains large volumes of beer, resulting in high beer losses. Freeman reports that sediment volumes are typically between 1 and 3%; however, amounts as high as 10% have been observed with high levels (300 g/hl) (15). Attempts to filter prematurely will result in "blinding" of the filter and cause short filtration runs because of incomplete settling of the clay particles. Bentonite is often used in conjunction with tannic acid or for that matter silica xerogel.

Polyvinylpolypyrrolidone

Polyvinylpolypyrrolidone (PVPP) is used to absorb polyphenolic materials. Part of PVPP's success as a beer stabilizer stems from the fact that it mimics the action of proteins by combining with polyphenols at a greater magnitude than proteins (41). PVPP is popular with many Continental European lager brewers because it is in compliance with Germany's Reinheitsgebot and because it doesn't remove foam-positive proteins like other fining agents. PVPP can either be added during maturation and removed during filtration or incorporated into the body feed of diatomaceous earth filtration (19). PVPP has not demonstrated any deleterious effects to beer attributes such as flavor or foam (48). In some countries, PVPP is legally banned.

While PVPP is an effective polyphenol adsorbent, PVPP is an expensive material in the context of treating beer on a commercial scale. Due to its expense, most brewers use regenerated PVPP that has been treated with basic and acidic solutions. However, recycling systems often involve large amounts of capital expenditures and are more suited for plants with an annual production ranging from 30,000 to 50,000 hl. Regenerated PVPP can be used many times for chillproofing beer (47).

> **Note | PVPP Used With Other Chillproofing Agents**
>
> PVPP is often used in combination with either silica gel either added from separate dosing vessels or by a combined PVPP/silica xerogel (3:7) preparation at a rate of 80 g/hl (5). They act instantaneously so can be injected into a beer stream on the way to the filter or even incorporated into the filter sheet material. This method is very effective.

PVPP can either be added to the beer during conditioning or dosed into the beer stream during filtration.

Added to Conditioning Tanks – The slurry of PVPP can be added to the beer during transfer from the fermenter to the conditioning tank. In this case the action is rapid, typically taking no longer than a few hours. Most of the PVPP-polyphenol complex is removed from the base of the tank after sedimentation followed by final removal of the suspended material during filtration.

PVPP is typically dosed in at rates between 15 and 50 g/hl. The dosage level for a typical lager with a 20% adjunct level is anywhere between 15 and 25 g/hl for single-use grades and 25 to 40 g/hl for regenerated material (41). Briggs *et al.* report dosage rates between 15 and 25 g/hl and rates between 30 and 50 g/hl for regenerated material (5). These dosage levels are usually sufficient to achieve a haze-free shelf life well in excess of 12 months (41).

At normal use rates, there is little danger that PVPP will affect beer attributes such as flavor or foam. However, some brewers have reported significant taste changes associated with PVPP treatment (49). At higher levels, there is some concern that excessive PVPP treatment tends to reduce color as well as hop bitterness since hop resins are phenolic materials (13).

Incorporated into Body Feed – The most common procedure for removing polyphenolic materials is to add the PVPP slurry to the beer stream along with the diatomaceous earth body feed. PVPP is also frequently incorporated into filter sheets for removing polyphenolic materials. Cartridge membrane filters are often employed downstream to trap any PVPP which bleeds through the plates.

Brewhouse Procedures Used for Stabilizing Beer

It should be noted that neither, stabilization or clarification or for that matter filtration are end-all measures for correcting mistakes made in the production process. Therefore, from the very beginning, it is necessary to perform stringent quality control paying attention from the selection of raw materials all the way through to conditioning, filtration, and packaging. Some of the techniques that a brewer can employ to achieve colloidal stability are as follows:

- Avoidance of high protein malt
- Use of well-modified malt to minimize problems with beta-glucans
- Use of low-polyphenol or polyphenol free malts
- Limit use of high protein adjuncts (e.g., barley and wheat-based adjuncts)
- Use adjuncts to dilute the amount of polyphenols coming from the malt
- Care in the use hop extracts and oils
- Maintenance of healthy yeast
- Correct mashing regimes
- Avoidance of over-sparging or over-raking
- Vigorous rolling boil to enhance precipitation of polyphenols and polypeptides
- Sufficient cold conditioning times (-1 or $-2°C$ for 3 days)
- Correct clarification and stabilization regimes
- Scrupulous limitation of oxygen ingress
- Correct filter aid selection

Conditioning Tanks

Conditioning tanks or commonly referred to as bright beer tanks are either horizontal or cylindroconical and are usually constructed of AISI 304 stainless steel, as shown in Figure 14.1. Horizontal tanks usually range in size of 100 to 500 hl while vertical cylindroconical tanks can be up to 6,500 hl in size (5). The greater ratio of surface area to beer depth for horizontal tanks provides a distinct advantage over vertical tanks in the conditioning of beer. Although horizontal tanks use more floor space per barrel of capacity, they offer quicker clarification, and the sediment has a shorter distance to fall, than in upright vessels. Horizontal tanks are usually located within the brewery in a temperature controlled room while cylindroconical tanks can either be located in the brewery or outdoors. Tanks are normally fitted with impellers for mixing.

Vessel Size

If the sole purpose of the tank is for conditioning, there is no restriction on the height of the tank. However, if the tank is also used for fermentation the height can't exceed 15 m (5). This has to do with the hydrostatic pressure on the yeast during fermentation. Tank diameters and the cone angle can vary but generally fall within the range of 3.5–4.75 m and 60 to 75°, respectively (5). The diameter to height ratio generally ranges from 1:1 to 1:5. If the tank is used solely for conditioning, a headspace volume of 5 to 10% is sufficient (5). Fermentation vessels usually need a headspace of 25%.

Figure 14.1
Bright Beer Tanks
Courtesy of JVNW Inc.

Temperature Control

Coolants

For temperature control of cylindroconical tanks the brewer can either use direct or indirect cooling:

1. Direct cooling with ammonia gas is accomplished with direct expansion and evaporation of the liquefied gas in the cooling jackets integrated in the tank wall. Ammonia is supplied by pumps from an ammonia gas/liquid separator. Direct expansion systems are more efficient and have no ozone depleting effect compared to fluorinated hydrocarbons. However, brewers typically prefer indirect cooling systems since direct expansion systems are more difficult to operate, more costly to implement, and because ammonia gas is a very corrosive, dangerous and pungent.
2. In indirect cooling a secondary coolant is used such as brine, ethanol, or propylene glycol in the cooling jackets integrated into the tank wall. Chilled water is unsuitable for use as a coolant since there is the requirement during conditioning to cool the beer to less than 0°C and to hold this temperature for several days.

Cooling Jackets

The coolant is circulated through the conditioning tanks (i.e., cooling jackets) and then returned to the plant. Cooling jackets are of several designs:

1. Pressed corrugated – widely used for secondary refrigerant systems
2. Dimple – very popular in beer and wine production
3. Inflated (quilted) – suited more to low volume primary refrigerant systems acting as an evaporator
4. Limpet – continuous coils welded to the vessel offer reduced heating/cooling times and a greater degree of accuracy in controlling product temperatures by allowing high pressures and high velocities of the heating/cooling medium

Most tanks will have two cooling jackets on the sides and an additional jacket on the cone.

References

1. Ambler, P.M. 1986. "Management of Cask Conditioned Beer (post fermentation)." *Brewers' Guardian* 115 (5).
2. Andrews, David. 1987. "Beer Off-Flavors –Their Cause, Effect & Prevention." *Brewers' Guardian* 116 (1).
3. Archibald, Hugh. 1993 "Conditioning, Filtration and Blending of Beer in Relation to Total Quality Management." *Brewers' Guardian* 122 (1).

4. Ballard, G. P. S. 1987. "Isinglass Types in Relation to Foam Stability." *Proceedings Institute of Brewing*, Australia and New Zealand Section, Adelaid.

5. Briggs, Dennis E., Boulton, Chris A., Brookes, Peter A., and Stevens, Roger. 2004. *Brewing: Science and Practice*. Cambridge, England: Woodhead Publishing Limited.

6. Broderick, Soyna. 2006. "Silica-Based Colloidal Stabilisation." *The Brewer and Distiller* 2 (6).

7. Campbell, I. 1987. "Microbiology of Brewing: Beer and Lager." *Essays in Agricultural and Food Microbiology*, edited by J.R. Norris and G.L. Pettipher. Chichester [West Sussex], England: John Wiley & Sons Ltd.

8. Coors, Jeffery H. 1977. "Cellar Operations." *The Practical Brewer*, edited by H. M. Broderick. Madison, Wisconsin: Master Brewers Association of the Americas.

9. Cox, R. H. 1973. "Conditioning and Chilling Processes." *Brewers' Guardian* 102 (3).

10. Delcour, Jan A. and Piet Van Loo. 1988 "Hydrolysable Tannins for Chillproofing Beers." *The New Brewer* 5 (2).

11. Fernyhough, R. and D. S. Ryder. 1990. "Customized Silicas – A Science for the Future." *MBAA Technical Quarterly* 27 (1).

12. Fernyhough, Robert, Ian McKeown, and Ian McMurrough. 1994. "Beer Stabilization with Silica Gel." *Brewers' Guardian* 123 (10).

13. Fix, George. 1989. *Principles of Brewing Science*. Boulder, Colorado: Brewers Publications.

14. Fix, George and Laurie Fix. 1997. *An Analysis of Brewing Techniques*. Boulder, Colorado: Brewers Publications.

15. Freeman, G. J. and M. T. McKechnie. 1995. "Filtration and Stabilization of Beers." *Fermented Beverage Production*, edited by A. G. H. Lea. New York, New York: Blackie Academic & Professional.

16. Grace, Michael. 1994. "Making Things Clearer." *The New Brewer* 11 (6).

17. Johnstone, J.T. and D.J. Whitmore. 1971. "Brewing Hygiene and Biological Stability of Beers." *Modern Brewing Technology*, edited by W.P.K. Findlay. Cleveland, Ohio: The Macmillian Press.

18. Harding, J.A. 1975. "Clarification & Stabilization." *Brewers' Guardian* 104 (9).

19. Hewson, J.R. 1985. "Preparation of Beer for Bottling and Canning." *The Brewer* 71 (851).

20. Hough, J.S. and A.L. Lovell. 1979. "Recent Developments in Silica Hydrogels for the Treatment and Processing of Beers." *MBAA Technical Quarterly* 162 (2).

21. Hough, J.S., et al. 1982. *Malting and Brewing Science*. Volume 2. London, England: Chapman-Hall, Ltd.

22. Knox, R.D. 1973. "Filtration and Stabilization." *Brewers' Guardian* 102 (4).

23. Knudsen, Finn B. 1977. "Fermentation Principles and Practice." *The Practical Brewer*, edited by H. M. Broderick. Madison, Wisconsin: Master Brewers Association of the Americas.

24. Kunze, Wolfgang. 1996. *Technology Brewing and Malting*, translated by Dr. Trevor Wainwright. Berlin, Germany: VLB Berlin.

25. Leather, R.V. 1994. "The Theory and Practice of Beer Clarification – Part 1 – Theory." *The Brewer* 80 (960).

26. Leiper, Kenneth A., Graham G. Stewart1, Ian P. McKeown, Tony Nock and Matthew J. 2005. "Optimising Beer Stabilisation by the Selective Removal of Tannoids and Sensitive Proteins." *Journal of the Institute of Brewing* 111 (2).

27. Lewis, Michael J. and Tom W. Young. 1995. *Brewing*. London, England: Chapman & Hall.

28. Linley, P.A. 1978. "Lager Fermentation and Maturation." *Brewers' Guardian* 107 (1).

29. MacDonald, J., et al. 1984. "Current Approaches to Brewery Fermentations." *Progress in Industrial Microbiology*, edited by M.E. Bushell. Guildford, United Kingdom: Elsevier.

30. Mackie, A. 1985. "The Production of Cask Conditioned Ales in Cylindroconical Fermenters." *The Brewer* 71 (849).

31. Martin, E. C. B. 1972. "Lager Fermentation, Storage, Filtration." *Brewers' Guardian* 101 (4).

32. Mathews, Tony. 1990. "Finings and Beer Clarification." *Brewers' Guardian* 119 (3).

33. Meier, Josef. 1987. "Keeping it Stable." *Brewers' Guardian* 116 (6).

34. Moll, Manfred, ed. 1995. *Beers and Coolers*. New York, New York: Marcel Dekker, Inc.

35. Munroe, James H. 1995. "Aging and Finishing." *Handbook of Brewing*, edited by William A. Hardwick. New York, New York: Marcel Dekker, Inc.

36. Munroe, James H. 1995. "Fermentation." *Handbook of Brewing*, edited by William A. Hardwick. New York, New York: Marcel Dekker, Inc.

37. Munroe, James H. 2006. "Aging and Finishing." *Handbook of Brewing*, edited by Fergus G. Priest and Graham G. Stewart. Boca Raton, Florida: CRC Press, Taylor & Francis Group.

38. Noonan, Gregory J. 1996. *New Brewing Lager Beer*. Boulder, Colorado: Brewers Publications.

39. Mathews, Tony. 1990. "Finings and Beer Clarification." *Brewers' Guardian* 119 (3).

40. Meier, J. 1993. "Modern Filtration." *Brauwelt International* 5.

41. O'Reilly, Jamie P. 1994. "The Use and Function of PVPP in Beer Stabilization." *Brewers' Guardian* 123 (9).

42. O'Rourke, Timothy. 1994. "The Requirements of Beer Stabilization." *Brewers' Guardian* 123 (8).

43. O'Rourke, Timothy. 195. "What Ever Happened to Maturation?" *The New Brewer* 12 (2).

44. Patino, Hugo. 1999. "Overview of Cellar Operations." *The Practical Brewer*, edited by John T. McCabe. Wauwatosa Wisconsin: Master Brewers Association of the Americas.

45. Pryor, Jim. 1999. Personal Communication. Vafac Inc.

46. Russell, Inge. 1995. "Yeast." *Handbook of Brewing*, edited by William A. Hardwick. New York, New York: Marcel Dekker, Inc.

47. Ryder, David S. and Joseph Power. 1995. "Miscellaneous Ingredients in Aid of the Process." *Handbook of Brewing*, edited by William A. Hardwick. New York, New York: Marcel Dekker, Inc.

48. Ryder, David S. and Joseph Power. 2006. "Miscellaneous Ingredients in Aid of the Process." *Handbook of Brewing*, edited by Fergus G. Priest and Graham G. Stewart. Boca Raton, Florida: CRC Press, Taylor & Francis Group.

49. Sobus, Michael T. 1988. "Beer Chillproofing Using Silica Hydrogel." *The New Brewer* 5 (2).

50. Thompson, G. J. 1994. "The Theory and Practice of Beer Clarification – Part 2 – Practice." *The Brewer* 80 (961).

51. Thompson, Donald, Karl Ockert, and Vince Cottone. 1999. "Craft Brewing." *The Practical Brewer*, edited by John T. McCabe. Wauwatosa, Wisconsin: Master Brewers Association of the Americas.

Chapter Fifteen

Beer Filtration

Although conditioning—maturation, clarification, and stabilization—plays an important role in reducing yeast and haze loading materials, a final filtration is needed in order to achieve colloidal and microbiological stability. The beer must be rendered stable so that visible changes do not occur during its shelf life.

Filtration Methods

Depth Filtration

Depth filtration removes particles from beer within the depth structure of the filter medium itself. The particles are either mechanically trapped in the pores or absorbed on the surface of the internal pores of the filtration medium. The filter media can be pre-made sheet filters or fine powder made of, for example, diatomaceous earth (DE), also known as kieselguhr, which is introduced into the beer and re-circulated past screens to form a filtration bed. Depth filtration is required when a significant quantity of suspended material needs to be removed from the beer. However, depth filtration is not as effective as surface filtration in removing suspended solids and microorganisms. Powder and sheet filters are examples of surface filters.

Surface Filtration

Surface filtration can be either absolute or nominal with a minimal depth capacity. Surface filtration consists of a thin membrane or a thin membrane covered with polypropylene or polyethersulfone in which particles are trapped in pores in the filter medium. Prior filtration with a depth filter is usually required to prevent clogging the surface of a cartridge membrane filter.

> **Note | Absolute versus Nominal Pore Ratings**
>
> An absolute pore size rating specifies the pore size at which an organism of a particular size will be retained with 100% efficiency under strictly defined test conditions. Absolute filtration is useful for critical applications such as sterilizing and final filtration when you require precise accuracy of filtration to limit risk. A nominal pore size rating describes the ability of the filter media to retain the majority of particulate at (60–98%) the rated pore size. Process conditions such as operating pressure, concentration of contaminant have a significant effect on the retention efficiency of the filters.

Single- or Double-Pass Filtration

The beer can undergo a single- or a double-pass filtration process. The double-pass filtration consists of two steps: a primary (rough) filtration, and a secondary (polish) filtration. Primary filtration removes the bulk of yeast and suspended material and the secondary filtration produces a brilliantly clear beer. Filtered beer is subsequently stored in a finishing tank. Double-pass filtration can be achieved with two sets of sheet filters of decreasing pore size or more commonly with a powder filter followed by a sheet filter. If sterile filtration is required the beer is filtered through a cartridge membrane filter too.

Powder Filters

Many brewers employ powder filters for single pass filtration and are able to produce beers that are suitable for packaging. In a powder filter, filtration is achieved with the use of filter aids that form a filter bed on stainless wire mesh inside a pressurized vessel; or, on cellulose sheets in a plate and frame filter; or, in the form of small holes in a candle filter. While the screens and/or cellulose do not filter at all, but act as septum for the filter aids, it is the applied "cake" of filter aid which does the filtration. The mechanics are very similar to those of sheet filters.

Powder filters are used because they are cost effective and successful in clarifying beer and also because they can be emptied, cleaned and put back in operation with little down-time. Long filtration cycles at high flow rates are possible with powder filters.

Types of Powder Filters

There are several types of powder filters, namely: (a) plate and frame filters in which the support is a cloth, (b) horizontal leaf or vertical leaf filters in which the support screen is a wire mesh, and (c) candle filters which support the filter aid on a long thin perforated rod.

Plate and Frame Filters

The plate and frame filter, as shown in Figure 15.1, has been the workhorse in breweries around the world for many decades. It is robust and reliable, consistently filtering beer to the specified standards. Plate and frame filters consist of a series of chambers enclosed within a metal frame. Between adjacent frames is a double-sided porous filter plate covered with a cellulose filter sheet folded at the top of the plate so that both sides of the plate are covered. The filter sheet acts as a trap for the filter aid, which otherwise might bleed through, thereby assuring, excellent clarity. Filter sheets are generally made with cellulose fiber, diatomaceous earth, perlite, and a resin for bonding to give dry and wet strength. The average pore size of filter sheets is between 4 and 20 microns (μm); therefore, plate and frame filters are readily pre-coated and less susceptible to malfunction. Each plate alternates with a frame with the entire system held together by a screw or hydraulic clamp mechanism.

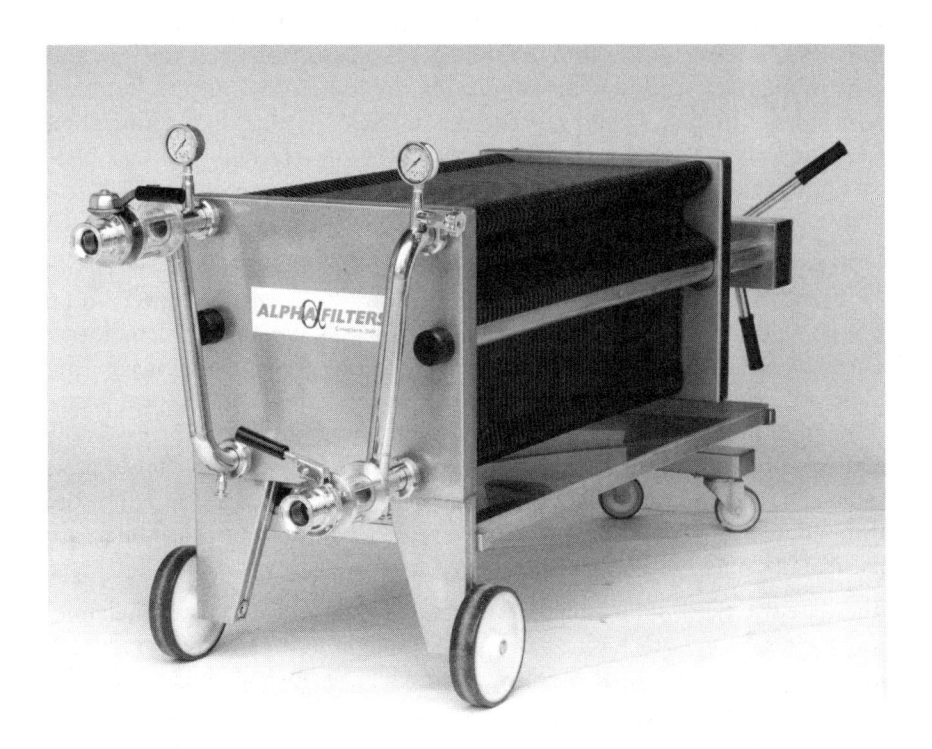

Figure 15.1
Plate and Frame Filter
Courtesy of Criveller Company

This type of filter is very similar in appearance to the sheet filter, except it has sludge frames.

The beer and filter aid enters the filter through the top and bottom inlet manifold, then flows into the frames and through the filter cake collected on the sheets or cloths. Thus, every plate and frame couple is a filter unit. The filtered beer leaves the filter through the top and bottom outlet manifold of the plate. At the end of filtration, de-aerated water is pumped into the filter to displace the residual beer, and the filter is opened to allow the filter cake to fall out onto a discharge screw. The DE can be dislodged from the sheets by high-pressure spraying. The sheets are washable and have a long life. The flow rate per unit area of 3 to 3.5 hl/m^2/hr and is the lowest of the three DE filter types (8).

Leaf Filters

Leaf filters have a series of stainless steel leaves that are arranged either vertically or horizontally inside a filter body. The filter leaf consists of a stainless steel mesh septum attached to a stainless steel support plate. Screen septums are made with openings ranging from 45 to 70 microns. Unfiltered beer enters the pressure vessel, passes through the filter cake established on the leaf, and exits through the hollow shaft connecting the leaves. In a vertical leaf filter both sides of the support are coated with filter aid, whereas in the horizontal leaf filter only the upper surface is used for filtration. The buildup of the filter cake adheres to a wire mesh screen instead of a cellulose sheet that is used with plate and frame filters.

Horizontal Pressure Leaf – Horizontal leaf filters, as shown in Figure 15.2, are widely used in Europe and are considered by many brewers to be technologically better than vertical leaf filters. This type of filter employs a number of stainless steel discs, or elements, rotating about a vertical axis. The filter aid is deposited on the upper surface of the leaves, and the beer passes through to the center column and then to a bright tank. The bed is very stable and will not become dislodged with low pressure fluctuations. However, as the bed lies on the top surface, only half the filtration area is utilized, so the throughput rate per unit of tank volume is reduced. Horizontal leaf filters are particularly suited to PVPP stabilization.

There are two types of horizontal leaf filters, manual and automatic discharge. Manual discharge, which is now common throughout the craft brew industry, implies that at the end of the filtration cycle, the operator must manually remove the bell of the filter, tilt the screen assembly, and proceed with the cake removal. With automatic discharge leaf filters, the filter cake is removed by water or air (back flush) and then spun off by rotating the leaves at a high speed. With automatic discharge systems, the majority of the cake can be discharged to the bottom of the tank semi-dry. The flow rate per unit of area is between 5 and 8 hl/m^2/hr (15).

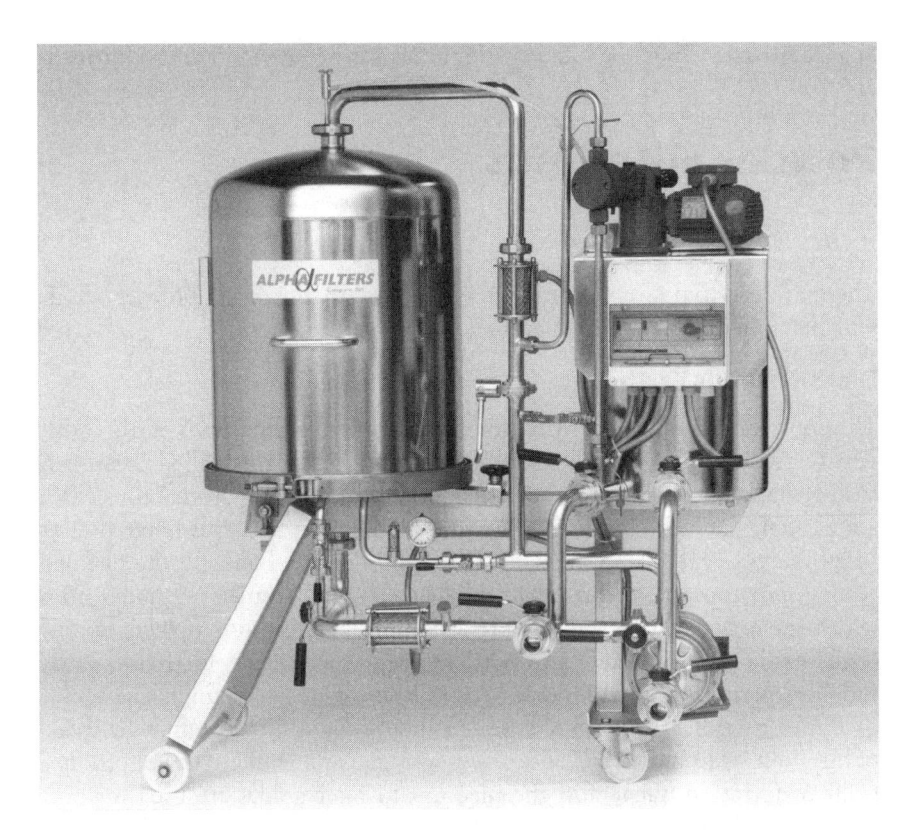

Figure 15.2
Horizontal Leaf Filter
Courtesy of Criveller Company

Vertical Pressure Leaf – Vertical leaf filters are widely established in North America, Australia, and many other countries. Unlike horizontal leaf filters, both sides of the plate can be utilized for powder support, thereby reducing the filter's overall size. Discharge can be achieved by mechanical vibration or by removing the outer shell and hosing off the plates.

Candle Filters

Candle filters utilize a series of candles, containing a number of rings that are hung from a rigid horizontal plate. The candles can be of porous ceramic, but they are usually perforated or fluted stainless tubes covered or surrounded by a stainless steel support of various types. The candles and support plate are housed within a vertical cylindrical vessel with a conical base. The beer and filter aid is fed into the base of the filter case, forming a cake on the outside of the candles. Beer flows into the candles and is collected through the dished end cavity. Due to the circular design, the increase in bed thickness during operation is less than with other filters, and the pressure drop increase occurs at a slower rate. The flow rate per unit of area is between 3.5 and 6 hl/m²/hr (15).

Powder Filter Aids

Types of Filter Aids

There are two types of filter aids used in conjunction with powder filters, which are diatomaceous earth and perlite.

Diatomaceous Earth

The most popular powder used for filtration is diatomaceous earth (DE), which consists of skeletal remains of single-celled plants called "diatoms" that contain silicon dioxide. The three classifications of diatomaceous earth are natural, calcined, and flux-calcined. The natural product is referred to as "diatomaceous earth" and the name "kieselguhr" is used only for the calcined grades of diatomaceous earth. Kieselguhr consists of a variety of grades from which the brewer chooses in order to accomplish clarification objectives. The term "grade" generally refers to particle size, which affects filter flow rates, filter bed permeability, and the degree of filtration (coarse to fine). The choice of a fine or smaller particle grade results in a brilliant filtration but the speed of filtration is less. Coarser grades give a rapid flow rate but might not produce brilliant filtration. Generally, the medium to finer grades are used for beer filtration. When choosing grades of kieselguhr primarily for microbiological considerations, it is necessary to rely on the tighter grades.

Perlite

Perlite is alumino-silicate rock that has been expanded by heat treatment and later crushed and graded. Perlite represents a low health risk but, because of its low density, it disperses easily in the air creating dust problems in the brewery. Perlite is less efficient than DE in removing submicron particles in filtration and is usually used for pre-coat applications.

Safety and Disposal

Diatomaceous earth is very dangerous when inhaled; it can lead to irritation of the lungs and even long-term lung damage. It is considered carcinogenic to human beings and can also cause silicosis, a progressive and sometimes fatal lung disease, following long exposure to silica dust. When handling diatomaceous earth, the handler must wear a face mask or respirator that has been fit-tested and approved as listed on the Material Safety Data Sheet (MSDS). The mask should seal tightly on the skin to prevent dust from entering. A painter's mask is insufficient, as it does not form a tight seal with the face. Use of uncalcined diatomaceous earth represents a moderate risk but it often contains undesirable traces of iron and other metals. Perlite is a lower risk, with only 60% of the density of diatomaceous earth, tends to disperse in the air very readily and spread as dust.

Powder Aid Filtration Process

Slurrying

Smaller breweries will usually empty the filter aid from the bags directly into the slurry tank adjacent to the filters. Larger breweries rely on dry-conveying the filter aid to the tank using screw or pneumatic systems. The wet slurry is then transferred from a centralized make-up system to one or more filter stations. The slurry tank should have adequate ventilation. It should also be equipped with a slow-moving gate stirrer to prevent vortex formation, which could entrain air and adversely affect filtration. The slurry medium can either be water or beer, although water is preferred since beer may contain oxygen. Also, any foam in the beer at the mixing stage could inhibit slurrying. The water should be de-aerated and preferably sterilized by heat or filtration for additional security. A carbon dioxide purge, about 30 minutes prior to use, is required to de-aerate the filter aid slurry since it contains considerable amounts of air (15).

The slurry takeoff point should be high enough from the bottom of the tank to safeguard against the discharge of lumps, pieces of paper bag, and other foreign objects into the dosing line. Care should be taken to ensure that the concentration of filter aid in the slurry is correct, and sampling of the slurry for a solids measurement is done on a regular basis. It is also recommended that the slurry be discharged and remade if not used up within 24 hours of makeup. However, the length of time the slurry is held depends on the level of particle fracture, which is a direct result of the speed of the agitator.

Pre-coating

In order to form a stable filter bed structure, it is first necessary to pre-coat the cloth or wire mesh screen with a number of DE layers of sequentially finer grades. The pre-coat is built up by circulating the beer through the powder filter and dosing it with DE until the emerging liquid is clear. This signals that the support mesh is fully bridged over or coated with DE. Some brewers will run without a coarse first pre-coat if the filter is a plate and frame type and fitted with filter sheets.

First Pre-coat – The first DE pre-coat, known as the bridging pre-coat, often has a higher percentage of coarse particles (perhaps of perlite or flux-calcined diatomaceous earth) than the second pre-coat. The first pre-coat is necessary to ensure efficient filtration of the early part of the beer run and, even more important, to guarantee the integrity of the filter throughout the run. The first pre-coat seldom performs any real filtration function, but it is essential as an interface between the filter septum and body feed, serving as a mechanical or retaining layer for subsequent body feed. About 600–800 g/m² of powder are used for this layer, and if DE it will represent about 75% of the total pre-coat (7). After the first pre-coat has been established, the operation is repeated with a finer powder to form a second pre-coat, the filtration pre-coat.

Second Pre-coat – The second pre-coat layer can be made with a single grade or with a mixture of kieselguhr grades. It is very important to have a uniform distribution of the pre-coats over the entire filter surface. Thinner regions or edges in the pre-coats cause unevenness in flow and possibly leakage. Some brewers use higher flow rates to ensure even deposition of the pre-coat; the flow rate is normally 50% higher than the rating of the filter (19). After the second pre-coat layer has been applied, the filtration system should be allowed to re-circulate for several minutes at a high rate of flow to achieve maximum compaction of the pre-coat. The total pre-coat layer will be 1.3–3 mm thick (7).

Third Pre-coat – Some brewers even apply a third pre-coat, one that is finer than the body feed grade (15). In this case the filter aid becomes a full, genuine, two-stage filtration that achieves superior clarity.

Body Feed

Beer filtration is started when the pre-coats are established and the re-circulating liquid is clear. To prevent clogging of small pores of the filter and to achieve extended filter runs; filter aid is continually metered into the unfiltered beer as "body feed."

Care must be taken to avoid large pressure shocks, which can severely crack the filter cake or drop the cake off the septum. Such an event would require removal of the cake and re-establishing the pre-coat. For this reason, it is good practice to have a buffer tank installed in the line upstream of the main delivery pump to the filter. A trap filter may be placed in-line downstream of the powder filter to safeguard the product against leakage of filter aid or the sudden breakthrough of part of the filter cake.

There is a continual increase in the pressure differential from inflow to outflow of the filter as the thickness of the bed increases. Filtration is stopped when the filter cake begins to reach a thickness that bridges the spaces between the septa in the filter, or when the differential pressure rises beyond a designated point. The filter cake is then removed and the filter cleaned and sterilized for the next run.

Body Feed Quantity – The quantity of DE body feed is often determined by trial and error, with amounts usually ranging from 40 to 200 g/hl (15).

Colloidal Stabilization

To achieve colloidal stability it is necessary to remove either protein, polyphenol, or both from the beer. During filtration, the most commonly used stabilizers for removing proteins is amorphous silica gel (e.g., Lucite®). Polyvinylpolypyrrolidone or PVPP (e.g., Polycar AT®) is typically used for removing polyphenols (15). The stabilization procedures typically used are:

- Silica gels with main beer filtration
- PVPP with main beer filtration
- PVPP-Impregnated sterilizing sheets

Silica Gels with Main Beer Filtration

A wide range of amorphous silicas are available. These provide stability by selective adsorption of proteins or protein/polyphenol complexes. The silica gel is mixed with cold, de-aerated water, and with or without body feed filter aid in the slurry tank. The slurry is added to the beer stream at a rate of 40 to 100 g/hl, depending on the silica grade (15).

PVPP with Main Beer Filtration

PVPP can be added in combination with kieselguhr in the slurry tank; but to achieve better control, it may be preferable to add the PVPP separately prior to kieselguhr filtration. Generally, a contact time of 5 minutes is sufficient for the reaction to proceed to completion. The rates for PVPP addition are generally between 15 and 50 g/hl, with rates from 25 to 35 g/hl being typical (15).

Regenerated PVPP Systems

In re-generable PVPP systems, the PVPP is dosed continuously into the bright beer stream, forming a filter bed on the screens of a dedicated horizontal leaf filter. The reusable grades of PVPP can be regenerated by re-dissolving the adsorbed polyphenols in a solution of sodium hydroxide. This system can only be justified economically if used in large-scale operations.

PVPP-Impregnated Sterilizing Sheets

PVPP can be incorporated into sterilizing filter sheets. PVPP finds its widest application in breweries where products with a long shelf-life are not required or in small breweries where the cost of a horizontal leaf filter/regeneration plant would be prohibitive.

Sheet (Pad) Filters

Sheet filters (see Figure 15.3) are classified as depth filters made of preformed sheets ranging in thickness from 3 to 4 mm. The sheets are of a simple construction, being made of ordinary cellulose fibers. Sometimes they also contain extra substances to increase their filtration efficiency, such as kieselguhr and/or perlite. Some are impregnated with PVPP or activated carbon too. As a general rule, the more filter aids the better the retention. An inert resin is added to provide wet strength; and—depending on the type of resin—also to create a positive charge or "zeta potential." This positive charge enhances the retention of small negatively charged particles/microorganisms. Sheet filters are available in a large variety of grades with flow rates ranging from 0.8 to 2.0 hl/m^2/hr, depending on the retentivity of the sheet.

In process filtration, sheet filters are inserted into filter presses between the plates in a sandwich formation. Take note that the beer does not pass from one end of the filter to the other, but it is distributed by the plates in such a way that the beer passes through one sheet only. The reason for having more than one sheet is merely to increase the rate at which the beer can be filtered: the more sheets the greater the flow rate, but the quality of the filtration

Figure 15.3
Sheet Filter
Courtesy of Schenk Filter Systems, Inc.

remains the same. Worth noting a less labor-intensive and more time saving way of using depth filter sheets are the so called lenticular-, stack- or modular filters: 3 names for 1 product—a pre assembled filter media pack inserted into special filter housings.

Sheet filters can be effective for final filtration though they are not suitable for high-solids filtration, as they do not have the depth capacity that is supplied by powder filters. Some sheet filters are fine enough to approach cartridge membrane filtration for microbiological stability. A 0.45 μm filtration is considered sterile for craft brewing (45).

As with sheet filters, or for that matter depth filters, they reach their end of service life when their inner matrix is loaded with "dirt" particles. Indicators are the rise of the differential pressure between inlet and outlet side; and the decreased flow rate.

Filtration Mechanisms

Three different filtration mechanisms are responsible for separation effects. The first separation mechanism is surface filtration, which filters out particles that are too large to enter the filter sheet. The second filtration effect is mechanical depth filtration, which deals with particles that are small enough to enter the sheet filter but too large to exit. The third filtration mechanism is adsorptive retention filtering. This type of filtration comes into play at the tighter levels of sterile filtration. The three mechanisms act interchangeably and do not appear on their own in practical filtration applications.

Operation

In craft breweries, both plate and frame and sheet filters are often mounted on the same chassis and separated by a changeover plate. This allows brewers to accomplish two filtration stages—rough and polish filtration—in a single step. Alternatively, some craft brewers use a two-stage filter sheet process, with rough, clarifying sheets in the first stage and tight, sterile sheets in the second stage.

It is important to maintain an even flow rate throughout the filtration cycle when using sheet filters, allowing the flow rate to gradually climb. If the recommended flow rate is exceeded, larger particles can be pushed deeper into the sheet, reducing its capacity to trap or adsorb smaller particles. Furthermore, pressure surges should be avoided during filtration since sudden pressure changes can partially destroy the bridging effect of the sludge deposits.

In certain applications sheet filters can be back-flushed and cleaned for a longer life using a hot water rinse (60–70°C) followed by sterilization. The hot water rinse will loosen and flush out retained beer matter in the filter sheet. Sterilization involves flushing heated water (at 82.5°C) in the direction of filtration for 30 minutes. Not sterilizing the pads can result in off-flavors, e.g., a taste similar to that of wet newspapers. Generally, chemical disinfection by oxidizing chemicals is not recommended.

Advantages and Disadvantages

The advantage of sheet filtration is that it requires no secondary filter equipment or materials. The disadvantages are the high labor cost of handling the sheets, the lack of automation, and the cost of sheet filters, which is about double the comparable cost for other filter aids. These pads may be regenerated after a filtration cycle but they cannot be regenerated indefinitely. Although filter sheets are regenerated by backwashing, they must be replaced every few months.

Cartridge Membrane Filters

Cartridge membrane filters or trap filters are classified as surface filters where the membrane is a uniform continuous size. The cartridge membrane filter is often pleated and encased in a plastic cartridge that is usually inserted into a single housing, in order to handle a meaningful flow. Cartridge membrane filter and housing are shown in Figure 15.4. They are available in many different pore sizes.

It is common to have these filters downstream as a final step after using powder filters and sheet filters. Cartridge membrane filters are often used as trap filters downstream of the main DE filters, to catch any bleed-though of the DE particles or yeast. The large commercial brewers usually install a duplex trap system, in which the second filter is switched automatically into service upon registration of a high-pressure drop across the first filter. The first filter is then regenerated with hot water while the second filter is in operation.

Filtration Mechanisms

Membrane filters, given the mechanisms of retention, work as either depth filters or surface (screen) filters, and in some cases has aspects of both. However, membrane filters are predominately used in the final filtration step because they assure the most reliable retention of particles and specific beer-spoiling microorganisms. Retention mainly takes place on the membrane surface by the sieving affect. Particles and microorganisms larger than the actual pore size of the membrane are effectively removed. The selection of the right pore size of the membrane is based on the potential beer-spoiling microorganism present and the size of the particles which need to be retained.

Membrane filters, are made from a film of synthetic polymer. These filters have an "absolute" micron rating. This means that particles larger than the

Figure 15.4
Cartridge Membrane Filter and Housing
Courtesy of Pall Corporation

stated pore size will be removed based on a percentage rating. The holding capacity of membrane filters, unlike depth filters, is limited, and it is not economical to use membrane filters as depth filters.

Operation

Multiple-stage cartridge filtration is generally practiced to limit particle loading on the final stage. The first stage should be designed for high solids capacity with a rating of between 10 and 20 µm used as a "trap filter" for kieselguhr or PVPP leakage. A membrane filter with this micron rating provides protection to the downstream filters. A cartridge filter with a micron rating of 6.0 will clarify beer to some extent, removing most chill haze particles and reducing the level of yeast. A micron rating of 3.0 is slightly better at chill proofing and is effective in removing mature and dead yeast, except for young yeast cells. A micron rating of 1.0 will remove all yeast cells and some bacteria, except Lactobacilli and Pediococci. Finally, a filter with a micron rating of 0.50 to 1.0 will impart good final beer clarity and remove most insoluble filter aids, such as diatomaceous earth (43). In addition, yeast carryover can be reduced, which will help maintain the character of the beer during storage. As mentioned, the pore size for sterile filtration is typically between 0.45 and 0.65 μm and usually results in a significant reduction of microbiological organisms.

Sterile Filtration

Cartridge membrane filters are effective for sterile filtration, or microbiological stabilization of beer after it has undergone depth filtration. For brewery use a pore size of 0.45 μm is necessary to retain most potential spoilage organisms, yeast, and inert particulate matter on the surface. Sterile filtration is usually achieved by filtering the beer through 2 or 3 grades of cartridge membrane filters. It is recommended that the first housing contains filters with a porosity of 10 μm followed by housings with 1 μm and finally with 0.45 μm filters. For small systems, this system can be reduced to a 2 stage system with 2 and 0.45 μm filters. Wood recommends a two-stage membrane process consisting of a 0.65 μm pre-filter and a 0.45 μm absolute filter for removing beer-spoiling microorganisms (47). Cartridge membrane filtration is the only sterile filtration method that will provide absolute sterility that can effectively eliminate 99 to 100% of undesirable microbial content. Pasteurization is unnecessary for sterile-filtered beer, so membrane filtration avoids the substantial energy cost and the potential flavor defects from heating the beer.

Advantages and Disadvantages

Cartridge membrane filters have the advantage of providing good pressure stability against momentary leakage of microbes from pressure surges, while depth filters (e.g., diatomaceous filtration) are significantly more susceptible to surges. Another advantage of most cartridge membrane filters is their ease of operation and the fact that when the filter is plugged up, it can easily be back-flushed by reversing the flow. When backwashing, the flow rate should be ramped slowly to avoid damaging the cartridge. While backwashing is very effective in extending the filter life, it will not restore the cartridge to its original performance. Cleaning can also accomplished by soaking the cartridge in hydrogen peroxide or enzymes (45).

However, in spite of their advantages, cartridge membrane filters are prone to blockage. Consequently, it is essential that the beer first undergo a satisfactory primary filtration to remove as much inert particulate material and as much microbial biomass as possible before reaching the cartridge membrane filter.

Deep-Bed Filtration

Deep-bed filters, as shown in Figure 15.5, consist of a stainless steel vertical frame with filter pads stacked one upon the other into the frame. Each conical filter pad has several layers with different degrees of permeability—very coarse on the inlet side and proceeding in stages to finer grades of filtration toward the outlet side. The step-by-step separation process is very efficient in providing good separation of haze particles, yeasts, and bacteria.

Feische reports that the deep-bed filter can be used as a pre-filter to protect the membrane filter, as a fine filter in front of a thermal treatment to reduce the PU's (pasteurizing units), or as a sheet filter replacement behind the DE filter (16). However, the deep bed filter is usually employed as a final sterile filter directly in front of the bottling or keg line to minimize recontamination out of the bright beer tanks. In this application, most breweries do not have a sheet filter operation, but go from DE filter into the bright beer tank and then through the deep bed filter into the bottle filler. Deep-bed filters can be safely used in front of the filling line because of their ability to withstand pressure shocks.

Figure 15.5
Deep Bed Filter
Courtesy of The Handtmann Group

Sterile Filtration

The term "sterile filtration" refers to the reduction of yeast and bacteria to levels that do not result in spoilage of the beer over its planned shelf life. Viruses do not survive the brewing process.

Sterile filtration has been used as an alternative to pasteurization for many years. It has the advantage over pasteurization in that the risk of flavor damage by heat is eliminated. However, many brewers report that there is little difference in flavor stability between sterile-filtered and properly pasteurized beer (37).

The brewer should set a specification for the maximum allowable concentration of yeast and bacteria in sterile-filtered beer since not all microorganisms are removed by sterile filtration. Opinions differ about what is the critical level for contaminants, and for beer it has been set at between 4 and 10 yeast cells per 12 ounce bottle. In the case of lactobacilli, it has been reported that between 1 and 3 organisms per 12 ounce bottle is acceptable. In other words, there is no universally agreed-upon minimum safety level for *Lactobacillus* sp., and use of the term "sterile filtration" can be misleading.

Sterile filtration, if overdone, has the net effect of stripping out much of the body, flavor, and even color from the beer. A filter tight enough to trap all bacteria also traps large molecules such as proteins, dextrins, and melanoidins. These substances are responsible for mouthfeel, head retention, color, and beer flavor. Thus, sterile filtration is fundamentally more compatible with light-bodied pale beers than with darker, fuller-bodied, craft-brewed beers. Some brewers report that removal of microorganisms via a sheet filter of 1 µm or less will significantly change the beer constituents by adsorption (20).

There are many plant and operating options that will achieve sterile-filtered products. They all rely on a cascade type approach, where sequentially finer stages of filtration are used ahead of packaging. Sterile filtration can be achieved in several ways: 1) powder filtration followed by sheet filtration, using depth filtration mechanisms; 2) powder filtration followed by sheet filtration and then cartridge systems that use membranes with an absolute cutoff for particles below the size of bacteria; and 3) powder filtration followed by cartridge systems that use depth filtration mechanisms, mainly as a means of protecting subsequent membrane cartridge filters. Typically, the bulk of the suspended particles are removed with a powder filter, and this is followed by a sheet filter that removes the residual material so as to reach haze specification prior to the cartridge membrane filter. The type of filter setup for sterile filtration will depend on the brewer's needs and on features such as capacity, ease of maintenance, cleaning, and sterilization.

References

1. Abec, Peter M. 1999. Personal Communication. ABEC Filtration Systems, Inc.

2. Anders, Thomas G. 1996. "Filtration: Types + Techniques." *The New Brewer* 13 (6).

3. Anders, Thomas G. 1999. Personal Communication. Scott Laboratories.

4. Ault, R.G. and R. Newton. 1971. "Brewing Hygiene and Biological Stability of Beers." *Modern Brewing Technology*, edited by W.P.K. Findlay. Cleveland, Ohio: The Macmillian Press.

5. Back, W., M. Leibhard, and I. Bohak. 1992. "Flash Pasteurization – Membrane Filtration." *Brauwelt International* 1.

6. Bennett, Keith. 1991. "The Importance of Filtration in Producing Quality Beers." *Brewers' Guardian* 120 (1).

7. Briggs, Dennis E., Boulton, Chris A., Brookes, Peter A., and Stevens, Roger. 2004. *Brewing: Science and Practice*. Cambridge, England: Woodhead Publishing Limited.

8. Candy, E. P. 1991. "Beer Filtration." *The Brewer* 77 (923).

9. Comton, John. 1977. "Beer Quality and Taste Methodology." *The Practical Brewer*, edited by H. M. Broderick. Madison, Wisconsin: Master Brewers Association of the Americas.

10. Coors, Jeffery H. 1977. "Cellar Operations." *The Practical Brewer*, edited by H. M. Broderick. Madison, Wisconsin: Master Brewers Association of the Americas.

11. Criveller, Bruno. 1999. Personal Communication. Criveller Company.

12. Daquino, Anthony. 1983. "Modern Applications for Centrifuges in Breweries." *MBAA Technical Quarterly* 20 (2).

13. DeYoung, Robert E. 1977. "Packaging – Bottling Operations." *The Practical Brewer*, edited by H. M. Broderick. Madison, Wisconsin: Master Brewers Association of the Americas.

14. European Brewery Convention (EBC). 1995. *Beer Pasteurisation, Manual of Good Practice*. Nürenberg, Germany: Getränke-Fachverlag Hans Carl.

15. European Brewery Convention (EBC). 1999. *Beer Filtration, Stabilization and Sterilization, Manual of Good Practice*. Nürenberg, Germany: Getränke-Fachverlag Hans Carl.

16. Feische, Michael. 1999. *Demands on a Modern Post-Filtration System*. Munich, Germany: Handtmann Filtration.

17. Fix, George. *Principles of Brewing Science*. 1989. Boulder, Colorado: Brewers Publications.

18. Fix, George and Laurie Fix. 1997. *An Analysis of Brewing Techniques*. Boulder, Colorado: Brewers Publications.

19. Freeman, G.J. and M.T. McKechine. 1995. "Filtration and Stabilization of Beers." *Fermented Beverage Production*, edited by A.G.H. Lea and J.R. Piggott. London, United, Kingdom: Blackie Academic & Professional.

20. Gaub, R. 1993. "Criteria for Fine and Sterile Filtration of Beer." *Brauwelt International* 5.

21. Haase, Nicolas. 1999. Personal Communication. Drinktec USA, Ltd./ A. Handtmann Filtration GmbH.

22. Hough, J.S., et al. 1982. *Malting and Brewing Science*. Volume 2. London, England: Chapman-Hall, Ltd.

23. Jany, Alex. 1999. Personal Communication. Handtmann Filtration.

24. Johnstone, J.T. and D.J. Whitmore. 1971. "Brewing Hygiene and Biological Stability of Beers." *Modern Brewing Technology*, edited by W.P.K. Findlay. Cleveland, Ohio: The Macmillian Press.

25. Knox, R.D. 1973. "Filtration and Stabilization." *Brewers' Guardian* 102 (4).

26. Kunze, Wolfgang. 1996. *Technology Brewing and Malting*, translated by Dr. Trevor Wainwright. Berlin, Germany: VLB Berlin.

27. Leeder, Geoff. 1993. "Features of a Modern Filter Line." *The Brewer* 79 (948).

28. Leeder, Geoff. 1995. "Beer Filtration from the Supplier's Perspective." *Brewers' Guardian* 124 (2).

29. MacDonald, J., et al. 1984. "Current Approaches to Brewery Fermentations." *Progress in Industrial Microbiology*, edited by M.E. Bushell. Guildford, United Kingdom: Elsevier.

30. Mailes, B.J., R.N. Ricketts, and A.L. Henderson. 1985. "Beer Filtration." *Brewers' Guardian* 114 (4).

31. Metz, David. 1995. *Prospero's Beer Bottling Handbook - Bottle Shop Operations for Small Brewers*. Muscatine, Iowa: Prospero Equipment Corporation.

32. Moll, Manfred, ed. 1995. *Beers and Coolers*. New York, New York: Marcel Dekker, Inc.

33. Munroe, James H. 1995. "Aging and Finishing." *Handbook of Brewing*, edited by William A. Hardwick. New York, New York: Marcel Dekker, Inc.

34. Murton, Dick. 1995. "Filtration: The Brewer's Viewpoint." *Brewers' Guardian* 124 (3).

35. Oechsle, Dietmar. 1997. "Cold Sterile Production of High Quality Beers by Means of Sheet Filtration." *The New Brewer* 14 (1).

36. O'Rourke, Tim. 1994. "The Requirements of Beer Stabilization." *Brewers' Guardian* 123 (8).

37. O'Rourke, Tim. 1996. "Strict Hygiene Regime is Essential for Trouble-Free Bottle Filling." *Brewers' Guardian* 125 (8).

38. Posada, J. 1987. "Filtration of Beer." *Brewing Science*. Volume 2, edited by J.R.A. Pollock. Reading, England: Academic Press.

39. Reed, Robert. 1989. "Advances in Filtration." *The Brewer* 75 (899).

40. Robinhold, Brian. 1997. "DE Filtering 101" *The New Brewer* 14 (2).

41. Stewart, G.G. and I. Russell. 1985. "Modern Brewing Biotechnology." *Food and Beverage Products*, edited by Murray Moo-Young. Oxford, England: Pergamon.

42. Tubbs, Jerry. 1998. "The Basics of Cartridge Filtration." *The New Brewer* 15 (1).

43. Tubbs, Jerry. 1998. "The Basics of Cartridge Filtration – Part II." *The New Brewer* 15 (3).

44. Thilert, Tom. 1987. "Beer Filtration." *The New Brewer* 4 (2).

45. Thompson, Donald, Karl Ockert, and Vince Cottone. 1999. "Craft Brewing." *The Practical Brewer*, edited by John T. McCabe. Wauwatosa Wisconsin: Master Brewers Association of the Americas.

46. Weaver, Mikoli. 1997. "Breaking the Bottling Barrier." *Brewing Techniques* 5 (3).

47. Wood, Grant E. 2006. "Clarification, Filtration, and Finishing Operations." *Fermentation, Cellaring, and Packaging Operations*. Volume 2, edited by Karl Ockert. St. Paul, Minnesota: Master Brewers Association of the Americas.

48. Woolley, D. A. 1981. "The Diatomaceous Earth Filter Types." *Brewers' Guardian* 110 (9).

Chapter Sixteen

Beer Carbonation

The next major process that takes place after filtration and prior to packaging is carbonation. Carbon dioxide not only contributes to perceived "fullness" or "body" and enhances foaming potential it also acts as a flavor enhancer and plays an important role in extending the shelf life of the product.

The level of dissolved carbon dioxide in beer following primary fermentation varies as a result of a number of parameters such as temperature, pressure, yeast, type of fermentation vessel, and initial wort clarity. Typically, carbon dioxide levels range from 1.2 to 1.7 volumes of carbon dioxide per volume of beer (v/v) for non-pressurized fermentations. Consequently, carbon dioxide levels need adjustment, unless the beer has undergone secondary fermentation. Common practice is to raise the carbon dioxide level between 2.2 and 2.8 v/v and possibly more prior for bottled and canned products. The carbon dioxide levels for kegged beer typically range from 1.5 to 2.5 volumes (2).

Principles of Carbonation

The time required to reach a desired carbon dioxide concentration depends on a number of physical factors. Temperature and pressure play an important role in determining the equilibrium concentration of carbon dioxide in solution. Increasing the pressure leads to a linear increase in carbon dioxide solubility in beer. Decreasing the temperature gives a nonlinear increase in carbon dioxide solubility in beer. Consequently, the equilibrium concentration cannot be attained without either increasing the pressure or decreasing the temperature. Thus, the closer the carbonating temperature is to 0°C and the higher the pressure, the greater the carbon dioxide absorption. The amount of carbon dioxide that dissolves is a function of time, with the rate decreasing exponentially as equilibrium is approached.

Carbon dioxide levels are stated as volumes of gas at standard temperature and pressure per volume of beer. Fixing the temperature and pressure at appropriate settings will bring about the desired carbon dioxide concentration. This relationship between pressure, temperature, and carbon dioxide volumes is given shown in the Carbon Dioxide Volume Table in Appendix B. To use this carbonation chart, look up the volume of CO_2 that you wish to dissolve in the beer, cross reference it to the temperature your beer is at, and this will tell you the gas pressure needed. For example, if you want 2.1 volumes of carbon dioxide in your beer and the temperature of your beer is at 0°C you need to adjust the psi to 8.0.

The rate of dissolution is influenced not only by temperature and pressure but also by the headspace (surface area) exposed to the gas, with the rate of dissolution being greater as surface area increases. Thus, carbon dioxide absorption can be increased by using very shallow tanks. Alternatively, the surface area can be increased by making waves or bubbles. The smaller the bubble size the greater the rate of absorption since more surface area is exposed with the smallest sized bubbles. It also follows that the smaller the bubble size, the slower the bubble travels up through the body of the beer and the longer the time allowed for absorption in a given travel.

In North America, carbon dioxide solubility in beer is usually measured in volumes of carbon dioxide per volume of beer (volumes of CO_2 = liters CO_2/liters beer) at standard temperature and pressure. One volume of CO_2 is equal to one liter of carbon dioxide dissolved in one liter of liquid, two volumes is equal to two liters of gas in one liter of liquid, etc. In Europe, carbon dioxide content in beer is expressed as percent by weight, in comparison with a given weight of beer. Gas volumes may be converted into percent by weight by using the following formula:

% CO_2 by weight = (gas volumes x 0.2)/specific gravity of beer

Percent of carbon dioxide by weight may be converted into gas volumes by the following formula:

gas volumes CO_2 = % CO_2 by weight × 5.1 x specific gravity of beer

Carbon Dioxide Levels in Beer

Typically, American lagers require carbon dioxide levels ranging from 2.5 to 2.8 volumes of carbon dioxide, while Continental lagers require carbon dioxide levels between 2.4 and 2.5. British ales with a high hop profile require carbon dioxide levels ranging from 2.2 to 2.4 volumes, and German weiss beers have levels more than 4.0 volumes. Cask-conditioned ales have extremely low carbonation levels, typically between 1 and 1.5 volumes of carbon dioxide. During packaging, carbon dioxide will be lost. To offset the losses, the beer should be carbonated 0.10 to 0.15 volumes higher than the desired carbonation.

Over-carbonation can promote gushing and have a "masking" effect on beer flavors. The positive flavor tones that result from dry-hopping or late kettle-hopping can be completely absorbed by excess carbon dioxide. If over-carbonation occurs, nitrogen gas (oxygen free) is used to reduce the excess carbon dioxide.

Methods of Carbonation

Secondary Fermentation

The traditional method involves carbonating the beer during secondary fermentation at low temperatures and under counter-pressure. The beer transferred to the conditioning tank should have at least 0.5 to 1.0°P of fermentable extract and be placed under pressure from 12 to 15 psi in conditioning tanks (16). While in the tank, the remaining extract ferments and creates sufficient carbon dioxide to saturate the beer to equilibrium. Munroe reports that if the pressure is too high during secondary fermentation, yeast growth may be affected and change the flavor characteristics of the beer (20).

An accurate, adjustable, relief valve and a dedicated pressure-regulating bung device are essential for this process. For the desired level of carbonation, the tank must be set at the correct pressure and temperature. Few brewers rely on this technique because of the difficulty in precisely controlling the carbon dioxide concentration from batch to batch, and the fact that the concentration may require some adjustment prior to packaging. Use of the technique requires precise control of temperature, degrees of Plato, and counter-pressure.

Kraeusening

Another carbonation technique involves priming the green beer with wort-primed beer, a process referred to as "kraeusening." As the second fermentation takes place, the excess carbon dioxide is bled off until the fermentation end point is obtained. This technique will produce a smoother beer with finer bubbles than other means of carbonation. It more completely bonds carbonic gas to the beer so that carbonation is less apparent. The technique is used by many commercial brewers to naturally condition beer. Kraeusen is usually sufficient to raise the carbon dioxide levels to acceptable levels. However, in North America, popular trend dictate higher levels of carbon dioxide.

Bottle-Conditioning

Before the advent of artificial carbonation, brewers carbonated beer by adding priming sugars (glucose, dextrose, and invert sugar) to the bottle for conditioning. This is a practice typical of both British and Belgian brewers. The amount of priming sugar needed for carbonation is easily determined, with 1 volume of carbon dioxide being generated for 3.7 g/l of added glucose. For example, to increase carbonation from 1.0 volume to 1.7 volumes, 1.3 grams of glucose is needed per liter, as is shown in the following formula:

priming addition of glucose (g/l) = (1.7 − 1.0)(3.7 g/l)

priming addition of glucose (g/l) = (0.7)(3.7 g/l)

priming addition of glucose (g/l) = 1.3 g/l

Mechanical Carbonation

Mechanical carbonation is accomplished either by in-line or in-tank techniques. Carbon dioxide may be purchased from suppliers of industrial gases. Alternatively, carbon dioxide may be recovered from fermentation vessels and then purified, liquefied, and stored until needed for carbonation. However, this collection system can be too expensive for most craft brewers. In general, the viability of collecting carbon dioxide depends on an alternative cost of purchased carbon dioxide, its availability, and the quantity used in the brewery. Some brewers report that use of mechanical carbonation actually has a greater influence on reducing acetaldehyde levels than does kraeusening the beer.

In-Line Carbonation

In-line carbonation involves injecting carbon dioxide into the beer through a fritted stainless steel diffuser (see Figure 16.1) between the outlet of the filter and the bright beer tank. It cannot be done upstream from DE filtration because carbon dioxide bubbles would disturb the filter bed (14). The diffuser creates very fine carbon dioxide bubbles, which go into solution readily as long as the beer is not saturated with carbon dioxide (3). Some brewers prefer the in-line method because it cuts down on pumping and lessens the chance for picking up air. In-line carbon dioxide measuring equipment is available to continuously indicate the level of carbon dioxide in the beer.

Control of this process may be hard to achieve under craft brewery applications. The best way to use in-line carbonation for a craft brewery

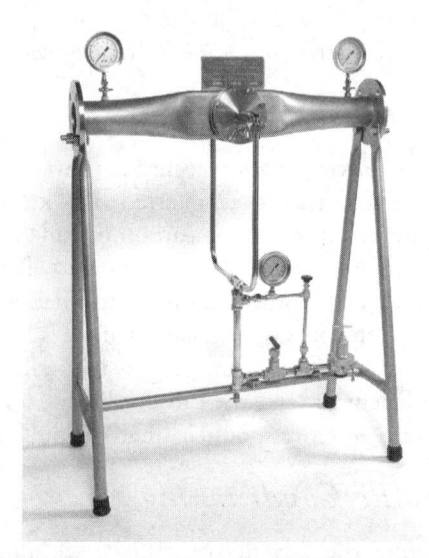

Figure 16.1
In-Line Carbonator
Courtesy of Wittemann Company

is to set up a portable system and mount it over the re-circulating pump (1). The beer is pumped from the bottom of the vessel and passed through the carbonator into the same tank. This process is continued, under pressure, until the desired level of carbon dioxide is reached.

In-Tank Carbonation

In-tank carbonation involves injecting carbon dioxide through either a submersed ceramic or sintered stainless steel stone (see Figure 16.2) at the bottom of the tank until a given backpressure is reached. The introduction of minutely divided gas directly into the bottom of a tank closely resembles the gas generating action of beer fermenting under pressure. Reportedly in-tank carbonation brings about more uniform carbonation and eliminates the stratification of gas content that is often prevalent with other methods. This method also aids in the removal of dissolved air inadvertently introduced into the beer during processing. Briggs *et al.* considers this method less efficient and more difficult to control than other methods (2).

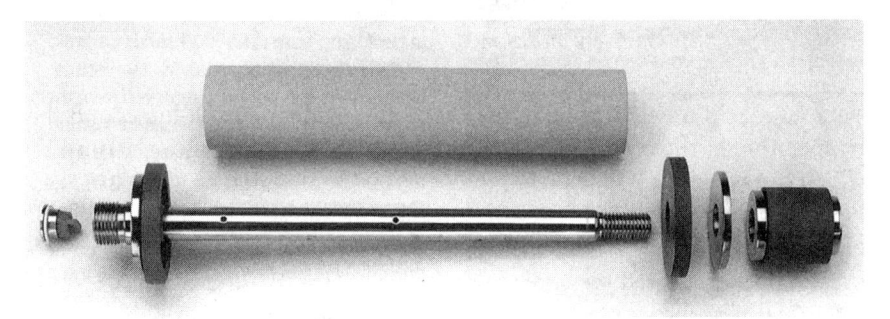

Figure 16.2
Check Valve, Stainless Steel Stone Holder, Ceramic Stone, and Nut
Courtesy of Zahm & Nagel Company, Inc.

Carbonation tanks are normally fitted with automatic carbon dioxide pressure regulators, which maintain a constant differential pressure and uniform gas flow, as shown in Figure 16.3. Unless these regulators become frozen, or fouled by substances ejected with foam, they maintain the beer at the predetermined pressure.

The first step is to determine the temperature of the beer and the desired carbonation level, in order to arrive at the required saturation pressure. The next step is to determine liquid head pressure of the beer above the stone since the liquid pressure affects the total internal head pressure. As a general rule, for every 2.3 feet of beer the pressure is equal to approximately 1 pound per square inch (psi). To obtain a more accurate pressure, multiply the total inches of liquid above the stone by the specific gravity of the finished beer, then divide by 27.684 in/psi (14).

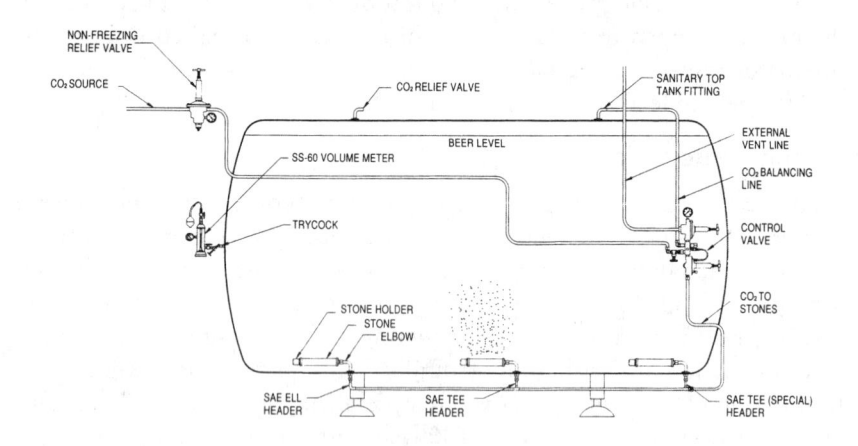

Figure 16.3
Carbonation Tank
Courtesy of Zahm & Nagel Company Inc.

For example, if the desired volume of carbon dioxide is 2.52 volumes and the temperature of the beer is 1.7°C, then the corresponding pressure is 10 (psi), which can be derived by referring to Appendix B. The total head pressure can be arrived at by adding the wetting pressure of the stone to the liquid head pressure of the tank. In this example, the wetting pressure of the stone is 5 psi and the liquid head pressure of the tank is 3.9 psi (9 feet/2.3 feet = 3.9 psi) for a total head pressure of 8.9 psi. The total head pressure is the minimum pressure at which the stone will begin releasing bubbles into the beer.

It is advantageous to start the flow of gas at a relatively low differential pressure, with relief value handle in a released position, for the purpose of removing dissolved air and expunging the air from the head space in the tank. After one-half to one hour, the relief valve closed and adjusted to the desired relief

Note | Measuring Carbon Dioxide Levels

Carbonation is measured either by using a Gehaltemeter or Zahm meter. Gehaltemeter is an analogue meter that can determine the carbon dioxide content in beer, either by directly sampling during production (for example from tanks and pipes) or after production (for example from kegs). The Zahm is a volume meter that also attaches to the tank to sample the beer. The temperature and temperature readings are checked against a chart (see Appendix B) that calculates the volume of carbon dioxide in the beer.

pressure. Then carbonation is started slowly, using a relatively low-pressure differential greater than 8.9 psi. As the pressure in the headspace of the tank increases, the carbon dioxide pressure is increased until the desired level of 10 psi is reached.

Carbonation time depends on the size of the carbonation stone and the tank configuration, but for many craft brewers it usually takes no longer than 12 hours. Some craft brewers believe that carbonation times more than 24 hours yield better results.

Safety Procedures

The brewer should only use pressure-rated vessels for carbonation and must know the maximum working pressure of the tanks. Pressure-rated tanks are certified by the American Society of Mechanical Engineers (ASME) certifying its ability to withstand normal operation pressures. Tanks are equipped with pressure relief valves as a precaution against over-pressurization. Relief valves should always be mounted directly to the highest point on the tank. It is very important that no beer or foam come in contact with the relief valve, as contact can affect the operation of the relief valve, causing it to fail to blow off at the set pressure. Pressure relief valves should be inspected and tested on a regular basis, and they should be regularly cleaned to avoid buildup that can lead to malfunction. To avoid buildup from beer or beer foam, never fill tanks completely.

References

1. Black, Bernard. 1989. "In-Tank/In-Line." *The New Brewer* 6 (2).
2. Briggs, Dennis E., Boulton, Chris A., Brookes, Peter A., and Stevens, Roger. 2004. *Brewing: Science and Practice*. Cambridge, England: Woodhead Publishing Limited.
3. Coors, Jeffery H. 1977. "Cellar Operations." *The Practical Brewer*, edited by H.M. Broderick. Madison, Wisconsin: Master Brewers Association of the Americas.
4. Eckhardt, F. 1993. *The Essentials of Beer Style* (5th edition). Portland, Oregon: Fred Eckhardt Associates.
5. Fix, George. 1989. "Beer Foam and CO_2 Levels." *The New Brewer* 6 (2).
6. Fix, George and Laurie Fix. 1997. *An Analysis of Brewing Techniques*. Boulder, Colorado: Brewers Publications.
7. Hough, J. S., et al. 1982. *Malting and Brewing Science*. Volume 2. London, England: Chapman-Hall, Ltd.

8. Koch, Gary S. 1999. Personal Communication. Zahm & Nagel Co., Inc.

9. Kunze, Wolfgang. 1996. *Technology Brewing and Malting*, translated by Dr. Trevor Wainwright. Berlin, Germany: VLB Berlin.

10. Lapp, Alan. 1999. Personal Communication. Zahm & Nagel Co., Inc.

11. MacDonald, J., et al. 1984. "Current Approaches to Brewery Fermentations." *Progress in Industrial Microbiology*, edited by M. E. Bushell. Guildford, United Kingdom: Elsevier.

12. Mallet, John. 1993. "CO_2: Its Dynamics in Beer." *The New Brewer* 10 (2).

13. Markowski, Michael R. 1999. Personal Communication. The Wittemann Company.

14. Meheen, Dave. 1994. "Dynamics of Tank Carbonation." *The New Brewer* 11 (6).

15. Munroe, James H. 1995. "Aging and Finishing." *Handbook of Brewing*, edited by William A. Hardwick. New York, New York: Marcel Dekker, Inc.

16. Patino, Hugo. 1999. "Overview of Cellar Operations." *The Practical Brewer*, edited by John T. McCabe. Wauwatosa Wisconsin: Master Brewers Association of the Americas.

17. Ruff, Donald G. and Kurt Becker. 1955. *Bottling and Canning of Beer*. Chicago, Illinois: Siebel Publishing Co.

18. Scheer, Fred. 1989. "Methods for Carbonating Beer." *The New Brewer* 6 (2).

19. Scheuermann, Mangel. 1989. "In-Tank/In-Line." *The New Brewer* 6 (2).

20. Slater, Gord. 1989. "Carbonation." *The New Brewer* 6 (2).

21. Thompson, Donald, Karl Ockert, and Vince Cottone. 1999. "Craft Brewing." *The Practical Brewer*, edited by John T. McCabe. Wauwatosa Wisconsin: Master Brewers Association of the Americas.

22. Warner, Eric. 1992. "Working with CO_2." *The New Brewer* 9 (5).

23. Weaver, Mikoli. 1997. "Breaking the Bottling Barrier." *Brewing Techniques* 5 (3).

Chapter Seventeen

Bottling

Once the final quality of the beer has been achieved, it is ready for bottling. The bottling of beer is one of the most complex aspects of brewery operations and the most labor intensive of the entire production process. The layout of the bottling line will depend on a number of factors but typically consists of a series of processes as shown below if non-returnable bottles are used.

- Sterilization of beer
- Bottle feeding
- Bottle rinsing
- Bottle filling
- Tunnel pasteurization
- Bottle labeling
- Case packing

If returnable bottles are used it will also include the following processes: depalletzing, decrating, bottle washing, bottle rinsing, and empty bottle inspection. These processes will not be discussed since non-returnable bottles are mainly used in the U.S. and European markets.

Sterilization of Beer

Absolute sterility of the bottled beer is essential given the fact that bottles are shipped over long distances, subject to varying temperature regimes, and often remain on the shelf for fairly long periods. Traditionally sterility of beer was accomplished by pasteurizing the beer in the bottle after filling and crowning by a process known as tunnel pasteurization. Alternatively, the beer can be sterilized prior to packaging either by flash pasteurization or sterile filtration.

Flash Pasteurization

Flash pasteurization has not been widely used by breweries in North America (though it is very popular with the dairy and juice industries), but in recent years it has grown in popularity in bottling beer. Europe and Asia, on the other hand, has adopted this process extensively. There are many hundreds of such systems in existence, most of which are used for kegs, and some for bottles and cans. Tunnel pasteurization is most commonly used with small-pack beer, either bottles or cans.

Flash pasteurization is a method of heat pasteurizing bulk beer prior to filling the bottles for the purposes of killing spoilage microorganisms. The basis for flash pasteurization is to heat the beer for a predetermined period of time at specific temperatures, thereby assuring the microbiological stability of the beer. In this process the product is handled in a controlled, continuous flow and subjected to a temperature, normally in the range of 71°C to 79°C, for a time period of 15 to 60 seconds (6). The amount of heat imparted to the product during the elevated temperature and time period is expressed in Pasteurization Units (PU's). A Pasteurization Unit (PU) is defined as a one-minute exposure to a temperature of 60°C. A PU is a measure of the lethal effect on microorganisms of the heat treatment. The aim is to attain the minimum degree of pasteurization necessary to inactivate beer spoilage organisms.

Flash pasteurization of beer typically uses a two- or three-stage plate heat exchanger with hot water as the heat exchange medium. The heat exchanger is designed so that a particular flow rate will achieve maximum efficiency. Consequently, the temperature—not the flow rate—must be adjusted to alter the number of PUs for a given beer since the heat exchanger is designed for a particular flow rate for maximum efficiency. The use of buffer tanks before and after the heat exchanger is essential to prevent interruptions of flow and pressure surges.

The advantages of flash pasteurization compared to conventional tunnel pasteurizers are: 1) kill-off is assured independent of the type of microbe, 2) less space is required, 3) lower capital expenditures, 4) operational costs are less (less coolant, steam, etc.), and 5) less of an impact on flavor than tunnel pasteurization. Beer composition and, more importantly, the oxygen content play major roles. Flavor changes may occur with flash pasteurization if dissolved oxygen levels are greater than 0.3 mg/l (or ppm) (6).

Sterile Filtration

Sterile filtration physically removes organisms from the beer as it passes through a cartridge membrane filter. It is essential the beer first undergo primary diatomaceous earth filtration and in some cases sheet pad filtration before it undergoes sterile cartridge membrane filtration. Briggs *et al.* recommends a series of filtration levels with progressively smaller micron sizes as follows (6):

- Kieselguhr filtration
- Cartridge membrane filter 1, 5µm pore size
- Cartridge membrane filter 2, 1µm pore size
- Sterilizing cartridge membrane filter 3, 0.45µm pore size

Some of the advantages of sterile filtration include: 1) lower operating costs compared to tunnel pasteurization, 2) improved non-biological stability, 3) better clarity of beer, and 4) more importantly sterile filtration avoids potential flavor deterioration effects of pasteurization and reportedly results in a fresher tasting beer.

Bottle Feeding

The first step in bottling begins with loading the empty bottles on the conveyor that feed into the unscrambling table either from a pre-pack or bulk-pack bottles.

Pre-Pack

Craft brewers often buy what is known as a "pre-pack," which includes printed six-pack carriers inserted in a printed case along with the bottles. For craft brewers, this is the most convenient and cost-effective way to go, even though it is a relatively expensive way to buy bottles.

In craft breweries, loading with pre-packed bottles is usually done manually at a rate of 80 to 100 bottles per minute (BPM). The table funnels the wide mass of bottles into a single stream. Several types of mechanisms prevent the bottles from bridging as they are funneled to a single line. These include mechanical joggers, reversing chains, and good inherent design.

On faster bottling lines with rates more than 100 BPM, the loading of bottles is automated. One method is to use an automatic unpacker/packer, as shown in Figure 17.1, which shows unpacker removing bottles from prepacked shipping containers. Fully automatic unpacker/packers are available at virtually any speed, with a single-head machine able to handle 20 cases per minute. The most common type lifts from 2 to 6 cases at a time, sending the empty case to the case packer. The process begins by conveying the cases of bottles to the unpacker. The unpacker then lifts the bottles out by their necks and places them onto a conventional unscrambling table to be fed into the machines via the conveyor. An advantage of pre-packed glass is that fewer machines are required to complete the packaging operation. On the other hand, pre-packed glass costs more than bulk-pack glass.

Figure 17.1
Case Unpacker/Packer
Courtesy of Krones, Inc.

Bulk-Pack

Another method, common with larger breweries and some craft breweries, automatically feeds bottles onto the conveyor from a palletized bulk pack. Bulk-pack bottles are full pallets of bottles with no cartons; rather, the bottles are separated by cardboard sheets and wrapped in plastic shrink. A de-palletizer removes the bottles from the pallet, usually a layer at a time, and places them on the unscrambling table.

Buying in bulk is not popular with most craft brewers because of the cost of a de-palletizer. Also, the warehouse space required for a bulk glass packaging operation is larger than that needed for a pre-pack operation. This is because more space is required for the bottling line itself, and additional space is needed to store cartons and baskets away from glass. The advantage in buying bulk glass is the savings in glass cost. Typically, a bulk glass operation is only suitable when the brewery has economies of scale justifying the investment.

Bottle Rinsing

Cleaning Treatment

Although the non-returnable bottles are virtually sterile when received by the manufacturer, they must undergo a relatively simple rinsing. The bottles often contain particles called "case dust," which cause foaming problems during filling since the dust particles act as nucleation sites. The cleaning is accomplished by turning the bottles upside down and spraying inside and outside several times before returning the bottles to the upright position for filling. Steam is usually jetted into the bottle followed by a purge of sterile air to ensure sterility by killing microorganisms.

Rinsing with a sanitizer in the rinse water is very effective too and a very inexpensive way to sterilize bottles. Ozone is a sanitizing agent that is gaining wide acceptance because of its effectiveness and short life.

Types of Bottle Rinsers

Bottle rinsers are used to sanitize and remove warehouse dust from bottles prior to filling. There are three types of bottle rinsers—twist, gripper, and rotary.

Twister Rinsers

Most craft breweries use twist rinsers. These use a set of belt drives to feed the bottles into a set of rails which twist and invert the bottle as it passes over a set of spray nozzles which spray into it. After spraying, the bottle is allowed to drain and is then re-inverted to return to the line right side up as it moves to the bottle filler. The longer the rinser's twist section, the faster it can run. Units of 14 feet in length can usually handle 150–250 bottles per minute (BPM). Each bottle size and shape requires a different "twister," but the cost of twist rinsers is quite low.

Gripper Rinsers

Gripper style rinsers can handle various bottle sizes and shapes without parts having to be changed and are generally used in larger lines. They can achieve very high speeds while requiring less floor space than a twist rinser of comparable capacity.

Rotary Rinsers

Rotary rinsers are compact as well, and are known for their smooth bottle handling. However, rotary rinsers are the most expensive option for rinsing. A rotary rinser can also be incorporated into a filler as part of a "monobloc" arrangement, as shown in Figure 17.2. Monobloc machines also combine a crowner with the rinser-filler on the same chassis. Generally, monoblocs that

Figure 17.2
Rinser/Filler Bloc
Courtesy of Krones, Inc.

combine a filler and crowner are better than separate units, since the bottle must be capped as soon as possible after filling to exclude air.

Bottle Filling

Filler Bowl Operations

The filling unit or filler bowl should be cleaned and sanitized before bottling. To prepare for filling, the beer lines, hoses, and filler should be cooled down using cold water. The water should be blown out before beer is introduced into the filler. The system is then pressurized, and beer is supplied to the filler bowl from the bright beer tank. Some brewers will vent the bowl pressurizing gas prior to the introduction of beer to lower the oxygen contact with the incoming product. It is better to fill the bowl from the bottom to reduce turbulence and air pickup. Brewers usually slightly over carbonate the beer to compensate for any loss of carbon dioxide during the filling operation.

During the filling operation, a pump is normally placed between the bright beer tank and the filler bowl to increase the pressure in the filler bowl relative to the bright beer tank. There are several types of pumps and control mechanisms for regulating the flow and pressure of the filler bowl. The simplest is a low-rpm centrifugal pump to increase the pressure by a fixed amount. A frequency controller can be added to a centrifugal pump to adjust the pressure. Positive-displacement pumps can also be used. It is important that a positive-displacement pump has a feedback control to regulate the pump speed based upon a signal from the filler bowl. The controlling signal can be from a liquid depth-sensing probe or from a pressure control. Although they operate at a higher rpm, centrifugal pumps do not knock carbon dioxide out of solution to any greater degree more than positive-displacement pumps.

> **Note | Beer Temperature**
>
> The temperature of the beer should fall in the range of − 1 to 0°C; at higher temperatures, foaming can be a problem when filling the bottles. It is important for the bright beer tank to have cooling jackets both above and below the manway. If there is a jacket only above the manway of the tank, the beer will start warming as soon as its level drops below the cooling jacket. The transfer line should be insulated to prevent the beer from warming if it is over a few feet from the bright beer tank to the filler bowl.

A better, simpler method of supplying beer to the filler is to do so directly from the bright beer tank without the use of a pump. This requires a tank rated as a "pressure vessel," since most bright beer tanks have pressure ratings of 15 psi or less. If the pressure in the bright beer tank is too high, the beer can pick up excessive carbon dioxide, which is more likely to occur toward the end of the bottling run. Excessive carbon dioxide will not only make the beer foamy, it will also result in problems with bottle filling. Foaming problems can develop if the beer is warm too.

The transfer line should be of sufficient size, and low restriction fittings and valves should be used in order to minimize foaming. Transfer lines should be designed with a minimum number of elbows. An air and foam vent (called a "lantern") is usually installed on the inlet of the filler to allow the operator to vent the air from the line before startup. A sight glass gives the operator visual assurance that the beer line has beer in it and not foam.

Types of Bottling Machines

The bottles are delivered on a conveyor belt, separated at a predetermined spacing by a separating device, and positioned on a lifting platform under the filling elements by a star wheel infeed device. Basically for the brewer, there are two types of bottling machines available:

1. Tandem fillers
2. Rotary fillers

Tandem Fillers

These are small long tube machines which are built to fill from one to 8 bottles at a time. Most common are the popular Meheen tandem machines (see Figure 17.3) with 4 spouts, although newer models are offered with 6 and 8 spouts. The main shortcoming of these machines is that because they fill bottles in tandem (side by side) and have a fixed overhead bridge that supports the fill valves and crowner heads, they are limited to one bottle size and type. However, one of the big advantages is their low cost making them ideal for small operations. Rated speeds of the four spout models is 20–30

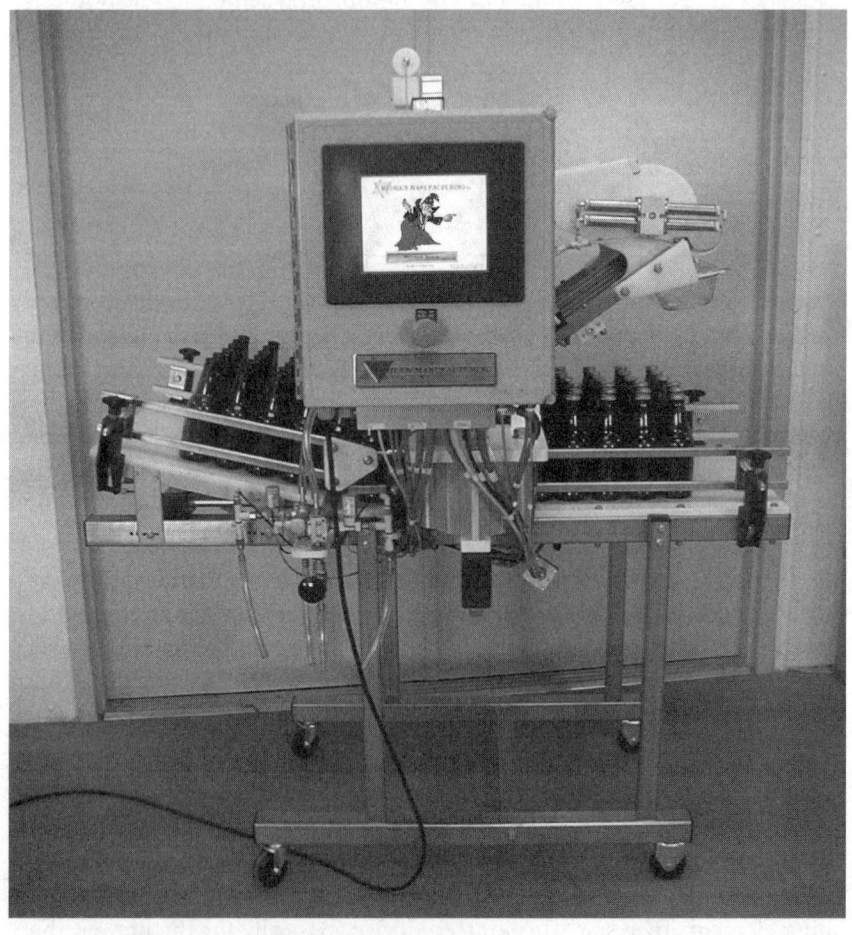

Figure 17.3
Tandem Bottle Filler
Courtesy of Meheen Manufacturing, Inc.

BPM, though some operators claim that better airs are obtained by limiting speed to 15–18 BPM.

Rotary Fillers

All high-speed machines are rotary machines as shown in Figure 17.4. These have a carousel with anywhere from eight to more than 100 valves and bottle platforms, and range in speed from 20 BPM to more than 1,000 per minute. These are available as short-tube (or vent tube), and long tube machines, as discussed below.

Figure 17.4
Rotary Bottle Filler
Courtesy of Criveller Company

Filling Valve Technology

There are two basic types of filling valves, short tube and long tube. Long or short tube refers to the length of the tube associated with the filling head. They work very differently, but both fill under counter pressure (the bottle is counter-pressurized to the same pressure as the filler bowl above the valves containing the carbonated beer) and the product flows by gravity into the bottle. Short tube fillers fill the bottle by running the beer down the sides of the bottle whereas with long tube fillers fill the bottle from the bottom up.

Short Tube Fillers

Short tube fillers have a vent tube about which exhausts the air or gas in the bottle as it fills with beer. The product flows in around the vent tube and over a small rubber umbrella spreader which fans the product to the walls of the bottle neck, where it cascades down the inside walls of the bottle. The tube serves to vent air or gas displaced by the beer from the bottle—no product flows through the vent tube in normal operation. In fact, the vent tube controls the fill level in the bottle—when the beer level reaches the end of the vent tube, the product stops flowing because air can no longer leave the bottle. Fill height adjustments are made by changing the length of the vent tube with some types having an adjustable plastic sleeve.

Older short tube fillers, lacking pre-evacuation or pre-purging, simply vent the air and gas from the bottle into the filler bowl headspace. This can increase air pick-up in the beer.

Long Tube Fillers

Long tube fillers use a long fill tube which extends nearly to the bottom of the bottle to fill the bottle from the bottom up. Unlike short tube fillers, they fill the bottle completely full, with correct fill height being assured by the displaced volume of the tube as it is withdrawn from the bottle. The long tube filler was the mainstay of the U.S. large breweries until the 1990s.

Bottle Evacuation, Counter-Pressure, and Filling

In pre-evacuation short tube fillers, the bottle is pressed airtight against the filler valve and air is evacuated and pressurized with carbon dioxide, followed by a second evacuation and pressurizing with carbon dioxide is what is called double pre-evacuation. Double pre-evacuation can reduce the air fraction in the bottle to about 1% (12). The beer flows into a virtually pure carbon dioxide atmosphere.

> **Note | Dissolved Oxygen in Beer**
>
> To successfully bottle the beer it is essential that the dissolved oxygen in the beer is less than 0.2 mg/l and in some cases less than 0.1 mg/l (6). If the air fraction is too high, it will cause oxidation haze and stale flavors in the bottled beer.

When the pressure in the bottle and the filler bowl are equalized, the bottle is filled to the desired height with beer. The beer is deflected by the deflecting cone against the bottle wall, and it flows down the sides of the bottle. The displaced carbon dioxide flows back into the filler bowl through the air pipe. The pressure in the bottle is bled off slowly to prevent the bottle from gushing when it comes off the spout. Some bottle fillers vent the gas displaced from the bottle during filling (a mixture of carbon dioxide and a small amount of air) into the atmosphere.

During filling it can be expected that some bottles will break—they are called "blow-outs" in the trade. About 0.1% breakage can be expected when bottling (16). Out-of-round bottles and crooked necks can cause breakage. On smaller fillers, the operator stops the machine and removes the broken glass by hand. High-speed machines have automatic high-pressure spouts to wash the broken glass off the filler platform, the vent tube, and the filling valve.

Post Bottle Filler Operations

Once the bottle has been filled and dropped away from the filler valve, it is necessary to clear the head space of oxygen prior to crowning. This is done by creating foam in the bottle neck, with the foam rising and displacing the gas in the head space. There are several methods by which this can be accomplished. The first is to arrange the filling parameters so that the bottle foams as it descends from the filling valve; the second is to add a fobber to the machine.

The first method does not require auxiliary equipment but has the disadvantage of uneven fobbing by the individual filling valves, resulting in uneven oxygen values in the packages.

Fobbers or jetters are devices that shoot a high-pressure jet of sterile water into the head space of the bottle neck. This excites the top layer of beer, causing the beer to foam before it enters the crowner. The most important parameter determining the amount of foam is the pressure of the water jet. Less important are water temperature and jet diameter. Jetting with liquid nitrogen is a recent technology that shows promise as a very effective system in removing oxygen from the head space. By using a jet of an inert material, some of the problems associated with water fobbers are eliminated.

In order to achieve low levels of air in the finished package, it is necessary to fob-out about 3 to 5 ml of beer (13). Lower values of fobbing result in significantly higher levels of air in the packaged product. Simply allowing the foam to crown over the bottle top is not as sufficient as fobbing.

Ideally, the foam should be creamy, with tiny bubbles. This is referred to as "tight" foam as opposed to the large "soda pop" bubbles that are formed at the outset of fobbing.

It is important for the foam to crest just above the lip of the bottle before the cap is closed around the bottle top. If the foam peaks and recedes prior to capping, air can be pulled back into the bottle. Also, active foaming helps remove air from under the cap before capping.

Crowning

After the bottle has been cleared of air, the next step is to cap the bottles as soon as possible with the crowner. The caps have a sprayed-on, hot-pressed PVC-based sealing insert or a cold-pressed sealing insert not containing PVC. The caps are conveyed to the crown hopper by means of a magnetic belt or a pneumatic crown feed, or they can be dumped manually directly into the crown hopper. Whatever method is used, the crown hopper should be kept only half full of caps. This is to lessen the possibility of crowns becoming packed and not feeding into the chute fast enough. In addition, when the caps become packed, the possibility of scratching the cap finish increases.

The closing element is lowered onto the bottle until the cap rests on the bottle mouth, bending the teeth of the cap against the upper edge of the mouth. For reliable crowning in the short term and for stability over time, there should be a maximum of 30 crowns per crowning head per minute (13).

The operation of the crowner should also be checked frequently. This is done by removing a bottle from each crowning head. A visual inspection of the crown surface for indentations and scratches should be made. If indentations are present, a more thorough inspection of the crowning head should be made to check for malfunctioning parts. A gauge plate with holes of various sizes is useful in determining whether the crowner is crimping to the proper value.

After crowning, the filled containers are conveyed past a fill-check unit. This unit can be adjusted to reject any container that has a low fill, a headspace not filled with beer foam, or a bottle that lacks a cap. The fill-check unit, if installed after the filler, must be installed downstream from the filler at a point where the beer is quiet and the foam has not yet started to settle.

Post Rinse

When a bottle comes off the filler, it receives a post rinse treatment to wash off the foam, and is then passed through a blower to remove the bulk of the water. If foam were left on the bottle, not only would it be unsightly, glue adhesion could be reduced. This could cause (depending on the glue) slippage and wrinkles on the bottle label. In addition, dried foam can mildew in damp climates if left under the lips of the crown. Water not blown off can cause labeling problems and lead to an unhygienic rust formation. If the brewery plans to use a tunnel pasteurizer, a post rinse is not necessary as the intense water spray in the tunnel pasteurizer will clean the bottle.

For efficient bottle shop operation, the water used to rinse the bottles should be relatively low in bicarbonates of calcium and magnesium. Bicarbonates in amounts exceeding 100 mg/l will tend to deposit white film on the bottles (19). The brewer can treat the water with lime (calcium hydroxide) by the precipitation and removal of calcium and magnesium salts.

Bottle Drying

Before labeling, it is absolutely essential that the bottles are dry and free of condensation. This is especially important with pressure-sensitive labels that use non-water-soluble glues. Wet glue, which is typically used as an adhesive in glue applications, is much less sensitive to residual water on the bottle. An air knife can eliminate virtually all of the surface moisture on the bottles after the post rinse.

Tunnel Pasteurization

An alternative to flash pasteurization and sterile filtration is tunnel pasteurization. Tunnel pasteurization is employed after bottles have been filled and crowned. The bottles are loaded at one end of the pasteurizer and passed under sprays of water as they move along the conveyor. The sprays are so arranged that the bottles are subjected to increasingly hot water until the pasteurization temperature is reached by the beer in the bottles (16). The bottles are then gradually cooled with water until they are discharged from the end of the pasteurizer. Temperature changes have to be made in stages to prevent the bottles from breaking. Bottle breakage is usually no more than 0.1 to 0.2% in the tunnel pasteurizer (12). If greater, it is usually due either to poorly made bottles or the lack of head space.

> **Note | Optimization of Beer Pasteurization**
>
> Factors such as bottle size, shape and material influence the specific processing conditions, such as residence time within the tunnel, to achieve appropriate results. In order to monitor the pasteurization process, i.e., to measure the lethal effect of the heat treatment on the microorganisms, the concept of Pasteurization Unit [PU] was introduced. It was defined that 1 PU is aggregated to the product when it is exposed to the temperature of 60°C for one minute.

Tunnel pasteurization has been implicated in causing off-flavors in beer. According to Hough, such changes are mainly attributed to the presence of oxygen, which can lead to cooked, biscuity flavors, especially when the dissolved oxygen content of the beer is high (greater than 0.3 mg/l) (10). Reportedly, tunnel pasteurization can also lead to destruction of the fine fresh-hop aroma.

Advantages and Disadvantages

Tunnel pasteurization, although reliable and simple to operate, is an expensive method of ensuring microbiological stability. Consequently, tunnel pasteurization of bottles is becoming less popular nowadays and is gradually being replaced by flash pasteurization and sterile filtration (12). However, tunnel pasteurization is the most reliable method of ensuring a sterile package because it treats the entire package. Both flash pasteurization and sterile filtration depend on sterile conditions during the filling and crowning, which is considered by many brewers technically difficult to achieve.

Bottle Labeling

Bottle labeling in the sequence of processes follows bottle filling, which may itself be preceded by tunnel pasteurization.

Types of Labelers

The principal types of labelers used in the brewing industry are pressure-sensitive, rotary, and tandem.

Pressure-Sensitive Labeler

These units are used in the wine industry and food industry, but are more typically used in the cosmetic, pharmaceutical, and soap industries. Pressure-sensitive or self-adhesive labels are also used in the craft brewing industry though not as much as paper labels. Pressure-sensitive labels are applied from a coiled strip of waxed carrier (also called "backing") paper. The labeling machine peels them off the carrier paper and transfers them to the bottle. One advantage of pressures-sensitive labelers is their reliability. Also, while higher in cost, they require less time for setup and cleanup than other types of labelers. They usually do not require a full-time operator, but rather only someone to change rolls and dispose of the rewind material. The big disadvantage is that the adhesives used are non-water-soluble, so labels may slide off if the bottles are not properly dried prior to labeling. Some craft brewers will apply the label at the beginning of the line in order to avoid having to apply it when the bottle is wet. This technique may work, but problems with scuffing and mold growing on the labels may result. Some brewers use label-indent bottles to prevent scuffing during packaging. Generally, 50–60 BPM is a high speed for pressure sensitive label applications.

Rotary Labeler

Rotary labelers are machines for which the bottles must be transferred from the bottle conveyor into the labeler carousel (see Figure 17.5). Rotary labelers operate at higher speeds than pressure-sensitive and tandem labelers. They are equipped to apply front, back, full wrap, and neck labels. Rotary labelers have two main advantages. The first is their ability to handle multiple bottles and multiple labeling configurations. The second is that they can be made virtually any size and can handle very high speeds. One big disadvantage with rotary labelers is that they are equipped for just one size and shape of label

Figure 17.5
Rotary Cold Glue Labeller
Courtesy of Criveller Company

and for one bottle size and type. If any of these is changed, the rotary labeler must be refitted, leading to down-time and additional costs.

Tandem Labeler

An in-line labeler is a type of machine for which the bottle travels in a straightforward motion and the label or labels are applied while the bottle is moving along the bottle conveyor. Tandem labelers cannot do front and back labels at the same time, and they are not well-suited for applying neck labels. Unlike other rotary labelers, the tandem labeler does not bond labels with a full width of glue, but rather with only two vertical strips, one at each side of the label. Tandem-applied labels, because they are not fully back-glued, can pucker on wet bottles or may be crooked or have corners flagged (edges lifted) compared to rotary applied labels. These limitations have increasingly contributed to the obsolescence of tandem labelers; nonetheless, their low cost and simplicity assures their continued use, particularly by craft brewers. Speeds of 50–80 BPM per machine are typical, assuming machines are in good repair. High-speed applications, such as 150–200 BPM or more, are better suited to rotary machines.

Case Packing

Case packing is done manually or by case packers. Two people standing at a discharge table can pack up to 100 BPM. This assumes that empty cases are within easy reach and that another person closes and stacks the case. For rates above 100 BPM, an automatic case erector and packer are required. A case packer has a linear collection table on which the bottles are automatically placed in ranks that match the packing pattern in the case. A conveyor supplies the packer with the empty shipping cases, and from the packer the conveyor carries the filled cases to the palletizing area.

Types of Case Packers

There are two types of case packers, "drop packers" and "pick and place packers." Drop packers accumulate and drop the bottles through flexible fingers that are inserted into each cell of the receiving case. Pick and place packers (see Figure 17.6) use pneumatic grippers that lift the bottles off the collection table and then lower them precisely through flexible fingers into the case. The pneumatic grippers release the bottles once they reach the bottom of the case.

The choice between drop packers and pick and place packers are application-specific and depend on speed, floor space, and case and bottle configuration.

Both types of case packers protect the bottle labels. However, manufacturers should be aware that if labels are applied with cold adhesive, the bottles require set-up time for full label adhesion before case packing or they can become skewed during handling.

Figure 17.6
Pick and Place Case Packer
Courtesy of A-B-C Packaging Machine Corporation

Case Sealing

Once the case is packed, it is sealed. In craft breweries, case sealing is done manually; in larger breweries, it is done automatically. The simplest automatic case sealer is a taper. The case taper machine has its own conveyor, with a device that closes the cases' flaps and seals them using a piece of wide, clear sealing tape. An operator manually palletizes the cases as they come off the sealer's conveyor.

Warehousing

If the palletized load of beer is to be warehoused, the temperature of the beer should not exceed 3°C. Care should be taken to avoid damage from fork trucks, and excessive heights should be avoided in stacking of pallets.

References

1. Andrews, David. 1987. "Beer Off-Flavors – Their Cause, Effect & Prevention." *Brewers' Guardian* 116 (1).
2. Ault, R.G. and R. Newton. 1971. "Brewing Hygiene and Biological Stability of Beers." *Modern Brewing Technology*, edited by W.P.K. Findlay. Cleveland, Ohio: The Macmillian Press.
3. Back, W., M. Leibhard, and I. Bohak. 1992. "Flash Pasteurization – Membrane Filtration." *Brauwelt International* 1.
4. Brenneman, Konie. 1999. Personal Communication. Krones, Inc., Franklin, Wisconsin.
5. Criveller, Bruno. 1999. Personal Communication. Criveller Company.
6. Briggs, Dennis E., Boulton, Chris A., Brookes, Peter A., and Stevens, Roger. 2004. *Brewing: Science and Practice*. Cambridge, England: Woodhead Publishing Limited.
7. DeYoung, Robert E. 1999. "Packaging – Bottling Operations." *The Practical Brewer*, edited by H. M. Broderick. Madison, Wisconsin: Master Brewers Association of the Americas.
8. Gaub, R. 1993. "Criteria for Fine and Sterile Filtration of Beer." *Brauwelt International* 5.
9. Johnstone, J.T. and D.J. Whitmore. 1971. "Brewing Hygiene and Biological Stability of Beers." *Modern Brewing Technology*, edited by W.P.K. Findlay. Cleveland, Ohio: The Macmillian Press.
10. Hough, J.S., et al. 1982. *Malting and Brewing Science*. Volume 2. London, England: Chapman-Hall, Ltd.
11. Huige, N., G. Sanchez, and J. Surfus. 1989. "Pasteurizer Operation and Control." *MBAA Technical Quarterly* 26 (1).

12. Kunze, Wolfgang. 1996. *Technology Brewing and Malting*, translated by Dr. Trevor Wainwright. Berlin, Germany: VLB Berlin.

13. Langbehn, Larry. 1999. Personal Communication. SMB Technik, Inc.

14. Machin, T.J. 1984. "The Kegging Line – A Brewer's Requirements." *The Brewer* 70 (832).

15. Markowski, Phil. 1996. "To Air is Human – Techniques and Technologies for Minimizing Air in Bottled Beers." *Brewing Techniques* 4 (2).

16. Metz, David. 1995. *Prospero's Beer Bottling Handbook - Bottle Shop Operations for Small Brewers.* Muscatine, Iowa: Prospero Equipment Corporation.

17. Muller, Gene. 1995. "Bottles, Bottles: Which Do You Choose?" *The New Brewer* 12 (6).

18. O'Rourke, Tim. 1996. "Strict Hygiene Regime is Essential for Trouble-Free Bottle Filling." *Brewers' Guardian* 125 (8).

19. Ruff, Donald G. and Kurt Becker. 1955. *Bottling and Canning of Beer.* Chicago, Illinois: Siebel Publishing Co.

20. Scheer, Fred. 1990. "Filtering and Cleaning." *The New Brewer* 7 (1).

21. Sharpe, F.R. 1984. "Evaluation and Control of Flavor in Beer." *Brewers' Guardian* 113 (1).

22. Thompson, Donald, Karl Ockert, and Vince Cottone. 1999. "Craft Brewing." *The Practical Brewer*, edited by John T. McCabe. Wauwatosa Wisconsin: Master Brewers Association of the Americas.

23. Weaver, Mikoli. 1997. "Breaking the Bottling Barrier." *Brewing Techniques* 5 (3).

Chapter Eighteen

Kegging

Kegs, another option in packaging beer, are used in bars and catering establishments where beer is served "on draught." Kegging involves filling carbonated pasteurized beer into sterile aluminum or stainless steel kegs of various sizes. Aluminum kegs are generally more popular than stainless steel kegs because they are lighter and more resistant to minor damage. Kegging fits into the cost structure for craft brewers with limited startup capital for bottling lines and low product output.

Keg Styles

Open-Keg Systems

Open-keg systems are characterized by their barrel shape and bunghole on the side by which the interior can be readily accessed without extracting the valve body or spear. The bunghole is closed by a wooded or plastic bung plug. Hoff-Stevens kegs and the almost-extinct Golden Gate kegs are two examples of open-keg systems. These kegs are frequently stored and shipped on their side because of their barrel shape, though they must be upright when served.

Closed-Keg Systems

Closed-keg systems are identified by their typically straight sides, with a rim, called a chime, on each side. The top chime has integrated handles for easy handling. The top also contains a concentric valve fitting in the center allowing for easy cleaning and filling by automated systems. Draught accounts prefer closed-keg systems over open-keg systems for several reasons including ease of tapping, ease of storage and handling, and the improved profit margin due to the reduction of beer loss that is commonly associated with older keg styles.

Unlike open-keg systems, closed-keg systems can be accessed only through the valve housing. All cleaning and filling is accomplished without extracting the valve. The keg can be filled inverted or upright. Closed-keg systems are less vulnerable to infection than open-keg systems because they restrict the access of air to the interior more than bung-style kegs do. The downside is that closed-system kegs require special machinery and equipment for washing, sterilizing, filling, and dispensing in order to reap the full benefit of this aseptic package. Closed-keg systems must use Sankey-style single-valves.

Sankey-Style Valve

The newer, more common single-valve keg (SVK) or "Sankey" keg (see Figure 18.1) has a valve arrangement that consists of a stainless steel, rod housing, called a combination fitting that is permanently installed into the top center of the keg and sealed with a spring-loaded check ball. The tapping device, or tavern head, fits into the lug housing of the valve. When the tap is opened, a probe extends and opens the check ball of the combination fitting. Carbon dioxide enters the keg and forces beer up the rod into the beer line and into the faucet. Virtually all large breweries, and many of the micros, are now using this combination.

The SVK and its integral valve and spear tube have evolved over the years into many sizes and varieties. Today's international standard SVK is the 50 liter (13.2 U.S. gal) stainless steel container, although sizes around the world range from the 2.5 liter (0.66 U.S. gal) SVK in Japan to the U.K. SVK barrel that holds almost 164 liters (43.3 U.S. gal). Regardless of the different sizes and valve styles, the SVK system provides a reliable, fully aseptic, returnable container.

Figure 18.1
Single Valve (SVK) Keg

Keg Racking Machines

In the brewing industry, kegs are usually filled with beer using a piece of equipment commonly called a racker. Empty kegs are positioned successively at one end of the racker and pass sequentially through a series of stations or heads where different operations are performed on the kegs. Typically, the kegs pass through a plurality of cleaning heads and a steam sterilizing head before they arrive at a filling head where the kegs are filled. Movement of the kegs between the heads is usually by means of a walking beam conveyor. After filling, the kegs move to a discharge platform where they are removed from the machine. There are two basic designs of keg racking machines—rotary and in-line.

Rotary Machines

On rotary machines kegs are handled simultaneously as they move around in a circular motion on a series of stations. Rotary machines are not suitable for frequent changes in keg sizes. These machines require less space than in-line machines, and on a like-for-like output basis, a rotary system has about half the number of mechanical components compared to an in-line system (7). Rotary systems are capable of kegging up to 2,000 kegs/hour (4).

In-Line Machines

With an in-line system (see Figure 18.2), empty kegs are positioned successively at one end of the racker and pass sequentially through a series of stations or heads where different operations are performed on the kegs. Typically, the kegs pass through a plurality of cleaning heads and a steam sterilizing head before they arrive at a filling head where the kegs are filled. Movement of

Figure 18.2
In-Line Kegging System
Courtesy of IDD Process & Technology Packaging, Inc.

the kegs between the heads is usually by means of a walking beam conveyor. In these systems, kegs are fed into the racker from a number of lanes. The main disadvantage of these multi-head machines is that they are difficult to maintain due to the huge number of valves. Production rates of 1,200 kegs/hour are achievable (4).

Sterilization of Beer

There are two common methods for sterilizing beer for kegging—flash pasteurization (see Figure 18.3) and sterile filtration. Both are done prior to packaging. The flash pasteurization system is considered more complex and requires higher capital investment than sterile filtration. It can lead to a flavor profile change if oxygen in excess of 0.05 mg/l (or ppm) is present during pasteurization or if over pasteurized. Sterile filtration tends to be more expensive than flash pasteurization both in terms of operating and processing costs. With flash pasteurization or sterile filtration, the keg packaging line,

Figure 18.3
Flash Pasteurizer
Courtesy of COMAC Group

and the keg must all be treated as a single clean and sterile system to optimize the reduction of spoilage organisms.

> **Note | Kegging versus Traditional Cask Ale**
>
> Compared with traditional cask ale, keg beer is more resistant to haze formation and microbial contamination. Consequently, the shelf life for keg beer is between 1 and 3 months compared with the 2 to 4 weeks for traditional cask ale.

Kegging Operations

Kegging lines range from systems so simple that washing, inspecting, and filling of each keg are done by hand, to fully automated systems that require virtually no human attendance. The output of a kegging operation ranges from a single-head machine giving an output of 15 kegs per hour, to multi-head machines with outputs more than 1,000 kegs per hour (see Figure 18.4).

The scope of a kegging operation is typically all-inclusive and can be divided into the following steps:

- Keg de-palletizing
- Keg inspection
- External washing
- Internal cleaning
- Keg filling
- Volumetric versus fill by weight measuring methods
- Cold storage

Keg De-palletizing

Kegs can only be transported on pallets and moved to the keg line by a forklift truck. For this reason, kegs are de-palletized by pushing the kegs together and lifting the layer with pneumatically operated grippers.

Keg Inspection

After the kegs are de-palletized, they are tested for internal pressure. Kegs that fail inspection are removed from the line for subsequent inspection and repair.

External Washing

Prior to washing the outside of the keg, it is inverted so that the valve is at the bottom. Before the kegs are cleaned and filled, the protective caps are

Figure 18.4
Complete Kegging Line
Courtesy of COMAC Group

removed from the fittings. The outside of the kegs are then blasted with water and detergent to remove the dirt and labels and rinsed in the tunnel as they are fed by the empty keg conveyor.

Internal Cleaning

The total wash cycle of the inside of the keg consists of a pre-rinse reusing the final rinse water, a detergent wash for removing biological and inorganic contamination, and a final rinse. The purpose of the final rinse is not for washing but for removing chemical contamination from the previous steps. The next step is to sterilize the inside of the keg with steam.

Kegs are always washed upside down. When keg washing is carried out correctly, the wash medium is forced up the inside of the spear tube under regulated pressures and flows to ensure that the medium cascades over the outer surface of the valve spear tube and down the internal walls of the keg. Experience has shown the ideal flow rates to be 7.5 and 49 LPM (2 and 13 U.S. GPM), respectively.

Washing Cycle

The internal washing of kegs has many variants and is influenced by the keg's valve style and shape, and the product to be processed. However, the following steps are commonly used in breweries worldwide for most kegs.

1. The keg is purged of residual beer and gas using CO_2 and/or N_2. After the air-assisted de-ullage purging, the keg can be pressurized to 2 BAR (30 PSIG) and a pressure loss check carried out to ensure that the keg valve is not leaking.
2. A warm water (50°C) pre-rinse is done, with either fresh or recovered water, for a preset time to rinse out any remaining residual product.
3. The keg is purged with air to remove the rinse water and remaining residual product.
4. A hot caustic wash (50–70°C, 2% v/v) is done for a preset time to loosen, break down, and remove proteinaceous materials from the keg's inner surfaces.
5. Caustic soaking time can be added to the sequence if organic materials in the keg are difficult to remove.
6. Depending on the product and the process water hardness, a phosphoric acid wash from 0.2 to 0.4% v/v at 50 to 70°C may be employed after the caustic wash. This will help to prevent calcium precipitants and beer stone from building up on the inner surfaces of the keg.
7. The keg is purged with air and the caustic is transferred to a holding tank where it is reheated and titration is monitored and adjusted.
8. A hot water rinse (50–70°C) for a preset time is done to remove caustic residuals from the keg's inner surfaces.
9. The keg is purged of the rinse water. This water is commonly recovered in a holding tank, where it is then used as the pre-rinse for the next keg to be washed.
10. The next step is to steam sterilize keg by raising the pressure to 1.4 BAR (20 PSIG) to achieve a minimum sterilizing temperature of 110°C.
11. The keg is held at the sterilizing temperature for a preset time to achieve complete sterilization of the inner surfaces and valve components of the keg.

There is little to be gained in using chemicals and rinse waters at titration levels and temperatures above the recommended levels. In fact, many keg valve elastomer seal failures and precipitants present in kegs (such as calcium and beer stone) are due to excessive levels of caustic chemicals and high temperatures.

Keg Filling

Kegs can be filled in the upright or the inverted position. However, all automatic kegging systems today fill kegs in the inverted position. The pros and cons of upright and inverted filling are as follows:

Upright

1. Upright filling is slower when introducing beer into the keg via the spear tube, due to the small cross-sectional area of the inner sections of the valve over that of the outer gas valve.
2. Upright filling is more likely to overfill a keg if an accurate metering or filling by weight system is not used. This condition creates hydraulic pressure in the keg and can lead to dispensing problems if the hydraulic pressure of the beer is too high when the keg is first tapped in the bar.

Inverted

1. It is more difficult to hydraulically over-fill an inverted keg because the valve spear tube is approximately 5 mm (0.2 inches) from the bottom of the keg. Beer will exit through the spear tube and immediately shut off the beer supply on any automatically controlled filler even without a metering device.
2. Inverted filling through the keg valve gas port is faster, but initially more turbulent than upright filling. Some equipment manufacturers have overcome this problem by introducing various methods of variable-flow filling and counter pressure control.
3. Inverted washing and upright filling machines require the keg to be turned upright between the end of the washing cycle and prior to the commencement of the filling cycle. This method of handling keg filling on modern high output systems proves to be cumbersome and inefficient.

Filling Sequence

A typical preparation and filling sequence for a 50 liter (13.2 U.S. gal) keg is as follows:

1. After washing, the inside of the kegs are usually sterilized, typically with steam; however, some brewers will use a no-rinse sanitizer or hot water between 93 and 97°C. If steam is used, the process begins by sterilizing the keg neck and filler connection head for a preset time prior to opening the keg valve.
2. The steam pressure in the keg is set to just above atmospheric pressure.
3. The keg is then purged of steam using an inert gas, such as CO_2, N_2, or a mixture of both to remove condensate.

4. Following the purging the keg is pressurized with gas to a counter-pressurization level that is dependent upon the method and rate of filling used, the beer temperature, and the volume and type of gas or gases in solution in the beer.
5. Filling of the keg is achieved with a beer "overfill" monitoring system, which can be combined with a metering system or a "fill by weight" system at the filling head.
6. After filling, the keg valve is closed, and the residual beer in the connection head of the machine is disposed to a drain or a beer recovery system. The keg valve and neck are in some instances rinsed with sanitizing water prior to being released from the head.

> **Note | Energy and Water Conservation Measures**
>
> There are a number of energy- and water-conservation measures that can be employed when kegging which are: 1) Recovering pre-rinse water and passing it through a heat exchanger to heat up incoming fresh water and cool down effluent prior to disposal. The effluent can also be used in the rinse section of an external washer for kegs prior to disposal. 2) Recovering the final rinse water and using it for the next keg pre-rinse. 3) Recovering the heated cooling water and using it as the final rinse water.
>
> Installing the above conservation measures into a keg line can reduce the system's water consumption by 50% and the steam water heating consumption by up to 90%.

Special attention to filling techniques, with a view to faster but quieter filling, greater fill accuracy, and less product loss is a necessity. Automatic keg packaging lines include a beer boost pump (a pump between the beer buffer tank and the filler) fitted with a variable frequency drive to vary the pump output to match the varying demand of a multi-filling head keg line. Small automatic and semiautomatic keg lines will generally suffice, along with a "flat curve" constant speed pump, to achieve the same near-constant pressure requirement for good filling conditions.

All automatic and semiautomatic machines are fitted with a sensor in the outlet pipework of the filling head to stop the flow of beer when detected by the sensor. This indicates that the keg has filled correctly or that there has been a leak detected across the keg valve and the filling systems connection head.

Other methods of fill-monitoring (in addition to the outlet sensor) are generally added to the filler to comply with fill regulatory requirements and product economy. All kegs are slightly oversized so as to afford headspace expansion.

Volumetric versus Fill by Weight Measuring Methods

As mentioned, the correct amount of beer in the keg can either be measured by a metering system or by a "fill by weight" system. The various types of metering systems include turbine, volumetric, magnetic flow, or mass-flow metering. In many counties; however, volumetric measuring methods do not meet fill regulatory requirements and brewers are required to use a "fill by weight" system for measuring the beer.

The "fill by weight" system works by measuring the gross (filled) weight of a keg after filling and the contents are calculated by subtracting a nominal tare (empty) weight of the keg from this figure. This approach however is very inaccurate since the weight of kegs having, in theory, the same nominal tare weight may vary by as much as +2 kg, due to non-standardized production, damage during use (some kegs may be in use for 25 or more years) and replacement of parts such as the valve mechanism forming part of the keg.

Another method weighs each keg individually at the filling head. The empty weight of the keg is first taken and the weighing system then zeroed. The keg is then filled until a predetermined weight, corresponding to the desired weight of the contents is reached. The filling is then stopped.

Cold Storage

Once palletized, the full kegs are transferred either by conveyor or forklift into cold storage and held at approximately 4 or 5°C to prevent the development of live beer spoilage organisms that might affect the desired flavor and color profile.

References

1. Ault, R.G. and R. Newton. 1971. "Brewing Hygiene and Biological Stability of Beers." *Modern Brewing Technology*, edited by W.P.K. Findlay. Cleveland, Ohio: The Macmillian Press.
2. Babcock Douglas R., Darl Holmer, and Andy Brewer. 1999. "Packaging – Draft Operations." *The Practical Brewer*, edited by John T. McCabe. Wauwatosa Wisconsin: Master Brewers Association of the Americas.
3. Back, W., M. Leibhard, and I. Bohak. 1992. "Flash Pasteurization – Membrane Filtration." *Brauwelt International* 1.
4. Briggs, Dennis E., Boulton, Chris A., Brookes, Peter A., and Stevens, Roger. 2004. *Brewing: Science and Practice*. Cambridge, England: Woodhead Publishing Limited.

5. Broughton, Barry. 1999. Personal Communication. MicroMatic USA, Inc.

6. DeYoung, Robert E. 1977. "Packaging – Bottling Operations." *The Practical Brewer*, edited by H.M. Broderick. Madison, Wisconsin: Master Brewers Association of the Americas.

7. Dunn, Alexander R. 2006. "Packaging Technology." *Handbook of Brewing*, edited by Fergus G. Priest and Graham G. Stewart. Boca Raton, Florida: CRC Press, Taylor & Francis Group.

8. Gunn, Jeff. 1999. Personal Communication. IDD Process & Packaging, Inc.

9. Gunn, Jonathan. 1999. Personal Communication. IDD Process & Packaging, Inc.

10. Hough, J. S., et al. 1982. *Malting and Brewing Science*. Volume 2. London, England: Chapman-Hall, Ltd.

11. Huige, N., G. Sanchez, and J. Surfus. 1989. "Pasteurizer Operation and Control." *MBAA Technical Quarterly* 26 (1).

12. Kunze, Wolfgang. 1996. *Technology Brewing and Malting*, translated by Dr. Trevor Wainwright. Berlin, Germany: VLB Berlin.

13. Lee, Graham. 1999. Personal Communication. IDD Process & Packaging, Inc.

14. Machin, T.J. 1984. "The Kegging Line – A Brewer's Requirements." *The Brewer* 70 (832).

15. Markowski, Phil. 1996. "To Air is Human – Techniques and Technologies for Minimizing Air in Bottled Beers." *Brewing Techniques* 4 (2).

16. O'Rourke, Tim. 1996. "Strict Hygiene Regime is Essential for Trouble-Free Bottle Filling." *Brewers' Guardian* 125 (8).

17. Sharpe, F.R. 1984. "Evaluation and Control of Flavor in Beer." *Brewers' Guardian* 113 (1).

18. Strandberg, Del. 1999. Personal Communication. Spartenburg Stainless Products, Inc.

19. Thompson, Donald, Karl Ockert, and Vince Cottone. 1999. "Craft Brewing." *The Practical Brewer*, edited by John T. McCabe. Wauwatosa Wisconsin: Master Brewers Association of the Americas.

Chapter Nineteen

Beer Spoilage Organisms

Microbial contamination can originate from a variety of sources in the brewing process. Raw materials, air, brewing water, additives, and even pitching yeast can act as a constant supply of contaminants. Residues remaining in brewhouse tanks, pipelines, valves, heat exchangers, and packaging equipment harbor microorganisms too that represent a potential source of recontamination. Some of the effects of contamination range from comparatively minor changes in beer flavor and fermentation performance to gross flavor and aroma defects, turbidity problems, abnormal attenuation rates, and reduced yeast crops.

A number of microorganisms have been reported to be beer spoilage microorganisms, among which bacteria, as well as so-called wild yeast and molds.

Microorganisms – Brewing Stages

From a microbiological standpoint, the brewing process can be broken down into various stages from malting followed by those that include wort production, fermentation, and subsequent beer processing. Microorganism contamination can occur during anyone of these stages. A summary of such microorganisms and the stages of the brewing process at which they are found are outlined in Table 19.1 and will be discussed in more detail further below. The spoilage character of a particular organism will depend on the stage at which it is found in the brewing process.

Bacteria

Beer is a poor and rather hostile environment for most microorganisms. Its ethanol concentration and low pH is lower than most bacteria can tolerate for growth. Furthermore, the high carbon dioxide concentration and extremely

Table 19.1

Occurence of Beer Spoilage Microorganisms at Different Stages of the Brewing Process

Brewing Stage	Beer Spoilage Microorganisms
Malting	Molds
Mashing and wort separation	*Pediococcus, Bacillus, Rahnella, Citrobacter, Klebsiella*
Fermentation	Wild yeasts, *Pediococcus, Selenomonas, Zympohilus, Rahnella, Obsesumbacterium*
Conditioning and packaging	*Pectinatus, Megasphaera*, Lactic acid bacteria
Contaminants of finished beer	Lactic acid bacteria, *Pectinatus, Megasphaera, Zymomonas, Micrococcus*
Dispense	Acetic acid bacteria, Lactic acid bacteria, Wild yeasts

low oxygen content makes beer a near to anaerobic medium. Beer also contains bitter hop compounds, which are toxic. Only a few bacteria are able to grow under such inhospitable conditions and are able to spoil beer. These bacteria include both Gram-positive and Gram-negative species as listed in Table 19.2.

Gram-Positive Bacteria

Gram-positive bacteria are generally regarded as the most threatening contaminants in the brewery because of their rapid growth rate and tolerance to high temperatures and low pH conditions. Most hazardous microorganisms are those belonging to the genera *Lactobacillus* and *Pediococcus* and are often referred to as lactic acid bacteria because of their propensity to produce lactic acid from simple sugars. The optimum pH for growth is about 5.5, but some strains can survive at pH levels as low as 3.5 (15). They are virtually present in every brewery and can cause a variety of problems. These bacteria spoil beer by producing extra-cellular slime (jellylike strands) and cause unpleasant off-flavors such as the sweet butterscotch or honey note provided by diacetyl. Lactic-acid bacteria are often contaminants from pitching yeast or from air. They may be the most significant infectious organism during fermentation and conditioning.

Lactobacillus spp.

The genus *Lactobacillus* is the largest genus of lactic acid bacteria and includes numerous species including *L. brevis, L. pastorianus, L. delbrueckii, L. lindneri, L. curvatus, L. casei, L. buchneri, L. coryneformis, L. plantarum, L. brevisimilis, L. malefarentans,* and *L. parabuchneri.*

Table 19.2
Beer Spoilage Bacteria

Gram-Positive Bacteria	Gram-Negative Bacteria
Lactobacillus spp. 　*L. brevis* 　*L. brevisimilis* 　*L. buchneri* 　*L. casei* 　*L. coryneformis* 　*L. curvatus* 　*L. delbrueckii* 　*L. lindneri* 　*L. maleferentans* 　*L. parabuchneri* 　*L. pastorianus* 　*L. plantarum* *Micrococcus* sp. 　*M. kristinae* *Pediococcus* spp. 　*P. damnosus* 　*P. dextrinicus* 　*P. inopinatus*	Acetic acid bacteria 　*Acetobacter* spp. 　　*A. aceti* 　　*A. ascendens* 　　*A. rancens* 　*Gluconobacter* sp. 　　*G. oxydans* Enterobacteriaceae 　*Citrobacter* spp. 　*Klebsiella* spp. 　*Obsesumbacterium* sp. 　　*O. proteus* 　*Rahnella* spp. *Megasphaera* sp. 　*M. cerevisiae* *Pectinatus* spp. 　*P. cerevisiiphilus* 　*P. frisingensis* *Selenomonas* sp. 　*S. lacticifex* *Zymomonas* spp. 　*Z. anaerobia* 　*Z. mobilis* *Zymophilus* sp. 　*Z. raffinosivorans*

Most lactobacilli bacteria are generally sensitive to hop constituents and their growth is impeded in hopped beers. However, some lactobacilli are hop resistant such as *L. brevis* making them serious wort contaminants (25).

Generally speaking, the lactic acid bacteria are the most troublesome of the beer spoilage bacteria, thriving during the oxygen-deprived stages of fermentation and conditioning. Priest believes that *Lactobacillus* spp. is more dangerous during conditioning and packaging (24).

Lactobacillus spp. can spoil beer by causing acidity, off-flavors, and turbidity. Some *Lactobacillus* spp. are homo-fermentative, resulting in the production of lactic acid; other species are hetero-fermentative and yield a wide range of products, including lactic acid, acetic acid, and diacetyl (28). The most important off-flavor and aroma associated with some strains of lactic acid bacteria is the sweet butterscotch or honey note provided by diacetyl and related vicinal diketones. Diacetyl is especially a problem in lighter-flavored beers where its characteristic aroma and taste is considered highly undesirable, and it is detected readily in lager beers.

Some strains are notable for their ability to produce extra-cellular slime, thereby giving rise to jellylike strands in beer. The strands disappear when the beer is briefly stirred, but later reform as chains upon the beer surface. However, the production of "rope" is not a serious problem with lactobacilli and is more often attributable to pediococci or acetic acid bacteria (26).

The only way to prevent lactic acid infections is by sanitation at every stage of the brewing operation. *Lactobacillus* spp. is found in dried yeast deposits in fermenters and in dried wort droplets on equipment that has not been cleaned thoroughly. Malt husks are another source of lactic acid bacteria. Consequently, it is usually safer to crush the malt in an area well away from the fermentation area to keep malt dust away from the wort and the beer.

L. brevis appears to be the most important beer spoiling *Lactobacillus* species and is detected at high frequency in beer and breweries. It is reported to cause superattenuation due to its ability to ferment dextrins and starch. Over half of the bacterial incidents in brewhouses are caused by this species.

L. pastorianus is generally considered to be the most typical of the beer spoilage *Lactobacillus* spp. Like other rod-shaped bacteria, *L. pastorianus* can cause a silky appearance to contaminated beer. The result of their growth is a marked acidification of the beer and the production of off-flavors, such as butterscotch and diacetyl. The main distinguishing characteristic of *L. pastorianus* is its ability to grow readily in strongly hopped beer. *L. pastorianus var. brownii* is a variant strain of *L. pastorianus* that is often encountered in brewing.

L. delbrueckii grows very quickly in unhopped wort and causes rapid spoilage of mash tun worts. It metabolizes glucose and yields only lactic acid. They are normally killed by the boil, but if the wort is stored at temperatures less than 60°C it provides an ideal growth medium for this bacterium that produces copious amounts of lactic acid (25). Since it cannot grow in either hopped wort or beer, it does not constitute a serious beer spoilage organism.

Pediococcus spp.

Beer spoilage caused by *Pediococcus* spp. is characterized by acid production and more notably the formation of high concentrations of diacetyl, accompanied by a reduction in yeast growth and low fermentation rates (5). Some strains may also form extra-cellular slime, thereby giving rise to jellylike strands in beer. They are found at many stages in the brewing process from wort to finished beer. *Pediococcus* spp. is particularly prevalent as a spoilage organism in beers fermented at low temperatures. Pediococci prefer lower temperatures for growth than do lactobacilli and therefore are less deterred by low storage and conditioning temperatures (26).

Several *Pediococcus* spp. are found in breweries with *P. damnosus* and *P. inopiantus* representing the most common beer spoilers. *P. acidilactici* and *P. pentosaceus* can grow during early stages of the mashing process but are destroyed during boiling or by hop bitterness.

Pediococcus spp. is one of the most feared because of the difficulty in removing it from the brewery. Contamination is most often from calcified trub deposits.

P. damnosus is one of the most common species found in beer (16). It can be found in pitching yeast and in fermenting beer. *P. damnosus*, unlike other strains of bacteria, fails to produce diacetyl (15). Some of its varieties (e.g., *var. diastaticus*) are super-attenuating and some are rope-producing (e.g., *var. viscosus* and *var. limosus*). They are extremely dangerous beer spoilage organisms. It has been suggested that *Pediococcus* growth during beer fermentation takes place primarily in the yeast layer at the bottom of the fermenter after active fermentation has ceased, and also possibly during yeast storage. Because it can grow in anaerobic conditions at relatively low temperatures, *P. damnosus* is prevalent in lager breweries, but it can also be a dangerous spoilage bacterium in naturally conditioned bottled ales and stouts. *P. damnosus* can be minimized by use of a freshly propagated batch of yeast every 10 fermentations or so, and by acid washing the yeast (31).

Gram-Negative Bacteria

Important Gram-negative contaminants in the context of beer brewing are acetic acid bacteria, *Zymomonas* spp., *Pectinatus* spp., and various Enterobacteriaceae. Several members of this group not only distort the fermentation process or produce undesired by-products but also have been reported to survive the fermentation process and to transfer into the finished product.

Acetic Acid Bacteria

The relevant acetic acid bacteria are rod shaped aerobic bacteria from the genera *Acetobacter* and *Glucanobacter*, which are both are capable of converting ethanol into acetic acid. *Acetobacter* is responsible for ropiness of beer whereas *Glucanobacter* causes a cider-like aroma and prefers high gravity environments. They are acid- and alcohol tolerant and unaffected by hop components. The most likely places to find acetic acid bacteria are the yeast propagation tank and bar taps. Like *O. proteus*, acetic acid bacteria can be transferred with the pitching yeast to the following fermentation.

Acetobacter **spp.** – *Acetobacter* spp. is hop-insensitive, non-sporing, and very acid-tolerant. They are particularly known in breweries for their ability to produce acetic acid and, therefore, vinegary off-flavors, turbidity, and ropiness. The surface contamination they cause is often apparent as an oily or moldy film.

Acetobacter spp. are either aerobic or micro-aerophilic (requires oxygen to survive) and develop best in worts exposed to oxygen—e.g., aerated wort at pitching and early fermentation, wort inadvertently aerated at racking, and cask-conditioned or open-fermented beers. As a result of their oxygen requirements, their growth in beer is always concentrated in the surface layer. Growth is inhibited in beer containing more than 8 to 10% alcohol. The greatest danger of infection with this bacteria lies with partially filled storage tanks and beer kegs (10). Like *O. proteus, Acetobacter* spp. can be transferred with the pitching yeast to the following fermentation. Commonly occurring brewery strains are *A. rancens, A. ascendens*, and *A. aceti*, including the subspecies *xylinum*.

Gluconobacter spp. – *Gluconobacter oxydans* (formerly *Acetomonas*) can produce ropiness in beer in the presence of limited supplies of oxygen. *G. oxydans* is active over the entire pH range of the brewing cycle and is not inhibited by isohumulone from the hops.

Zymomonas spp.

Zymomonas spp. is a major anaerobic beer-spoiling bacterium that is responsible for off-flavors like acetaldehyde and rotten apple. It is particularly a problem in the U.K. brewing industry because of the continued practices of beer priming and cask maturation, and also due to the higher temperatures used in brewing ales. *Zymomonas* in primed beer is capable of fast growth; only a few cells per liter can produce spoilage within days. Contamination by *Zymomonas* in the brewery is very often restricted to the packaging stage, although in some cases it has been traced back to the fermentation stage.

Z. anaerobia – *Zymomonas anaerobia* is described as the most dangerous of *Zymomonas* spp. since it can cause spoilage within a relatively short time. *Z. anaerobia* rapidly converts glucose and fructose (some strains attack sucrose as well) into alcohol and carbon dioxide, causing considerable turbidity and an unpleasant odor of rotten apples. No acetic acid is produced, and the bacteria die off very quickly at the end of fermentation. Anaerobic conditions and the presence of priming sugars (as in cask and keg beers) are ideal conditions for growth of these bacteria, and in warm weather complete beer spoilage can occur within 48 hours.

Z. mobilis – *Z. mobilis* grows under micro-aerophilic conditions. Spoilage in beer can occur within a few hours; however, the organism is comparatively rare. *Z. mobilis* may be introduced by dirty keg fillers, by the brushes at cask washing, or from soil and dust during brewery reconstruction. Infection is normally confined to traditional cask-conditioned ales, i.e., sugar-primed, unpasteurized beer (15).

Pectinatus spp.

Pectinatus cervesiiphilus and *P. frisingens* are two common beer spoilage organisms. They are strictly anaerobic rod shaped and produce a range of off-flavors like hydrogen sulfide, acetic acid, and propionic acids in wort and packaged beers. They can survive in low alcoholic beer and have become more important with the increasing production of non-pasteurized or flash pasteurized beer. Infections are normally encountered at the bottling stage.

P. cervesiiphilus has proved to be very tolerant towards hop bitterness and is capable of causing beer spoilage by production of butyric acid, hydrogen sulfide and the development of turbidity.

Enterobacteriaceae

Members of the Enterobacteriaceae associated with spoilage (*Obesumbacterium, Klebsiella, Rahnella,* and *Citrobacter*) are related to those of *Escherichia coli*. For this reason, they are often referred to as coliform. They can grow in the presence or absence of oxygen, which in the context of brewing increases the number of locations where spoilage can occur. They are all tolerant of hop resins and can ferment in a wide range of sugars but cannot utilize ethanol (5). The cooled wort is an ideal medium for their growth especially if the cooled wort is left unpitched for any length of time (25). They can also grow during the early stages of fermentation. Although they are active over a wide temperature range, most enterobacterial species are inhibited below pH 4.4. Hough reports they are not commonly encountered during the later stages of brewing since the pH of the beer is too low for growth and the ethanol content is greater than 2.0% (v/v) (15).

Their main effect, if allowed to grow in wort, is the production of dimethyl sulfide (DMS) and the production of phenolic flavor compounds.

These organisms gain access into breweries via the shoes of operators, the water supply, unwashed hands, dirty equipment, and even airborne dust particles. However, the source is most often the rinsing water. Once they have developed in an area which cannot easily be cleaned, it is extremely difficult to free the plant from infection. Low-pressure steaming is often ineffective as a sanitizing procedure because the bacteria are thermophilic and protected by condensate and dead bacteria.

Obesumbacterium proteus – *Obesumbacterium proteus* is a very common Gram negative beer spoilage bacterium and poses a big risk in breweries. Formerly known as *Hafnia protea*, *O. proteus* develops only during the early stages of fermentation. It is generally considered a wort spoiler by producing diacetyl and dimethyl sulfide, and it may impart a very noticeable parsnip odor to the wort. *O. proteus* can reduce nitrate to nitrite, which results in a decrease in fermentation rate, a reduction in yeast crop, a slowing in the rate of pH fall, and an increase in both beer color and vicinal diketone content.

O. proteus is sufficiently resistant to survive long enough into fermentation to be passed on via the yeast collected for re-pitching. It is a noted contaminant of pitching yeast in ale breweries (18). Almost all pitching yeasts in commercial breweries are contaminated with *O. proteus* to some extent. Low levels of contamination are usually considered normal and acceptable, since the presence of only a small number of these bacteria have little effect on the flavor of the finished beer.

Breweries that acid-wash their yeasts periodically are able to hold down the numbers of these bacteria. Acid-washing is necessary because otherwise, *O. proteus* will continue to multiply from one pitching to the next, until it reaches a point where it adversely affects the flavor or the fermentation itself. The best protection against the spread of this type of infection is a vigorous main fermentation to bring about a rapid decrease of pH, as growth rarely proceeds below a pH of 4.40.

Wild Yeast

Wild yeast is any yeast other than the pitching yeast. Wild yeast can be isolated at all stages of the brewing process from raw materials, wort, pitching yeast, and fermenting beer, through to the packaged product and the dispense system. Wild yeast can produce unintended flavors because of differences in ester, fusel alcohol, and diketone production. They are particularly known for producing phenolic or medicinal notes. In the presence of air, some wild yeast can grow rapidly and form a film on the surface of the beer, which can cause haze. Other effects may include primary yeast fermentation and separation difficulties, significantly lower terminal gravities, and a higher, alcohol content in the finished beer. The lower terminal gravities are due to the ability of wild yeast to ferment sugars (such as maltotetraose and dextrins) not used by the primary yeast.

Sources of contamination can be wind-blown dust particles, malt, clothing, and brewery equipment. Wild yeast infection is usually more of a problem for brewers not having a pure culture yeast propagation system or utilizing multiple strains for different styles of beer.

The wild yeast identified as beer-spoilers are traditionally divided into *Saccharomyces* and the non-*Saccharomyces* yeast genera such as *Brettanomyces, Torulopsis, Pichia,* and *Candida.*

Control of wild yeast can be difficult for two main reasons: first, they are difficult to detect; second, unlike bacteria, they are not susceptible to acid-washing. Wild yeast cells cannot be detected microscopically because of their similarity to pitching yeast and the fact that they usually occur in small numbers in proportion to pitching yeast. To ensure that the pitching yeast is free from wild yeast, the brewery needs a regular regime of introducing pure culture yeast. Fining agents have limited use since wild yeast do not flocculate well and do not have a strong negative charge (15).

Saccharomyces spp.

Saccharomyces spp. in general is considered to be the most widespread and the most hazardous and for that matter the greatest threat since they are similar to production strains. The majority of the *Saccharomyces* spp. brewing contaminants detected belong to *S. cerevisiae* but also other *Saccharomyces* spp. have been reported. *S. diastaticus* is capable of fermenting complex sugars—maltotetraose and often dextrins—that brewer's yeast cannot handle, and has been associated in cases of superattenuation in beer (15). *S. diastaticus* is known for producing medicinal off-flavors. *S. pastorianus* is can grow in wort and produce off-flavors in the product. Common offenders of beer hazes include *Saccharomyces cerevisiae var. turbidans* and *var. ellipsoideus*, *S. carlsbergensis*, and *S. pastorianus*. The haze-forming strains of *S. cerevisiae* and *S. carlsbergensis* are usually non-fining yeasts, and their main effect on flavor is to produce a thin-drinking beer.

Certain strains of yeast, including *S. cerevisiae* and *S. carlsbergensis*, have developed the ability to secrete polypeptide "zymocin" that are capable of killing sensitive strains of yeast over a particular range of pH values (5). Under certain circumstances, these killer strains eliminate the culture strain and dominate fermentation rather than simply compete with culture yeast for available nutrients.

Non-Saccharomyces spp.

There are many non-*Saccharomyces* yeasts found in the brewery but most cannot compete with brewing yeasts. Non-*Saccharomyces* yeasts encountered in the brewery include representatives of the following genera—*Brettanomyces, Torulopsis, Pichia,* and *Candida*.

Brettanomyces spp.

Less frequently encountered than the genus *Saccharomyces*, but more troublesome as a producer of acetic acid off-flavor is the genus *Brettanomyces* (6). Cider-like tones and clove/medicinal aromas will be noted in the infected beer. The anaerobic conditions generated by active fermentation prevent further growth of this yeast; however, it is transferred with the pitching yeast to the following fermentation, where it continues growth. *Brettanomyces* spp. may take as long as six to eight weeks before they begin to produce off-flavors. This can be a serious problem in packaged beer, which may be on the shelves for 30 to 90 days. On the other hand, some *Brettanomyces* strains (e.g., *B. bruxellensis* and *B. lambicus*) are beneficial in Belgian lambics and barley wines used for the production of esters (10).

Torulopsis, Pichia, and *Candida* spp.

Turbidity may also arise from species of *Torulopsis*, which may consist of very small cells and are therefore slow in sedimenting (15). Breaking up of the film or pellicle produced by *Pichia* and *Candida* spp. may also produce turbidity, of which *Pichia membranaefaciens* and *Candida vini* are frequently recorded (6). *Pichia* and *Candida* species will oxidize ethanol in a vessel exposed to air to produce acetic acid.

Molds

Molds are non-chlorophyll-bearing plants that range in size from a single spore to large cell aggregates. Commonly occurring genera are *Mucor, Penicillium, Aspergillus, Cladosporium, Geotrichum,* and *Rhizopus*. Most molds are able to grow well at ordinary temperatures, the optimum range being from 25 to 30°C, although some species can grow at 35°C or even higher temperatures and others at much lower temperatures. Molds are normally aerobic organisms and can grow over a wide pH range, although most species prefer an acid pH.

Malt and adjuncts are sources of mold largely as a result in the poor handling of these ingredients during the manufacture process. The most widely recognized defect ascribed to the growth of mold on malts is that of gushing. Gushing occurs in bottled beers where on opening there is a sudden loss of carbon dioxide that causes foaming.

Mold contamination can occur when the air contains considerable quantities of mold spores too. In breweries, mold growth normally occurs on damp walls, floors, and ceilings; in wooden vessels, dirty casks, and bottles; and almost anyplace contaminated with beer residues.

Microbiological Quality Assurance

All breweries need quality assurance to maintain confidence in the beer they produce. It is required for a variety of functions from checking the quality of the raw materials, through monitoring beer production and packaging operations, to checking final product quality.

There are essentially two approaches to microbiological testing, the conventional techniques that involve inoculating a solid or liquid medium with a brewery sample, and after incubation, examining for the presence of absence of growth. General-purpose media for the cultivation and identification of microorganisms can be prepared from wort or beer but commercial media provide for consistency and ease of use. Although these traditional methods will continue to have a place in the brewery, they are very time consuming and do not yield immediate results. Minimizing the time needed to detect microorganisms can lead to significant savings through improved consistency of product quality. As a result, this has led to the development of rapid microbial detection and identification techniques, such as the ATP bioluminescence method, which is fast, easy to perform and yields reliable data.

References

1. Allen, Fal. 1994. "The Microbrewery Laboratory Manual – Part II: Bacteria Detection, Enumeration, and Identification." *Brewing Techniques* 2 (5).

2. Allen, Fal. 1994. "The Microbrewery Laboratory Manual – Part III: Wild Yeast Detection and Remediation." *Brewing Techniques* 2 (6).

3. American Society of Brewing Chemists (ASBC). 1992. "Yeast-3.c Triphenyltetrazolium Chloride for Identification of Respiratory Deficient Cells." *Methods of Analyses*, 8th ed., The Society, St. Paul, Minnesota.

4. Ault, R.G. and R. Newton. 1971. "Spoilage Organisms in Brewing." *Modern Brewing Technology*, edited by W.P.K. Findlay. Cleveland, Ohio: The Macmillan Press.

5. Briggs, Dennis E., Boulton, Chris A., Brookes, Peter A., and Stevens, Roger. 2004. *Brewing: Science and Practice*. Cambridge, England: Woodhead Publishing Limited.

6. Campbell, I. 1987. "Microbiology of Brewing: Beer and Lager." *Essays in Agricultural and Food Microbiology*, edited by J.R. Norris and G.L. Pettipher. Chichester [West Sussex], England: John Wiley & Sons Ltd.

7. Campbell, I. 1987. "Wild Yeasts in Brewing and Distilling." *Brewing Microbiology*, edited by F.G. Priest and I. Campbell. Essex, United Kingdom: Elsevier.

8. Dowhanick, Terrance M. 1990. "Is There a Need for Microbiological Quality Control in a Microbrewery." *Brewers Digest*. December.

9. Ernandes, J. R., J. W. Williams, I. Russel, and G. G. Stewart. 1993. "Respiratory Deficiency in Brewing Yeast Strains – Effects on Fermentation, Flocculation, and Beer Flavor Components." *J. Am. Soc. Brew. Chem.* 51.

10. Fix, George J. 1989. *Principles of Brewing Science*. Boulder, Colorado: Brewers Publications.

11. Fix, George J. 1990. "Banishing Bacteria." *The New Brewer* 7 (1).

12. Fix, George and Laurie Fix. 1997. *An Analysis of Brewing Techniques*. Boulder, Colorado: Brewers Publications.

13. Gilliland, R.B. 1981. "Brewing Yeast." *Brewing Science*. Volume 2, edited by J.R.A. Pollock. London, England: Academic Press.

14. Giorgio, Greg. 1990. "QC: Fact of Fiction?" *The New Brewer* 7 (1).

15. Hough, J. S., et al. 1982. *Malting and Brewing Science*. Volume 2. London, United Kingdom: Chapman-Hall, Ltd.

16. Annual Meeting American Society Brewing Chemists. 1968.

17. Lawrence, D. R. 1988. "Spoilage Organisms in Beer." *Developmental Food Microbiology*. Volume 3. London, United Kingdom: Elsevier.

18. Lewis, Michael J. and Tom W. Young. 1995. *Brewing*. London, England: Chapman & Hall.

19. Mackie, T.J. and J. E. McCartney. 1953. *Handbook of Practical Bacteriology*. Edinburg, United Kingdom: Churchill Livingstone.

20. Maule, D. R. 1980. "Propagation and Handling of Pitching Yeast." *Brewers Digest* 55 (2).

21. Middlekauff, James E. 1995. "Microbiological Aspects." *Handbook of Brewing*, edited by William A. Hardwick. New York, New York: Marcel Dekker, Inc.

22. Morris, Rodney L. 1994. "Simple Detection of Wild Yeast and Yeast Stability." *Brewing Techniques* 2 (3).

23. Piesley, John G. and Tomas Lom. 1977. "Yeast – Strains and Handling Techniques." *The Practical Brewer*, edited by H. M. Broderick. Madison, Wisconsin: Master Brewers Association of the Americas.

24. Priest, F.G. 1987. "Gram-Positive Brewery Bacteria." *Brewing Microbiology*, edited by F.G. Priest and I. Campbell. Essex, United Kingdom: Elsevier.

25. Priest, Fergus G. 2006. "Microbilogy and Microbiological Control in the Brewery." *Handbook of Brewing*, edited by Fergus G. Priest and Graham G. Stewart. Boca Raton, Florida: CRC Press, Taylor & Francis Group.

26. Rainbow, C. 1981. "Beer Spoilage Microorganisms." *Brewing Science*. Volume 2, edited by J.R.A. Pollock. London, United Kingdom: Academic Press.

27. Richards, M. Wallerstein Lab Communication. 33 (11), 1970.

28. Russell, Inge. 1995. "Yeast." *Handbook of Brewing*, edited by William A. Hardwick. New York, New York: Marcel Dekker, Inc.

29. Simpson, W. J. 1989. "Instant Assessment of Brewery Hygiene Using ATP-Bioluminescense." *Brewers' Guardian*. 118 (11).

30. Simpson, W.J. 1991. "Rapid Microbiological Methods in the Brewery." *Brewers' Guardian*. 120 (6).

31. Simpson, W.J. and J.R. M. Hammond. 1990. "A Practical Guide to the Acid Washing of Brewer's Yeast." *Ferment*, 3 (6).

32. Van Vuuren, H.J.J. 1987. "Gram-Negative Spoilage Bacteria." *Brewing Microbiology*, edited by F.G. Priest and I. Campbell. Essex, United Kingdom: Elsevier.

Chapter Twenty

Wastewater and Solid Waste Management

The beer brewing process generates large amounts of wastewater effluent and solid wastes that must be disposed of or treated in the least costly way to meet strict discharge regulations set by government entities.

Brewery wastewater typically has a high biochemical oxygen demand (BOD) from all the organic components (sugars, soluble starch, ethanol, volatile fatty acids, etc). Brewery wastewater usually has temperatures ranging from 25°C to 38°C. The pH levels can range between 2 and 12 and are influenced by the amount and type of chemicals used in cleaning and sanitation (e.g., caustic soda, phosphoric acid, nitric acid, etc.). Nitrogen and phosphorus levels are mainly dependent on the raw material and the amount of yeast present in the effluent.

Solid wastes consist of spent grains, spent hops, diatomaceous earth used for filtering, waste yeast, and trub. Broken glass, grit, paper, wood, and bottle caps that are classified as solid wastes too.

Treatment is normally provided either at the brewery or at a municipal wastewater treatment plant. Some of techniques for treating wastewater effluent at the brewery include "equalizing" the wastewater in holding tanks, screening, pH correction, sedimentation, and biological treatment (i.e., aerobic and anaerobic).

Brewery Wastewater

Wastewater is one of the most significant waste products of brewery operations. Even though substantial technological improvements have been made in the past, it has been estimated that approximately 3 to 10 liters of waste effluent is generated per liter of beer produced in breweries (7). The quantity of brewery wastewater will depend on the production and the specific water usage. Some of the biggest water uses in a brewery include bottle and keg washing, cooling, pasteurization, cleaning-in-place (CIP) equipment, line and filler flushing water. Smaller breweries or for that matter older breweries will typically discharge proportionately larger volumes of water compared to larger more modern breweries.

Characteristics of Brewery Wastewater

Wastewater is mostly water by weight. Other materials make up only a small portion of wastewater, but can be present in large enough quantities that may require some pretreatment before discharging the wastewater into the sewage system. Some of these materials found in brewery wastewater are:

- Spent grains
- Spent hops
- Pre- and after-runs from diatomaceous earth filtration and filling
- Waste yeast
- Trub
- Weak wort discharge
- Acid and alkaline solutions used for cleaning and disinfecting
- Rejected beer in the packaging area
- Returned beer
- Residual beer from returned bottles, casks, and kegs
- Ancillary materials used in packaging area e.g., label pulp, grit, broken glass, conveyor lubrication, and label glue

The composition of brewery effluent can fluctuate significantly throughout the day as it depends on various processes that take place within the brewery (raw material handling, wort preparation, fermentation, filtration, CIP, packaging, etc). For example, bottle washing produces a large volume of effluent that, however, contains only a minor part of the total organics discharged from the brewery. Effluents from fermentation and filtering are high in organics and biochemical oxygen demand (BOD) but low in volume. Effluent pH averages about 7 for the combined effluent but can fluctuate from 2 to 12 depending on the use of acid and alkaline cleaning and sanitizing agents. Effluent temperatures average about 30°C.

Wastewater Quality

Wastewater quality may be defined by its physical and chemical characteristics.

Physical Characteristics

Physical parameters include color, odor, temperature, and turbidity. Insoluble contents such as total solids (TS), oil and grease also fall into this category. Total solids may be further subdivided into total suspended solids (TSS) and total dissolved solids (TDS). TSS is the solids in water that can be trapped by a filter including a wide variety of organic and inorganic material. TDS is the amount of dissolved substances, such as salts or minerals, in water remaining after evaporating the water and weighing the residue.

Chemical Characteristics

Chemical parameters associated with the organic content of waste-water include biochemical oxygen demand (BOD), chemical oxygen demand (COD), total organic carbon (TOC), and total oxygen demand. Inorganic chemical parameters include salinity, hardness, pH, acidity and alkalinity, as well as concentrations of ionized metals such as iron and manganese, and anionic entities such as chlorides, sulfates, sulfides, nitrates, and phosphates.

Solid Wastes

The most cost-effective method for significantly reducing effluent load of brewery wastewater is to separate the solid wastes from the wastewater itself. The equipment necessary includes holding vessels, tanker trucks that can haul away the material, pumps, and dedicated piping or hoses for transfer. Typical solid wastes include spent grains, trub, spent yeast, diatomaceous earth slurry from filtration, and packing materials.

Spent Grains

Beer production results in a variety of residues, such as spent grains, which have a commercial value and can be sold as byproducts for livestock feed. The nutritional value of spent grain is much less than that of the same amount of dried barley, but the moisture makes it easily digestible by livestock. Farmers often mix it with silage to help prolong its storage time. Mushroom growers also use spent grain mixed with wood or grasses as an excellent growing medium. Autolyzed yeast, trub, spent hops and even used diatomaceous earth may be added to spent grains too. A range of other uses for spent grains has been proposed, as a source of biogas, but this may be more than most small brewers are willing to take on. Composting is another choice and, in many cases, may be a more viable option than biogas production. Some breweries use compost to enrich the soil of their own brewery herb and vegetable gardens. Another option is reusing spent grain in products like bread mixes or pizza dough. Spent grains are seldom dried commercially given the additional cost for drying-facilities and energy though they may be partially dried.

Trub

Trub is slurry consisting of wort, hop particles, and unstable colloidal proteins coagulated during the wort boiling. The trub suspension can be treated in different ways, added to the spent grains or sent directly to the sewer system.

To minimize the BOD/COD load and the extract loss, discharge of trub to the sewer system should be avoided.

Spent Yeast

In brewing, surplus yeast is recovered by natural sedimentation at the end of the fermentation and conditioning. Only part of the yeast can be reused as new production yeast. Spent yeast is very high in protein and B vitamins, and may be given to livestock as a feeding supplement. However, live yeast may harm livestock and should be killed before release by the brewery. Heating is the most common method of killing yeast. The stirred mixture can be brought up to 55–60°C and held for 12 hours for a complete kill (12). Yeast can also be used in pharmaceutical industry applications too.

Typically 2–4 kg of yeast (10–15% dry matter) is produced per hectoliter of beer. The yeast suspension contains yeast and beer and has a very high COD value (180,000–220,000 mg/l). It is important to collect as much surplus yeast as possible to avoid high COD discharge to the sewer system.

Diatomaceous Earth (DE) Slurry

Diatomaceous earth slurry from the filtration of beer also constitutes a very large category high in SS and BOD/COD. Different methods for regeneration are under development, but presently they are not capable of totally replacing new diatomaceous earth. Therefore, the disposal of diatomaceous earth is limited to landfills or used as a soil amendment. The spent diatomaceous earth, with the yeast cells it contains, is a valuable nitrogen-rich soil improver and plant fertilizer for farmers. Moreover, it also improves the water retention capacity of sandy soils without adversely affecting the ground-water. Sometimes the used filter-aid has been mixed with the spent grains, but this is not always acceptable. Utilization in building materials is also possible in the manufacture of bricks, asphalt or concrete. One can expect about 500 g of diatomaceous earth slurry per hl of filtered beer.

Packaging Materials

Other solid wastes include label pulp from the washing of returnable bottles, broken glass, cardboard, bottle caps, and wood that is usually disposed of at sanitary landfills.

Brewery Wastewater Treatment

Wastewater from a brewery may be discharged into either a waterway (rivers, streams, or lakes), directly into a municipal sewer system, or into the municipal sewer system after the wastewater has undergone some treatment.

If discharged into a waterway, the brewery will more than likely be subject to the Clean Water Act's National Pollution Discharge Elimination System (NPDES) permit program. These permits regulate what individual facilities can discharge into a waterway setting limitations on wastewater pollutant parameters such as organic load, suspended solids, pH, temperature, and chlorine.

If the brewery discharges wastewater directly into a public sewer system, the publicly owned treatment works (POTW) will usually set limits on the composition, volume, rate of flow, temperature, and pH of the wastewater effluent. For small breweries wastewater treatment may not be required. However, in the case of larger breweries some treatment may be needed given the negative impact the load and/or chemicals could have on the municipal sewer systems, not to mention the wastewater user fees levied by the POTW.

Most municipalities levy a range of fees to encourage breweries to reduce load/chemical charges into the municipal sewer system. For small breweries their water and sewage costs are typically minimal, and wastewater costs are not typically a major financial concern. Obviously, wastewater costs are a bigger concern for large breweries if only for reasons of scale. Breweries can exert a large demand on municipal wastewater treatment systems because of the high levels of organic materials in the wastewater, and may therefore be incur substantial fees assessed by the POTW.

The fees can include application fees, industrial permit fees, volume discharge-limit fees to expensive surcharge fees that are assessed when certain wastewater pollutant parameters (e.g., BOD, SS, nitrogen, etc.) exceed those set by the POTW. These fees are expected to increase given growing environmental concerns and aging municipal sewer systems. Consequently, breweries should always consult the POTW about wastewater fees as well as limiting factors such wastewater pollutant parameters (e.g., rate of flow, etc.) and municipal treatment capacity that may affect them.

Wastewater treatment aims at the removal of unwanted components in order to provide safe discharge into the waterway or sewer system. This can be achieved by using physical, chemical, and biological means, either alone or in combination. Some breweries do not carry out any treatment on-site, but those that do typically treat the wastewater with the objective of reducing the temperature to a moderate level, to restrict the pH to a specified range, and to reduce the BOD, COD and SS levels to below specified levels as set by the POTW.

Physical Treatment

Physical treatment is for removing coarse solids and other large materials, rather than dissolved pollutants. It may be a passive process, such as sedimentation to allow suspended pollutants to settle out or float to the top naturally. Other typical physical treatments include flow equalization, screening, and grit removal. Physical treatment is necessary to enhance the operation and efficiency of subsequent treatment steps.

Flow Equalization

Flow equalization is a technique used to consolidate wastewater effluent in holding tanks for "equalizing" before introducing wastewater into downstream brewery treatment processes or for that matter directly into the municipal sewage system. This method levels out operation parameters such as flow, pollutant levels, and temperature over a period of time. Holding tanks are usually equipped with agitators or aerators where mixing of the wastewater is desired and to prevent suspended solids from settling to the bottom of the unit. In many cases, flow equalization might be sufficient to meet local regulations.

Screening

Typically, the wastewater is first screened to remove glass, labels, and bottle caps, floating plastic items and spent grains. Manual bar screens may be adequate to smaller breweries; however, mechanical screens are normally used for much larger operations. These screens may be of many types, for instance, hyperbolic bar screens or screens of woven stainless steel mesh. The choice of aperture will affect the quantity and quality of the screening captured.

Grit Removal

After the wastewater has been screened, it may flow into a grit chamber where sand, grit, and small stones settle to the bottom. The grit and screenings removed by these processes must be periodically collected and trucked to a landfill for disposal or are incinerated. Not removing the grit that washes off the brewery floor could cause abrasive problems on pipes and pumps and sludge handling equipment.

Gravity Sedimentation

With the screening completed and the grit removed, wastewater still contains dissolved organic and inorganic constituents along with suspended solids. The suspended solids consist of minute particles of matter that can be removed from the wastewater with further treatment such as sedimentation or chemical flocculation.

Gravity sedimentation is one of the most frequently used processes in wastewater treatment. Many wastewaters contain settleable suspended solids that can be removed by gravity sedimentation. Settled solids (primary sludge) are normally removed from the bottom of tanks by sludge rakes that scrape the sludge to a central well from which it is pumped to sludge processing units. Scum is swept across the tank surface by water jets or mechanical means from which it is also pumped to sludge processing units.

In other cases where suspended materials do not settle readily, upstream unit processes are used to convert colloidal (non-settleable suspended solids) and soluble pollutants into settleable suspended solids. Suspended solids removal is important because of the pollutants associated with the removed solids, such as organics, nutrients (nitrogen, phosphorus), and heavy metals. Suspended diatomaceous earth and some yeast may be removed at this stage too. The wastewater effluent may also pass through oil traps.

> **Note | Dissolved Air Flotation**
>
> Dissolved air flotation (DAF) is an alternative to conventional gravity sedimentation as a way in treating brewery wastewater effluent. The DAF is a liquid-solid separation process in which microscopic air bubbles become attached to solids and then transported to the surface to form a floating blanket. Here a slowly rotating arm skims the top layer of thickened sludge. This treatment is different from conventional sedimentation tanks where suspended solids settle to the bottom under the influence of gravity. To improve solids separation, chemicals are added to the waste streams to achieve coagulation and flocculation of suspended solids (TSS).
>
> DAF has several advantages over conventional sedimentation which are:
> - More compact design
> - Shorter start-up time
> - Better performance
> - Lower chemical dose
> - Shorter flocculation time
> - Thicker sludge

Chemical Treatment

Among the chemical treatment methods, pH adjustment and flocculation are some of the most commonly used at breweries in removing toxic materials and colloidal impurities.

pH Adjustment

The acidity or alkalinity of wastewater affects both wastewater treatment and the environment. Low pH indicates increasing acidity while a high pH indicates increasing alkalinity (a pH of 7 is neutral). The pH of wastewater needs to remain between 6 and 9 to protect organisms. Alkalis and acids can alter pH thus inactivating wastewater treatment processes. Federal law, 40 CFR Sec. 403.5(3) (2), requires that the pH of the wastewater stream be adjusted to greater than 5.0 prior to discharge into the municipal sewer system (12). The upper range is set by the POTW and is usually less than pH 11.5.

The most common method in adjusting pH is by the addition of an acid or alkali to the wastewater at a range stipulated by federal regulations and/or the POTW. To raise pH, basic compounds such as sodium hydroxide (caustic soda), calcium oxide (lime), calcium hydroxide (hydrated lime), or sodium carbonate (soda ash) are used. To lower pH, sulfuric or hydrochloric acid may be used. Small breweries can implement a simple one- or two-tank batch system with above-ground or below-ground holding tanks. Larger plants usually have a central collection plant and up to two or three mixing and adjustment tanks, each outfitted with pH probes, mixers, chemical injectors, and sometimes aerators. When the pH of the wastewater in the tank reaches the required range, it is pumped out for further treatment (e.g., flocculation and biological) on-site or diverted to the POTW.

Flocculation

Flocculation is the stirring or agitation of chemically-treated water to induce coagulation. Flocculation enhances sedimentation performance by increasing particle size resulting in increased settling rates. If employed flocculation precedes the sedimentation process and usually consists of a rapid mix tank or in-line mixer, and a flocculation tank. The wastewater is initially mixed while a coagulant and/or a coagulant aid is added. A rapid mix tank is usually designed for a detention time of 15 seconds to several minutes. After mixing, the coagulated wastewater flows to a settling tank where slow mixing of the waste occurs. The slow mixing allows the particles to agglomerate into heavier, more settleable solids.

The chemicals used in the coagulation/flocculation process include alum and lime as well as a range of polymers (large molecules that can trap other substances). If lime is used, the resulting sludge has enhanced agronomic value.

Biological Treatment

After the brewery wastewater has undergone physical and chemical treatments, the wastewater can then undergo an additional biological treatment. Biological treatment of wastewater can be either aerobic (with air/oxygen supply) or anaerobic (without oxygen), which are discussed in more detail in the following sections. Generally, aerobic treatment has been applied for the treatment of brewery wastewater and recently anaerobic systems have become an attractive option.

Adoption of two-step anaerobic/aerobic wastewater treatment systems is increasingly adopted by breweries worldwide. This technique has the benefits of much reduced footprint, substantial electricity savings, and generation of biogas which can be used in boilers or for power generation.

Biological treatment systems use microbes which consume, and thereby destroy, organic compounds as a food source. The microbes use the organic compounds as both a source of carbon and as a source of energy. These microbes may also need supplemental nutrients for growth, such as nitrogen and phosphorus, if the waste stream is deficient in these nutrients. Aerobic microbes require oxygen to grow, whereas anaerobic microbes will grow only in the absence of oxygen. Facultative microbes are an adaptive type of microbe that can grow with or without oxygen.

The success of biological treatment is dependent on many factors, such as the pH and temperature of the wastewater, the nature of the pollutants, the nutrient requirements of the microbes, the presence of inhibiting pollutants, and variations in the feed stream loading. Certain compounds, such as heavy metals, may be toxic to the microorganisms and must be removed from the waste stream prior to biological treatment.

Although anaerobic technology is the most economical bioprocess due to lower operating costs for aeration and sludge processing, it is not cost effective for small breweries. In the case of small breweries, aerobic technology is the best option and at present aerobic treatment is still widely used both by small and large breweries.

Aerobic Wastewater Treatment

Aerobic biological treatment is performed in the presence of oxygen by aerobic microorganisms (principally bacteria) that metabolize the organic matter in the wastewater, thereby producing more microorganisms and inorganic end-products (principally CO_2, NH_3, and H_2O).

The bacteria in aerobic systems are generally more tolerant to pH and temperature swings than anaerobic systems. The pH of the system should range from 7.0 to 7.5. Aerobic bacteria are less sensitive to alkaline pH than to acidic pH, and systems operating below pH 6 may encounter problems.

Aerobic treatment utilizes biological treatment processes, in which microorganisms convert nonsettleable solids to settleable solids. Sedimentation typically follows, allowing the settleable solids to settle out. Three options include:

1. Activated sludge process
2. Attached growth (biofilm) process
3. Lagoons

Activated Sludge Process

In the activated sludge process, the wastewater flows into an aerated and agitated tank that is primed with activated sludge. This complex mixture containing bacteria, fungi, protozoans, and other microorganisms is referred to collectively as the biomass. In this process, the suspension of aerobic microorganisms in the aeration tank, are mixed vigorously by aeration devices which also supply oxygen to the biological suspension. Aeration devices commonly used include submerged diffusers that release compressed air and mechanical surface aerators that introduce air by agitating the liquid surface. Sufficient air must be provided to supply the biomass with the oxygen necessary for respiration. If too little air is introduced into the aeration tanks, the biomass will use anaerobic respiration to metabolize the organic matter, producing foul odors and poor effluent quality. Hydraulic retention time in the aeration tanks usually ranges from 3 to 8 hours but can be higher with high BOD wastewaters.

After the wastewater and biomass have been aerated for a sufficient period to allow the soluble BOD to be incorporated into the cells of the biomass, the mixture flows to the final, or secondary settling tanks. Since these tanks are not aerated or agitated, the biomass is allowed to settle. The remaining effluent, which by this point in the process appears quite clear, is ready to be discharged into the sewer system. A portion of the settled sludge (biomass) is reintroduced into an aeration tank to serve as the inoculum for the next batch, and the excess sludge disposed of.

The activated sludge process performs well as long as the system is designed and operated correctly. However, the process requires constant operator supervision, laboratory testing and monitoring to assure performance, and consistent maintenance of mechanical equipment. The process also produces a large amount of waste sludge that must be stabilized and disposed of (up to 0.65 lbs. of sludge for each lb. of BOD treated), and requires a great deal of energy to process the wastewater. The activated sludge system is very inefficient for lower than design flows and is not tolerant of shutdown periods or highly varying flows. The overall operational cost of the activated sludge process can be very high as compared to attached-growth systems.

To ensure biological stabilization of organic compounds in activated sludge systems, adequate nutrient levels must be available to the biomass. The primary nutrients are nitrogen and phosphorus. Lack of these nutrients can impair biological activity and result in reduced removal efficiencies. As a result, nutrient supplements (e.g., phosphoric acid addition for additional phosphorus) have been used in activated sludge systems.

Attached Growth (Biofilm) Process

The second type of aerobic biological treatment system is called "Attached Growth (Biofilm) Process" and deals with microorganisms that are fixed in place on a solid surface. This "attached growth type" aerobic biological treatment process creates an environment that supports the growth of microorganisms that prefer to remain attached to a solid material.

The material that the microorganisms attach themselves to is called "media." This media can be any solid material that the microorganisms are able to attach to, either natural or synthetic. The natural material used in wastewater treatment will be stone or rock, while the synthetic material is usually plastic or textile fibers.

Trickling Filter Process

In the trickling filter process, the wastewater is sprayed over the surface of a bed of rough solids (such as gravel, rock, or plastic) and is allowed to "trickle down" through the microorganism-covered media. The inert medium develops a biological slime that absorbs and biodegrades organic pollutants. Air flows through the filter by convection, thereby providing the oxygen needed to maintain aerobic conditions. At intervals surplus sludge sloughs away, and is collected in a settling tank. For BOD-rich wastewater, several filters may be operated in sequence. These systems occupy a large space and have a limited capacity for treating wastewater, thus are liable to "pond" if overloaded (3).

Biofiltration Towers

A variation of a trickling filtration process is the biofiltration tower or otherwise known as the biotower. The biotower is packed with plastic or redwood media containing the attached microbial growth. Biological degradation occurs as the wastewater is sprayed into the top of the tower and as it trickles down it passes over the media. Treated wastewater collects in the bottom of the tower. If needed, additional oxygen is provided via air blowers countercurrent to the wastewater flow. Biotowers have capacities about ten times those of trickling filters covering the same area (3).

Rotating Biological Contactor Process

The rotating biological contactor process consists of a series of plastic discs attached to a common shaft. The discs are partially submerged in a trough of continuously flowing wastewater. As the discs rotate, a film of microorganisms growing on the discs consume oxygen from the air and substrate from the wastewater. In this way, organic materials (substrate) are removed from the wastewater.

Lagoons

These are slow, cheap, and relatively inefficient, but can be used for various types of wastewater. They rely on the interaction of sunlight, algae, microorganisms, and oxygen (sometimes aerated). Algae grow within the lagoons and utilize sunlight to produce oxygen, which is in turn used by microorganisms in the lagoon to break down organic material in the wastewater. Wastewater solids settle in the lagoon, resulting in effluent that is relatively well treated, although it does contain algae.

Advantages and Disadvantages of Aerobic Treatment

The advantages of aerobic wastewater treatment include:
1. Low capital expenditures compared to anaerobic systems.
2. These systems are fairly easy to maintain in terms of pH, flow, and temperatures than anaerobic systems.
3. The aerobic process does not produce H_2S or methane gases, which are unpleasant, and problematic in an urban environment.

The disadvantages of aerobic wastewater treatment include:
1. Aerobic systems, as efficient as they are, need constant power in terms of aerators and agitator pumps to ensure efficient BOD/COD degradation.
2. Large amounts of sludge (biomass) are generated, requiring some method of disposal.
3. These systems are less suitable for high-load wastewater.
4. Aerobic treatment systems require large holding vessels, which may pose a problem for breweries located in metropolitan areas.

Sludge Treatment and Disposal

In general, aerobic treatment systems like the activated sludge system produce relatively large quantities of sludge which requires disposal. The sludge can undergo a dewatering treatment either by reconsolidated centrifugation, vacuum filtration or in a pressure filter.

Brewers usually dispose of sludge in a landfill or may arrange to have it spread on farmland as a soil conditioner and fertilizer. It can also be used for restoring and re-vegetating areas with poor soils due to construction activities, strip mining or other practices. Federal regulation (40 CFR Pert 503) defines minimum requirements for such land application practices, including contaminant limits, field management practices, treatment requirements, monitoring, recordkeeping, and reporting requirements.

Reportedly dried sludge can be used successfully as a high protein animal feed supplement. The nutritional value of the sludge is further increased by its high vitamin content. It has also been found that dried sludge is usable as a replacement for soybean meal and fish meal in poultry feeds.

Anaerobic Wastewater Treatment

Anaerobic wastewater treatment is the biological treatment of wastewater without the use of air or elemental oxygen. Anaerobic treatment is characterized by biological conversion of organic compounds by anaerobic microorganisms into biogas which can be used as a fuel—mainly methane 55–75 vol% and carbon dioxide 25–40 vol% with traces of hydrogen sulfide (3).

Anaerobic systems are best regarded as preliminary treatments for highly polluted wastewater flows from a brewery followed by an aerobic treatment for weak wastewater. Combined anaerobic/aerobic treatment of brewery effluent has important advantages over complete aerobic treatment especially regarding a positive energy balance, reduced sludge production, and significant low space requirements.

Several types of anaerobic digestors can be used for brewery wastewater treatment including Upflow Anaerobic Sludge Blanket (USAB) reactor and the Fludized Bed Reactor (FDR). However, the USAB reactor clearly accounts for most of the full-scale systems used in breweries (1).

Upflow Anaerobic Sludge Blanket

In the Upflow Anaerobic Sludge Blanket (UASB) reactor, the wastewater flows in an upward mode through a dense bed of anaerobic sludge. This sludge is mostly of a granular nature (1–4 mm) having superior settling characteristics (> 50 m/h). The organic materials in solution are attacked by the microbes, which release biogas. The biogas rises, carrying some of the granular microbial blanket. Towards the top of the vessel the rising mixture meets a three-phase separator. The gas is collected and taken from the vessel and the granular microbial material, which had been carried up by the gas bubbles, settles back into the body of the reactor. The liquid flows up and over one or more weirs and leaves the vessel. After passing through a de-foamer and being scrubbed the biogas, now chiefly methane may be stored in a gas-holder or flared off. This type of reactor has a compact design and is capable of handling high COD loads (10–15 kg of COD/m^3/day).

Fluidized Bed Reactor

In a Fludized Bed Reactor (FBR), wastewater flows in through the bottom of the reactor, and up through a media (usually sand or activated carbon) that is colonized by active bacterial biomass. The media provides a growth area for the biofilm. This media is "fluidized" by the upward flow of wastewater into the vessel, with the lowest density particles (those with highest biomass) moving to the top. A control system is used at the top of the reactor to remove excess biomass, and thus control the height of the expanded media bed. The high biomass maintained within the FBR bed makes it appreciably more efficient for water treatment than many other types of biological systems, and allows reactors to be considerably smaller (13). Successful treatment of wastewater with low COD (<1,000 mg/L) at low operating temperatures and short retention times (<6 hours) are possible with this design.

Advantages and Disadvantages of Anaerobic Treatment

The advantages of anaerobic wastewater treatment include:

1. Anaerobic systems typically produce a fraction of the sludge generated by aerobic systems since more carbon is directed towards methane production than towards bacterial biomass synthesis (2).
2. The energy required for wastewater treatment is less since no aeration is required.
3. Anaerobic systems are capable of providing high treatment efficiencies for high-strength brewery waste—typically COD levels at 8,000 to 12,000 mg/l (2). These systems are also typically more effective than aerobic systems in removing organic material.
4. A byproduct of the anaerobic degradation of pollutants is the production of a methane-rich biogas (0.3–0.5 m^3/kg COD removed) which can be used to supplement or replace natural gas for fueling plant boilers or engine generators (3).
5. Anaerobic systems require a fraction of the nitrogen and phosphorous addition for biomass growth that an aerobic system does.
6. Because anaerobic systems require fewer nutrients and electrical input and produce less sludge than aerobic systems, they have inherently lower operating costs.

The disadvantages of anaerobic wastewater treatment include:

1. Anaerobic systems are sensitive to temperature with microorganisms requiring a temperature of 35°C (12).
2. Anaerobic systems work optimally within a pH range of 6.8–7.4. If the pH is below 6, anaerobic digestion can stop completely (12).
3. Anaerobic systems are not effective at reducing COD levels much below 1,500 to 2,500 mg/l, consequently are often run in tandem with aerobic systems (2).
4. May be more sensitive to the adverse effect of toxins.
5. These systems have the potential for odor problems.
6. Aerobic systems require more initial capital outlay and expertise than aerobic systems.
7. Start-up times are slow (4–10 weeks) because the anaerobic microorganisms multiply slowly (3).

References

1. Batston, D.J., J. Keller, and L.L. Blackall. 2004. "The Influence of Substrate Kinetics on the Microbial Community Structure in Granular Anaerobic Biomass." *Water Research.* 38(1390).

2. Brauer, John. 2006. "Wastewater Treatment: Back to Basics." *Brewers' Guardian* 135 (1).

3. Briggs, Dennis E., Boulton, Chris A., Brookes, Peter A., and Stevens, Roger. 2004. *Brewing: Science and Practice*. Cambridge, England: Woodhead Publishing Limited.

4. Davis, M. L., and Cornwell, D. 1991. *Introduction to Environmental Engineering*. 2d ed. New York, New York. McGraw-Hill.

5. HDR Engineering Inc. 2002. *Handbook of Public Water Systems*. 2d ed. Hoboken, New Jersey: John Wiley & Sons, Inc.

6. Huige, Nick J. 2006. "Brewery By-Products and Effluents." *Handbook of Brewing*, edited by Fergus G. Priest and Graham G. Stewart. Boca Raton, Florida: CRC Press, Taylor & Francis Group.

7. Kanagachandran, K, and R. Jayaratne. 2006. "Utilization Potential of Brewery Waste Water Sludge as an Organic Fertilizer." *Journal of the Institute of Brewing*. 112(2).

8. Kunze, Wolfgang. 1996. *Technology Brewing and Malting*, translated by Dr. Trevor Wainwright. Berlin, Germany: VLB Berlin.

9. Ockert, K. 2002. Practical Wastewater Pretreatment Strategies for Small Breweries. *MBAA Technical Quarterly* 39(1).

10. Watson, C. 1993. Wastewater Minization and Effluent Disposal of a Brewery. *MBAA Technical Quarterly* 30.

11. Lewis, M. J. and Young, T. W. 2002. *Brewing*. New York, New York: Kluwer Academic/Plenum Publishers.

12. Porter, Fred and Karl Ockert. 2006. "Environmental Engineering." *Brewing Engineering and Plant Operations*. Volume 3, edited by Karl Ockert. St. Paul, Minnesota: Master Brewers Association of the Americas.

13. U.S. Environmental Protection Agency. 1993. *Nitrogen Control Manual*. EPA/625/R-93/100, Washington, D.C.

14. U.S. Environmental Protection Agency. 2004. *Primer for Municipal Wastewater Treatment Systems*. EPA 832-R-04-001. Washington, DC.

Chapter Twenty One

Beer Styles

Although beers are brewed from similar materials, beers throughout the world have distinctive styles. Their uniqueness comes from the mineral content of the water used, the types of ingredients employed, and the difference in brewing methods. In a strict sense, there are two classical beer styles—ales and lagers. However, in addition to ales and lagers, there are other classical beer styles such as wheat beers, porters, stouts, and lambics—to name a few—that merit differentiation.

Traditionally, ales are most associated with Britain, Ireland, and Scotland. British variations include mild, bitter, and pale ales; India Pale Ale; brown ale; old ale; and barley wines. Today, ales are produced throughout the world. The ale family also includes Belgian specialty beers, German specialty beers, and American ales. Ales tend to have a fruity aroma and palate, and often a complex flavor varying considerably among ales in bitterness, color, sweetness, and harshness.

Lagers are a whole family of beers ranging from the light delicate Pilsner beers, to the dark, aromatic Munich types identified as "dunkels" and the strong lagers known as bocks. There is a tremendous variety of German and Continental style lager beers. Lager is mostly known to people in the United States as a golden-colored beer that is heavily carbonated and served chilled to accentuate its clean, refreshing flavor. Characteristically, lagers are low in esters and VDKs and are lightly hopped, with the important exceptions of Pilsner and German export. Lager beers, like ales, are available to the beer drinker pasteurized or not, filtered or unfiltered, on draft or bottled.

With any beer style, there are no hard rules, and variations within the styles are to be expected concerning flavor, ingredients, and methods of brewing. Brewmasters each have their own interpretation of what they consider appropriate for the style.

The descriptions of classic beer styles below are not comprehensive and definitive but are regarded merely as descriptions of the styles' flavor notes and the ingredients.

American Beer Styles

Ales

Amber Ale

American amber ales range in color from amber to coppery brown and are characterized by American-variety hops used to produce high hop bitterness, flavor, and medium to high aroma. A citrusy hop character is common, but not required. Amber ales have a moderate to strong malt flavor with a moderate caramel character. Malt and hop bitterness are usually balanced and mutually supportive. Fruity esters can be moderate to none and diacetyl, if present, is barely perceived. Similar to American pale ales, but is usually darker in color, more caramel flavor, more body, and usually balanced more evenly between malt and bitterness.

Ingredients – Pale ale malt, typically American two-row, as well as medium to dark crystal malts are typically used in brewing American amber ales. American hops, often with citrusy flavors, are common but others may also be used. Water can vary in sulfate and carbonate content.

Commercial Examples – North Coast Red Seal Ale, Pyramid Broken Rake, Avery Redpoint Ale, McNeill's Firehouse Amber Ale, Mendocino Red Tail Ale

Vital Statistics
Original gravity: 1.045–1.060
Apparent extract/final gravity: 1.010–1.015
IBUs: 25–45
Color SRM: 10–17
Alcohol by volume: 4.5–6.2%

India Pale Ale

American India pale ales range in color from medium gold to medium reddish copper. The style is characterized by prominent hop bitterness and a full, flowery hop aroma, although the malt backbone will support the strong hop character and provide the best balance. India pale ales possess low to medium maltiness and body, and are generally clean and malty sweet although some caramel or toasty flavors are acceptable at low levels. Fruity-ester flavors and aromas are moderate to very strong and diacetyl, if present, is perceived at very low levels.

Ingredients – Pale ale malt, typically American two-row, as well as medium to dark crystal malts are typically used in brewing American India pale ales. Water character varies from soft to moderately sulfate.

Commercial Examples – Russian River Blind Pig IPA, Stone IPA, Victory Hop Devil, Sierra Nevada Celebration Ale, Dogfish Head 60 Minute IPA, Anchor Liberty Ale, Harpoon IPA

Vital Statistics
Original gravity: 1.056–1.075
Apparent extract/final gravity: 1.010–1.018
IBUs: 40–70
Color SRM: 6–15
Alcohol by volume: 5.5–7.5%

Pale Ale

American pale ales range in color from deep golden to deep amber and are characterized by a moderate to high hop flavor, often showing a citrusy American hop character. Hop flavor and bitterness often lingers into the finish. American pale ales have medium body and low to moderate maltiness, and may optionally show small amounts of specialty malt character (bready, toasty, biscuity). Caramel flavors are usually restrained or absent. Fruity esters should be moderate to strong.

Ingredients – Pale ale malt, typically American two-row are typically used in brewing American pale ales. Specialty grains are often used to add character and complexity, but generally make up a relatively small portion of the grist. Grains that add malt flavor and richness, light sweetness, and toasty or bready notes are often used too. Water is generally low in carbonates with varying sulfate content.

Commercial Examples – Sierra Nevada Pale Ale, Great Lakes Burning River Pale Ale, Bear Republic XP Pale Ale, Full Sail Pale Ale, Firestone Pale Ale, Left Hand Brewing Jackman's Pale Ale

Vital Statistics
Original gravity: 1.044–1.050
Apparent extract/final gravity: 1.008–1.014
IBUs: 28–40
Color SRM: 6–14
Alcohol by volume: 4.5–5.5%

Lagers

Dark

American dark beers are copper to dark brown in color, light to medium in body, and have low hop aroma and flavor. They are essentially colored versions of American standard lagers with minimal, or sometimes no, roasted or chocolate-like characters contributed by the darker malts. American dark lagers are usually not as dark as their German counterparts. They are usually a bit heavier than the lightest of the American lagers. Any aroma is usually indicative of its high adjunct grain recipe, i.e., corn.

Ingredients – Ingredients used include American six-row malt, 20 to 40% corn adjuncts, and for color crystal, Munich or black malts. Color can also be artificially derived from the addition of caramel syrup. Cluster hops are used for bittering and Cascade and Willamette hops for aroma.

Commercial Examples – Dixie Blackened Voodoo, Shiner Bock, San Miguel Dark, Beck's Dark, Saint Pauli Girl Dark, Warsteiner Dunkel, Heineken Dark Lager

Vital Statistics
Original gravity: 1.044–1.056
Apparent extract/final gravity: 1.008–1.012
IBUs: 8–20
Color SRM: 14–22
Alcohol by volume: 4.2–6.0%

Diet/Light

Diet/light lager is loosely based on the Czech Pilsner style and has a caloric value lower than conventional beer. It is extremely pale, with no malt flavor or aroma and a light, watery body compared to regular beers. Hop bitterness is usually below the threshold, and no hop flavor or aroma is detected. Esters or diacetyl are not desirable, but some dimethyl sulfide flavor and aroma is acceptable. Often, low-cal beers will have just as much alcohol as their American full-calorie counterparts.

Ingredients – Corn is the usual adjunct, often amounting to 50 to 65% of the total grain bill in brewing diet/light lagers. Clusters is the hops most often used. Low-calorie beers are often processed with added enzymes to break down non-fermentable carbohydrates (dextrins to maltose) so attenuation is higher than normal beers.

Commercial Examples – Bitburger Light, Sam Adams Light, Heineken Premium Light, Miller Lite, Bud Light, Coors Light, Old Milwaukee Light, Amstel Light

Vital Statistics
Original gravity: 1.028–1.040
Apparent extract/final gravity: 0.998–1.008
IBUs: 8–12
Color SRM: 2–4
Alcohol by volume: 2.8–4.2%

Premium

Premium beer is light in body, with low malt flavor and aroma. Hop bitterness is low to medium, but it is usually just above the taste threshold, with hop flavor and aroma barely detectable. Color is very pale to deep gold. Original gravities also tend to be higher than standard beers. Many Canadian, Australian, Mexican, and U.S. lagers fall into this category. Esters or diacetyl are not desirable in these beers.

Ingredients – Premium beers are usually brewed with a smaller proportion of adjuncts than American standard (25 to 30% as opposed to 60 to 65%). Sometimes two-row malt is used to supplement American six-row, and some brewers even use two-row malt exclusively in their premium brands. Usually, more rice is used than corn or other cereal grains to give the beer a "crisp" taste, with there often being no hint of corn found in the flavor of standard light lagers. Cluster hops are used for bittering; Cascade and Willamette, and sometimes Saaz or Hallertauer hops, are used for aroma. The hopping rates are often higher than for standard brands.

Commercial Examples – Full Sail Session Premium Lager, Miller Genuine Draft, Corona Extra, Michelob, Coors Extra Gold, Heineken, Beck's, Stella Artois, Red Stripe

Vital Statistics
Original gravity: 1.044–1.048
Apparent extract/final gravity: 1.010–1.014
IBUs: 6–15
Color SRM: 2–6
Alcohol by volume: 4.3–5.0%

Standard

Standard lagers are the most common style of beer produced in Canada and the United States. They are pale to deep gold in color, run the gamut from sweet to dry, and are lightly hopped, light-bodied, and highly carbonated. This style has low malt aroma and flavor. Hop bitterness is barely noticeable, with very low flavor and aroma, though Canadian light lagers have a bit more hop character. Esters or diacetyl are not desirable in standard lagers.

Ingredients – Ingredients used in brewing standard lagers include American six-row malt and adjuncts such as rice, corn, sugar syrups, and wheat. Adjuncts can make up as much as 65% of the total grain bill. Clusters hops are generally used for bittering, and Cascade and Willamette hops for aroma. Soft water is generally used for this style to avoid harsh mineral tastes. Low mashing temperatures are desirable to keep the body thin.

Commercial Examples – Pabst Blue Ribbon, Miller High Life, Budweiser, Kirin Lager, Grain Belt Premium Lager, Molson Golden, Labatt Blue, Coors Original, Corona Extra, Foster's Lager

Vital Statistics
Original gravity: 1.040–1.050
Apparent extract/final gravity: 1.004–1.010
IBUs: 8–15
Color SRM: 2–4
Alcohol by volume: 4.2–5.3%

California Common (Steam Beer)

This style of beer has its origin in the California Gold Rush. It is considered neither an ale nor a lager but rather a distinct style in itself. It is brewed with lager yeast but at ale fermentation temperatures. California common has the roundness and cleanness of a lager, but some of the complexity of ale. It has a clean malt character, a light fruitiness, and a noticeable to intense hop bouquet, with a residual sweetness of crystal malt. This style of beer is straw-like in color. California common beer is often referred to as "steam beer," but only in reference to beer made by Anchor Brewing Company in San Francisco, since it is that company's registered trademark.

Ingredients – California common is produced from two-row pale ale and crystal malts. It is hopped with Cascade and is heavily dry-hopped with Northern Brewer. Either ale or lager yeast is used, though brewers generally prefer lager yeast.

Commercial Examples – Anchor Steam, Barbary Coast Gold Rush

Vital Statistics
Original gravity: 1.048–1.054
Apparent extract/final gravity: 1.011–1.014
IBUs: 30–45
Color SRM: 10–14
Alcohol by volume: 4.5–5.5%

Stouts

American style stouts are generally a jet-black color, although some may appear very dark brown. They are characterized by a moderate to strong aroma of roasted malts, often having a roasted coffee or dark chocolate quality. Burnt or charcoal aromas are low to none. Hop aroma is medium to high, often with a citrusy or resiny American hop character. Hop bitterness may be moderate to high. The perception of esters is generally low and diacetyl should be negligible. Medium to dry finish, occasionally with a light burnt quality.

Ingredients – Common American base malts with varied use of dark and roasted malts, as well as caramel-type malts. Adjuncts such as oatmeal may be present in low quantities. American hop varieties are typically used such as Chinook hops.

Commercial Examples – Rogue Shakespeare Stout, Sierra Nevada Stout, North Coast Old No. 38, Bar Harbor Cadillac Mountain Stout, Avery Out of Bounds Stout, Mad River Steelhead Extra Stout

Vital Statistics
Original gravity: 1.050–1.075
Apparent extract/final gravity: 1.010–1.022
IBUs: 35–75
Color SRM: 30–40
Alcohol by volume: 4–5%

Wheat Beers

There are really no parameters for this beer style since it is so new; interpretations by brewers vary. In general, wheat beers have light grain flavors and aromas characteristic of wheat. They are light to medium in body and are usually pale straw to gold in color, although dark versions do exist. Fruitiness and esters are common in this style. American wheat beers are similar to German wheat beers, but without the spicy/phenolic character of the German beers. Hop aroma and flavor can vary, with bitterness usually ranging from low to medium.

Ingredients – Wheat malt is used in proportions ranging from 25 to 70% of the grist, though generally 40 to 60% is most common. Typically, U.S. wheat beers are brewed with smaller proportions of wheat malt than is found in German wheat beers (50–75%). Dextrin malt and the paler caramel malts are also used. Hopping rates are generally very low, but again this varies from brewery to brewery. American wheat beer does not use the traditional German weizenbier/weissbier yeast, but rather conventional ale yeast or lager yeast. The reason is to avoid the typical spicy/phenolic character of German wheat beers, which many beer drinkers find objectionable.

Commercial Examples – Boston Beer Climax Wheat Beer, Redhook Brewery Redhook Hefeweizen Beer, Wachsett Brewing Summer Breeze Beer, Bell's Oberon, Pyramid Hefe-Weizen, Widmer Hefeweizen, Sierra Nevada Unfiltered Wheat Beer, Anchor Summer Beer

Vital Statistics
Original gravity: 1.040–1.055
Apparent extract/final gravity: 1.008–1.013
IBUs: 15–30
Color SRM: 3–6
Alcohol by volume: 4–5.5%

Belgium Beer Styles

Ales

Flanders Brown

The brown ales of Flanders are a distinctively regional style in Belgium. These beers, deep copper to brown in color, are faintly to strongly tart with a dry, assertive, lactic character. They have a very complex caramel/nutty/slight chocolate malt character, with flavors sometimes reminiscent of olives, raisins, or spices. The unique fermentations lead to a fruity, spicy complexity with a vinous nature. Hop flavor and aroma do not make an impression, though the bitterness can be assertive.

Ingredients – Flanders brown is brewed primarily from pale ale malt (two- and six-row malts) and Vienna malt. These malts comprise at least 80% of the malt bill, with the balance of the grist coming from corn grits. Liefmans Oud Bruin is brewed with Pilsner, Munich, Vienna, and caramel malts, along with roasted barley. The hops used in brewing Flanders brown are mainly East Kent Goldings, though Liefmans uses Whitebread Goldings, Tettnanger, and Saaz.

Commercial Examples – Liefman's Goudenband, Liefman's Odnar, Liefman's Oud Bruin, Ichtegem Old Brown, Riva Vondel

Vital Statistics
Original gravity: 1.040–1.074
Apparent extract/final gravity: 1.008–1.012
IBUs: 20–25
Color SRM: 15–23
Alcohol by volume: 4.0–8.0%

Pale Ale

Belgian-style pale ales are similar to British pale ales but are more spicy and aromatic both in malt and yeast character. Belgium pale ales are amber to copper in color and characterized by low, but noticeable, hop bitterness, flavor and aroma. Light to medium body and low malt aroma are typical boasting sweetish to toasty malt overtones. May have an orange- or pear-like fruitiness though not as fruity/citrusy as many other Belgian ales. Diacetyl should not be perceived.

Ingredients – Pilsner or pale ale malt contributes the bulk of the grist with (cara) Vienna and Munich malts adding color, body, and complexity. Noble hops, Styrian Goldings, East Kent Goldings or Fuggles are commonly used. Yeasts prone to moderate production of phenols are often used but fermentation temperatures should be kept moderate to limit this character.

Commercial Examples – De Koninck, Speciale Palm, Dobble Palm, Russian River Perdition, St. Pieters Zinnebir, Brewer's Art House Pale Ale, Avery Karma, Eisenbahn Pale Ale

Vital Statistics
Original gravity: 1.044–1.054
Apparent extract/final gravity: 1.008–1.014
IBUs: 20–30
Color SRM: 3.5–12
Alcohol by volume: 4.0–6.0%

Red Ale

This West Flanders style known as the "burgundies of Belgium" is distinctively red in color, thin but firm in body, and tart with a wide range of fruitiness. The red color comes, in part, from the use of Vienna malt; but it is also derived from aging in the brewery's uncoated oak tuns, which creates caramel flavors, tannins, and acidity.

Ingredients – Red ale primarily consists of pale ale malts (two- and six-row varieties) and Vienna malt, with the balance of the grist coming from corn grits. The hop varieties used are primarily East Kent Goldings and Brewers Gold. A number of yeast cultures and *Lactobacillus* strains are used to ferment this beer.

Commercial Examples – Rodenbach Klassiek, Rodenbach Grand Cru, Bellegems Bruin, Duchesse de Bourgogne, New Belgium La Folie, Petrus Oud Bruin, Southampton Flanders Red Ale, Verhaege Vichtenaar, Monk's Cafe Flanders Red Ale, New Glarus Enigma, Panil Barriquée, Mestreechs Aajt

Vital Statistics
Original gravity: 1.048–1.057
Apparent extract/final gravity: 1.002–1.012
IBUs: 10–25
Color SRM: 10–16
Alcohol by volume: 4.6–6.5%

Saison

Saisons are the summer and harvest specialties for French-speaking Belgium. The characteristics of the style are a powerful effervescence and a unique fruitiness, often with citric notes and a pungent sourness accented with aroma hops. This ale is distinctively bitter but not assertive. The style is crisp, tart, and refreshing. Most of these beers have a distinctive orange color, though they can range in color from light to amber.

Ingredients – Saison is brewed predominantly with pale ale malts, though sometimes with Munich or crystal malts. Brewers have also been known to use a small portion of spelt (a variety of wheat), raw oats, or raw rice in brewing Saisons. Some versions are spiced with orange, coriander, and licorice root. Saisons are often heavily hopped, usually with Belgian or British varieties, and notably with East Kent Goldings. The use of hard water adds to the body and mouth-feel of this style of beer, and it enhances the extraction of flavors from the grains.

Commercial Examples – Saison Dupont Vieille Provision, Fantôme Saison D'Erezée – Printemps, Saison de Pipaix, Saison Regal, Saison Voisin, Lefebvre Saison 1900, Ellezelloise Saison 2000, Saison Silly, New Belgium Saison, Pizza Port SPF 45, Ommegang Hennepin

Vital Statistics
Original gravity: 1.048–1.065
Apparent extract/final gravity: 1.002–1.012
IBUs: 20–35
Color SRM: 5–14
Alcohol by volume: 5.0–7.0%

Strong Golden Ale

Belgian golden ales are strong, golden in color, full of fruit, and somewhat hoppy, with low hop flavor and aroma. Usually they are very effervescent. References to the devil are often a trademark of these beers.

Ingredients – Strong golden ale is brewed using two-row summer malted barley. The primary hop varieties used for this style are Saaz and Styrian Goldings.

Commercial Examples – Duvel, Gouden Carolous, Leffe Blond, Affligem Blond, La Trappe (Koningshoeven) Blond, Grimbergen Blond, Val-Dieu Blond, Straffe Hendrik Blonde, Brugse Zot, Pater Lieven Blond Abbey Ale, Troubadour Blond Ale

Vital Statistics
Original gravity: 1.070–1.095
Apparent extract/final gravity: 1.005–1.016
IBUs: 22–35
Color SRM: 3–6
Alcohol by volume: 7.5–10.5%

Trappist Ales

Trappist ale is beer that is brewed only in Trappist monasteries, or under license of a Trappist monastery. Its appellation is thus based on origin rather than referring to a style. These beers are identified as "trappiste" and are made by only five breweries in Belgium and one in the Netherlands. The five abbeys in Belgium are Orval, Chimay, Rochefort, Westvleteren, and Westmalle. There are commercial versions, not brewed under the control of monks that are referred to as "bière d'abbaye" or "abbey" beer. The ingredients and brewing methods for abbey beers are similar to those for the Trappist beers. Although widely varying in character, Trappist beers are generally regarded as relatively strong, malty, and fruity, with a unique Belgian spiciness and a slight acidity that sets them apart from all other ale traditions.

Two sub-styles of Trappist beer exist. They are generally termed "double" and "triple," often expressed in their Flemish spelling as "dubbel" or "tripel." The sub-style dubbel is reddish-brown in color and very lightly hopped, with a powerful malt component expressed more as a complex fruitiness with a dark fruit character and a slightly nutty taste. It is medium- to full-bodied, with low bitterness. The sub-style tripel is gold or amber ale in color and medium- to full-bodied, with delicate aromatic hop characteristics and a light citric fruitiness. These beers have a light, malty nose and subtle alcohol undertones, with a sweet finish. They are more alcoholic than the dubbel, but are drier and much lighter in color than British ales of similar alcoholic strength, e.g., barley wine.

Ingredients – Trappist ales are brewed primarily with two-row Pilsner or pale ale malts. Small amounts of Cara-Vienne, Cara-Munich, and Special B are added to impart raisiny and plumy flavors. Sugar in the form of "candi sugar" (99% sucrose) is another important ingredient. Trappists are generously hopped, which serves to emphasize aroma more than bitterness. The important hop varieties used in brewing trappist ales are Styrian Goldings, East Kent Goldings, Saaz, Hallertauer, Tettnanger, and Northern Brewer. These beers tend to be dominated by unique yeast strains with up to five different strains being incorporated during primary and secondary fermentation.

Commercial Examples (Dubbel) – Westmalle Dubbel, Affligem Dubbel, St. Bernardus Pater 6, La Trappe Dubbel, Corsendonk Abbey Brown Ale, St. Feuillien Brune, Grimbergen Double, Chimay Premiere (Red), Pater Lieven Bruin, Duinen Dubbel, New Belgium Abbey Belgian Style Ale, Stoudts Abbey Double Ale, Russian River Benediction, Flying Fish Dubbel, Lost Abbey Lost and Found Abbey Ale, Allagash Double

Commercial Examples (Tripel) – Westmalle Tripel, La Rulles Tripel, St. Feuillien Tripel, St. Bernardus Tripel, Chimay Cinq Cents (White), Watou Tripel, Val-Dieu Triple, Tripel Karmeliet, Affligem Tripel, Grimbergen Tripel, La Trappe Tripel, Witkap Pater Tripel, Corsendonk Abbey Pale Ale, Bink Tripel, New Belgium Trippel, Dragonmead Final Absolution, Allagash Tripel Reserve, Unibroue La Fin du Monde, Victory Golden Monkey8

Vital Statistics – Dubbel
Original gravity: 1.050–1.070
Apparent extract/final gravity: 1.012–1.016
IBUs: 18–25
Color SRM: 14–18
Alcohol by volume: 6.0–7.5%

Vital Statistics – Tripel
Original gravity: 1.060–1.096
Apparent extract/final gravity: 1.008–1.020
IBUs: 20–25
Color SRM: 3.5–7
Alcohol by volume: 7.0–10.0%

Lambics

Lambic is a type of wheat beer that has the unique distinction of using unmalted wheat and being spontaneously fermented. The lambic family of beers includes many different sub-styles, such as gueuze, faro, kriek, and framboise.

Lambic

This is sour wheat beer made from the wild yeasts of the Senne Valley in Belgium. Although no one is sure of its origin, the most likely explanation is that the name derives from that of Lembeek, a small town southeast of Brussels. Lambic is golden yellow to light amber in color, light- to medium-bodied, almost flat, pungently sour, and has earthy (horsy and mousy) aromas and fruity notes. Some acetic character is acceptable, but excessive amounts are undesirable. The hop bitterness can be undetectable to very low. Lambic begins with a single basic wort. Lambic is sold either when it is young (called "vos" or "foxy" Lambic) or old ("vieux" Lambic). Vos lambics are highly variable products; they are frequently sold straight from the cask in specialized places in Brussels and continue to change from day to day with age. The vos lambic is less than one-year old and is usually a hazy, very dry, rusty color, with very little carbonation. It can be quite sharp and lactic. Some brewers no longer sell lambic in this form because there is little or no demand for the product. The vieux lambic is two to three years old and becomes clearer and has very interesting flavors from an earthiness to a wide range of esters. Basically, color is light gold to amber. As long as the beer is in the cask, it is called "lambic." Much of the lambic is not sold "straight," but in a variety of blended forms such as gueuze and fruit lambics.

Ingredients – One of the basic requirements of lambic is that at least 30 to 40% of the grain bill be unmalted wheat—usually a soft, white variety. The malt used is usually a mixture of two- and six-row barley. Crystal and specialty malts are not used, but some brewers add proportions of corn, rice, or even rye.

Traditionally, aged hops (from 1 to 3 years) were used to brew lambic in order to reduce the flavor and aroma of hops, as well as the bitterness. Assertive hops flavors do not combine well with the tartness of wheat beers or the intentionally sour notes of lambic beers. Modern lambic brewers use mostly British rather than Belgian hops. Preferred hop varieties are East Kent Goldings, Fuggle, Brewers Gold, Northern Brewer, and Hallertauer.

Commercial Examples – Girardin Unblended Lambic, Cantillon Grand Cru Bruocsella

Vital Statistics
Original gravity: 1.040–1.054
Apparent extract/final gravity: 1.004–1.010
IBUs: 1–10
Color SRM: 3–7
Alcohol by volume: 5.0–6.5%

Gueuze

A more widely available member of the lambic family is gueuze (spellings vary). Gueuze is the result of blending young and old lambic beers, with the blend then undergoing secondary fermentation. This version is known as gueuze-lambic, especially if served from a cask. If blended and bottle-conditioned, it is generally identified simply as gueuze. Gueuze undergoes secondary fermentation in the bottle and, like champagne, has a sparkle, a toasty aroma, and enduring length. The taste is dry, tart, and fruity, yet sweeter than regular lambic, and several have a toasty aroma reminiscent of some champagnes. The hop bitterness can be undetectable to very low, and with age the gueuze will become drier and more complex.

Ingredients – The proportion of young lambic blended with old lambics can range from 15 to 70%. The young lambic adds a lactic character, while the old lambic adds aroma and depth. The gueuze is then bottled and conditioned at the brewery between 5 and 24 months, or even longer, at temperatures between 10 and 15°C. Gueuze is bottled in very strong champagne-style bottles (complete with punt) which are stacked on their sides to mature through the summer. Not all gueuzes, however, are produced using traditional methods. Some brewers blend lambic with a top-fermented beer, then bulk-ferment, filter, and artificially carbonate the beer before bottling. Gueuze produced in this fashion tends to be sweet, with a fuller mouthfeel and relatively little complexity.

Commercial Examples – Boon Oude Gueuze, Boon Oude Gueuze Mariage Parfait, Drie Fonteinen Oud Gueuze, De Cam Gueuze, Mort Subite (Unfiltered) Gueuze, De Cam/Drei Fonteinen Millennium Gueuze, Cantillon

Gueuze, Oud Beersel Oude Gueuze, Hanssens Oude Gueuze, Lindemans Gueuze Cuvée René, Girardin Gueuze (Black Label)

Vital Statistics
Original gravity: 1.040–1.060
Apparent extract/final gravity: 1.000–1.006
IBUs: 1–10
Color SRM: 3–7
Alcohol by volume: 5.0–8.0%

Fruit Lambics

Fruit lambics are produced by racking young lambic into casks filled with macerated fruit. If the fruit is cherries, the beer is referred to as a "kriek," if raspberries a "framboise," if strawberries an "aardbien," and if grapes a "druiven." In general, lambics made with fruits are very sweet beers, sometimes to the point of being cloying, but they often go well with fruit-based deserts. In Belgium, if a lambic is used as the base for a fruit beer, this will be indicated on the label. It should be noted, however, that not all fruit beers are lambics.

Ingredients – Kriek, the most traditional lambic, has a deep ruby to garnet color, offset with a pink, delicate head. Kriek has the acidity of champagne and a sharp, dry flavor, and the bouquet and flavor of cherries, with almond notes from the stones. The preferred cherry for kriek is the Schaerbeek, though many brewers now substitute Moreno or Northern cherries. In the traditional method, the cherries are dried, like plums, while still on the tree because it is believed that the fruity flavors are thereby strengthened, and the fermentable sugars concentrated.

Most framboise beers are quite sweet, though the Cantillon brewery produces a tart version called *Rosé de Gambrinus* that is based on the traditional kriek style.

Commercial Examples – Boon Framboise Marriage Parfait, Boon Kriek Mariage Parfait, Boon Oude Kriek, Mort Subite Kriek, Cantillon Kriek, Cantillon Lou Pepe Kriek, Cantillon Vigneronne (Muscat grape), Cantillon Lou Pepe Framboise, Cantillon Rose de Gambrinus, Cantillon St. Lamvinus (merlot grape), De Cam Oude Kriek, Drie Fonteinen Kriek, Girardin Kriek, Hanssens Oude Kriek, Oud Beersel Kriek

Vital Statistics
Original gravity: 1.040–1.060
Apparent extract/final gravity: 1.000–1.010
IBUs: 1–10
Color SRM: 3–7 (varies w/ fruit)
Alcohol by volume: 5.0–7.0%

Wheat Beers

Beers that originate around Leuven and Hoegaarden are traditionally known as "white" beers because of their very pale color and a degree of cloudiness. They are Belgium's dessert beers, with a tangy, faintly acidic aftertaste, and are sharply refreshing, with hints of orange, honey, and spices. They typically are yellow-white in color, with a dry finish and slightly noticeable hop bitterness. As with most Belgium styles, carbonation is on the "spritzy" side. Today, white beers are stronger and maltier than Berliner weisse, but not as acidic. The traditional Berliner weisse was at least as strong as the white beers.

Ingredients – White beers are brewed with unmalted wheat, malted barley, and, on occasion, oats. In brewing traditional white beers, the grain bill consisted of 55% malted barley, 40% unmalted wheat, and 5% unmalted oats. Unmalted wheat gives white beers firmness of body and graininess, more than their German counterparts. Traditionally, two-row pale malts have been the malt of choice, but some brewers have used six-row malts, presumably for their additional diastatic power.

White beers are often spiced with coriander (which is ground fresh before use) and orange peels from both the Curacao orange and bitter orange. The Curacao orange is grown primarily in Spain, Italy, and North Africa, while the bitter orange is grown in the United States. These beers are also spiced with cumin, cardamom, anise, and black pepper. The use of spices is typical of Belgium brewing. Spices are usually added at the knockout of the boil or in the last 15 minutes in order to retain as many of the aromatics as possible.

Generally, white beers are lightly hopped with low alpha hops such as Styrian Goldings, Saaz, Hallertauer, or East Kent Goldings. Some sourness is preferred to boost the contribution of the orange and the hops. Some brewers introduce sourness by inoculating the beer with a *Lactobacillus* culture after primary fermentation. This is followed by a pasteurization to arrest the bacteria's action.

Commercial Examples – Hoegaarden Wit, St. Bernardus Blanche, Celis White, Vuuve 5, Brugs Tarwebier (Blanche de Bruges), Ommegang Witte, Allagash White, Blanche de Bruxelles, Avery White Rascal, Unibroue Blanche de Chambly, Sterkens White Ale, Bell's Winter White Ale, Hitachino Nest White Ale

Vital Statistics
Original gravity: 1.044–1.050
Apparent extract/final gravity: 1.006–1.010
IBUs: 10–17
Color SRM: 2–4
Alcohol by volume: 4.8–5.2%

British Beer Styles

Ales

Barley Wine

Traditionally, British brewers used the term "barley wine" to describe their strongest ale, brewed from the first mash runnings, where a single mash produced multiple beers of descending gravity and alcohol. Barley wine is the richest and strongest of British ales, with wine-like alcohol content. They usually vary in color from bronze to mahogany, though there are some golden versions. They are malty, heavy, and full-bodied, and they have lots of fruitiness that is usually balanced with a high rate of hop bitterness and low aroma, both of which may diminish during aging. The high terminal gravities and residual sweetness balances the hop bitterness. Their flavor can be reminiscent of fortified wine or even brandy having a vinous (sometimes sherry-like) aroma and flavor. American versions of Barley wine tend to have more hop aroma than the traditional British versions, and are usually more bitter. The effect of extremely high gravities can make for a very estery beer. Barley wines often have little head retention.

Ingredients – Ingredients include pale ale malt or American two-row malt, which make-up the majority of the malt bill (80–100%), along with caramel, dextrin, and amber malts. Dark malts should be kept below 2% of the total grain bill. Some commercial brewers will add sugar or malt extract to the kettle to achieve higher gravities.

East Kent Goldings is the preferred hops for barley wines, but Fuggle, Challenger, Northdown, and Styrian Goldings are also appropriate. American commercial brewers are more likely to use Chinook, Centennial, and Cascade hops.

Commercial Examples – Fuller's Golden Pride, Young's Old Nick, Anchor Old Foghorn Barleywine Style Ale, Sierra Nevada Bigfoot Barleywine

Vital Statistics
Original gravity: 1.080–1.120
Apparent extract/final gravity: 1.018–1.030
IBUs: 35–70
Color SRM: 8–22
Alcohol by volume: 8.0–12%

Bitter

Bitter, a principal style of ale sold in Britain, is usually served on-tap and is usually called "bitter" or "best bitter." Its taste is rather mild to assertive, with medium or even low alcohol content. The beer is usually amber, often with a reddish tinge, but it can be golden in color. The dominating flavor of this beer is hop bitterness, which is accentuated by the low original gravity. Hop aroma is medium to high, but this is secondary. As opposed to mild and pale ale, Bitter should certainly be dry. Some styles will have a rich creamy head, while others are less carbonated. They are very similar to pale ales, and sometimes are identical. Traditionally, pale ales were bottled, while bitters were in casks or kegs. Nowadays, even this separation is no longer in use. Today, the major difference between pale ale and a bitter is the name. Bitters are generally available in three strengths: ordinary, special and extra special bitter (or ESB). Pale ales are usually around the ESB strength, though some fall into the area of special bitter.

Ordinary bitter is the mildest form. It is dark gold to medium copper-brown, with medium bitterness and low carbonation. Grain and malt tend to predominate over hop flavor and bitterness (although there are exceptions), with enough hop aroma for balance and to add interest. It has a light to medium body. Low diacetyl and fruity esters are acceptable, but should not be overpowering. Special bitter is similar to ordinary bitter, but stronger and more robust, with a more evident malt flavor and hop character. Extra special bitter is a full-bodied, robust, copper-colored beer with a maltier, more complex flavor than either ordinary or special bitter. Maltiness should be evident with medium to high hop bitterness, flavor, and aroma.

Special bitter is more robust than ordinary bitter that is gold to copper colored with medium bitterness. It has a medium body and medium residual malt sweetness.

Ingredients – Ingredients of bitter include pale ale, crystal, black patent, and chocolate malts. Specialized brewing sugars are extremely useful for bitter production, especially in brewing special and extra special bitter, to enhance the malt and to add complexity. The sugars are usually used in amounts not exceeding 15%. Northern Brewer, Fuggle, and Brewers Gold are good bittering hops. Fuggle, East Kent Goldings, and Hallertauer are the choices for dry hopping. Bitter is brewed with water that is extraordinarily hard.

Commercial Examples (Ordinary) – Fuller's Chiswick Bitter, Tetley's Original Bitter, Adnams Bitter, Young's Bitter, Greene King IPA, Boddington's Pub Draught, Brakspear Bitter, Oakham Jeffrey Hudson Bitter (JHB), Brains Bitter

Commercial Examples (Special) – Fuller's London Pride, Timothy Taylor Landlord, Greene King Ruddles County Bitter, Adnams SSB, Young's Special, Shepherd Neame Masterbrew Bitter, Goose Island Honkers Ale, Brains SA, Black Sheep Best Bitter, Rogue Younger's Special Bitter

Commercial Examples (Extra Special) – Fullers ESB, Adnams Broadside, Shepherd Neame Bishop's Finger, Young's Ram Rod, Samuel Smith's Old Brewery Pale Ale, Bass Ale, Whitbread Pale Ale, Shepherd Neame Spitfire, Gale's Hordean Special Bitter (HSB), Marston's Pedigree, Ushers 1824

Particular Ale, Black Sheep Ale, Vintage Henley, Mordue Workie Ticket, Morland Old Speckled Hen, Greene King Abbot Ale, Bateman's XXXB, Hopback Summer Lightning, Shipyard Old Thumper, Alaskan ESB, Geary's Pale Ale, Cooperstown Old Slugger

Vital Statistics (Ordinary)
Original gravity: 1.032–1.040
Apparent extract/final gravity: 1.007–1.011
IBUs: 25–35
Color SRM: 4–15
Alcohol by volume: 3.2–3.8%

Vital Statistics (Special)
Original gravity: 1.040–1.048
Apparent extract/final gravity: 1.008–1.012
IBUs: 25–40
Color SRM: 5–16
Alcohol by volume: 3.8–4.6%

Vital Statistics (Extra Special)
Original gravity: 1.048–1.060
Apparent extract/final gravity: 1.010–1.016
IBUs: 30–50
Color SRM: 6–20
Alcohol by volume: 4.6–6.2%

Brown Ale

Brown ale was traditionally associated with northeastern England, especially Newcastle-upon-Tyne, though today it is found throughout England. Generally, brown ales are sweeter, fuller-bodied, more reddish brown, and stronger than their relative, mild ales. Some esters and fruitiness are present, and hop aroma and bitterness are usually in the low range but can be higher. Brown ales are less bitter than pale ales, but usually are maltier and have a higher alcohol level. Roast malt tones may sometimes contribute to the flavor and aroma profile.

In Great Britain, brown ale is generally a bottled product, although some cask-conditioned versions may be found. Two separate sub-styles of brown ale exist: London or southern brown ale, and Northern or Newcastle brown. The London brown ale is sweet in palate with malt and caramel flavors, and is light to dark brown in color and low in hop flavor. The Newcastle brown ale is deep amber to reddish brown in color, with some noticeable sweetness and bitterness.

Ingredients – Ingredients used in brewing brown ale include pale ale malt as the base malt, followed by crystal malt, black patent, and chocolate malt.

Some brewers also add sugar. Fuggle and East Kent Goldings are the hops of choice in brewing brown ales.

Commercial Examples – Newcastle Brown Ale, Samuel Smith's Nut Brown Ale, Tröegs Rugged Trail Ale, Riggwelter Yorkshire Ale, Wychwood Hobgoblin, Alesmith Nautical Nut Brown Ale, Goose Island Nut Brown Ale, Samuel Adams Brown Ale

Vital Statistics
Original gravity: 1.033–1.042
Apparent extract/final gravity: 1.011–1.014
IBUs: 10–20
Color SRM: 19–350
Alcohol by volume: 2.8–4.1%

India Pale Ale (IPA)

India Pale Ale (IPA) was traditionally a high-gravity, heavily hopped beer brewed for export to India during the colonizing period of the British Empire. Today, India Pale Ale is a stronger variant of ordinary pale ale, but is usually hoppier, with a slightly higher final gravity than most pale ales. Usually, IPAs are maltier and fruitier than pale ales, with a dominant overpowering hop bitterness that ranges in color from amber to copper. The fruity, estery aroma may be a little more pronounced than other ales because it is slightly higher in alcohol. A touch of oak is appropriate in this style.

Ingredients – Ingredients used to brew IPAs include two-row pale ale as the base malt, along with small amounts (usually no more than 10%) of crystal or Carastan malts to enhance malt flavors and increase color. Various amounts of sugar are also added. Small amounts (1–2%) of Carapils or dextrin malt will result in a frothy head. The use of adjuncts is common, with American brewers using up to 15% flaked maize or rice to help produce higher alcohol levels.

Traditionally, East Kent Goldings and Fuggle were used for India pale ales, which resulted in a smooth, rounded flavor. The hops were usually added to the cask (dry hopping) before shipping. Many U.S. brewers use Chinook, Galena, Brewers Gold, and Centennial for bittering and use Tettnanger, East Kent Goldings, Cascade, or Fuggle for aroma enhancement. Calcium and sulfates ions found in the water are some of the reasons for IPA's success.

Commercial Examples – Meantime India Pale Ale, Freeminer Trafalgar IPA, Fuller's IPA, Ridgeway Bad Elf, Burton Bridge Empire IPA, Summit India Pale Ale, Brooklyn East India Pale Ale, Samuel Smith's India Ale, Hampshire Pride of Romsey IPA, Goose Island IPA

Vital Statistics
Original gravity: 1.048–1.075
Apparent extract/final gravity: 1.010–1.017
IBUs: 40–60
Color SRM: 8–19
Alcohol by volume: 4.6–7.5%

Mild Ale

Mild ale, originally a London style, is quite common in the Midlands, where it is still identified as something of a regional favorite. Mild always applies to draft ales. Mild ale is lower in alcohol than its relatives and not particularly robust, but it is flavorful and light-to medium-bodied. Hop aroma is absent or subdued, with little bitterness, which allows malt to predominate. Mild ales are relatively sweet and darker in color than pale ale. Mild has a very light nutty flavor.

Ingredients – Ingredients used in brewing mild ales include pale ale malt as the base malt, or mild malt (if available), followed by crystal (7–12%), black patent, and chocolate malts. The grist for dark milds typically includes a small amount of black patent or chocolate malt for color more than for flavor development. Specialty malts typically constitute from 3 to 5% of the total grain bill. Adjuncts used include brown sugar and corn sugar as well as cereal grains such as corn, wheat, flaked barley, and oats. Fuggle is the preferred hop followed by East Kent Goldings. Willamette has been an acceptable replacement for Fuggle in the United States. A low to medium attenuating yeast strain can help conserve body and sweetness in the finish. Moderately hard water with some carbonate content is acceptable given the use of darker malts.

Commercial Examples – Moorhouse Black Cat, Gale's Festival Mild, Theakston Traditional Mild, Highgate Mild, Woodforde's Mardler's Mild, Sainsbury Mild, Brain's Dark, Greene King XX Mild, Banks's Mild, Coach House Gunpowder Strong Mild

Vital Statistics
Original gravity: 1.030–1.038
Apparent extract/final gravity: 1.008–1.013
IBUs: 10–25
Color SRM: 12–25
Alcohol by volume: 2.8–4.5%

Old Ale

Old ale is primarily an English style that was named "old" because it was kept a long time before drinking. Most are full-bodied and tawny, with nutty malt sweetness. High original gravities lend a fruity character and body to old ales.

Color is usually light amber to very dark red. They are a high-alcohol version of pale ale, though generally not as strong or rich as barley wines.

Ingredients – Ingredients used in brewing old ale include pale ale malt as the base malt, and crystal, chocolate, black, and amber malts. Adjuncts are often used, including torrified corn and sugar. Old Peculier is made from pale ale and crystal malts, torrified wheat, caramel, and three different sugars. Old ale is usually hopped with East Kent Goldings, Fuggle, or Northern Brewer for bittering, and with Styrian Goldings or Fuggle for dry hopping. The yeast should result in an estery-fruity flavor, and usually more than one strain of yeast is used to ferment this beer. Typically, hard water is used to produce old ales.

Commercial Examples – Gale's Prize Old Ale, Burton Bridge Olde Expensive, Marston Owd Roger, Greene King Olde Suffolk Ale, Theakston Old Peculier, Harviestoun Old Engine Oil, Fuller's Vintage Ale, Fuller's 1845, Fuller's Old Winter Ale, Harvey's Elizabethan Ale, Young's Winter Warmer, Samuel Smith's Winter Welcome, Great Divide Hibernation Ale, Founders Curmudgeon, Cooperstown Pride of Milford Special Ale, Coniston Old Man Ale

Vital Statistics
Original gravity: 1.060–1.090
Apparent extract/final gravity: 1.015–1.022
IBUs: 30–60
Color SRM: 10–23
Alcohol by volume: 6.0–9.0%

Pale Ale

Pale ale, first brewed in Burton-upon-Trent and Tadcaster, represents the very best in British beer. The beer has a medium body, low to medium maltiness, a bronze or copper color, and is generously hopped, with a dry crisp taste and little sweetness. Pale ale is distinguished by its light nuttiness of malt character, and its estery overtones followed by lingering hop bitterness. Low caramel flavor is allowable. Pale ale tends to be more hoppy and higher in alcohol than bitter. Some pale ales, especially those fermented in Yorkshire stone squares, have a noticeable diacetyl flavor component that is often regarded as a desirable feature.

Ingredients – The base malt used in brewing pale ales is two-row pale ale malt. Crystal malt, typically at the rate of 5% of the total grist, is added to provide extra body and residual sweetness and to intensify the copper hue. Bitters usually require more crystal malts than pale ales. Some brewers add wheat malt, no more than 5% of the total grist, to improve the beer's head retention, which can very desirable with low-carbonated draught beer. British brewers will often use flaked corn (10–15%) to dilute nitrogen. American craft brewers do not usually employ adjuncts in brewing pale ales. Traditionally,

British brewers also used cane and invert sugar; more recently, the trend is to use syrups, which consist mainly of glucose and maltose.

Hops are important in developing flavor in pale ale. East Kent Goldings is the traditional hop used for bittering and dry hopping in the cask, giving the beer a delicate floral/grassy character. Saaz, Hallertauer, Wye Progress, Target, Challenger, and Styrian Goldings are also used by British brewers. American brewers are more likely to use Cascade, Chinook, Eroica, Tettnanger, Willamette, and Galena.

Ale yeast is generally a mixture of two or more strains. Using such a combination is often essential to ensure both a rapid fermentation and high attenuation. Another important characteristic of ale yeasts, which is partly a function of fermentation temperature, is the production of esters responsible for the fruity flavor and aroma of ales.

The main flavor ions in the brewing water of most British ales are sulfate and chloride. The ratio of sulfate to chloride influences the overall mouthfeel, palate fullness, and dryness of the beer. The higher the sulfate to chloride ratio is the dryer the beer. Generally, the ratio ranges from 10:1 up to 20:1 for pale ale, whereas a Pilsen has a ratio of 1:1 and dark lagers have a ratio of 5:1, which makes the beers much softer. Pale ales benefit from the use of water with high calcium content relative to bicarbonate. The traditional mashing method using pale ale malt is single-infusion at around 65 to 68°C. Heavily hopped, high-gravity IPAs are usually mashed at higher temperatures, while low-gravity bitters are mashed at the lower end of the range. Some brewers consider U.S. and Canadian pale malts to be too light in flavor for pale ales. Step-infusion mashing may be effective when brewing pale ales.

Commercial Examples – Royal Oak Pale Ale, Whitbread Pale Ale, Samuel Smith's Old Brewery Pale Ale, Young's Special London Ale, Anchor Liberty Ale, Bass Ale, Whitbread Pale Ale, Shepherd Neame Spitfire, Black Sheep Ale, Greene King Abbot Ale, Vintage Henley, Geary's Pale Ale, Ushers 1824 Particular Ale, Great Lakes Moondog Ale, Cooperstown Old Slugger

Vital Statistics
Original gravity: 1.032–1.040
Apparent extract/final gravity: 1.007–1.011
IBUs: 25–35
Color SRM: 4–14
Alcohol by volume: 3.2–3.8%

Porter

Porter was the principal beer style in Britain, and especially in London, during the country's greatest period of industrial and economic growth. Porter is a heavy beer of pronounced bitterness, reddish-brown to a very dark brown, but is usually lighter in body and malt character than stouts. Generally, porters are a medium-bodied counterpart to stouts, with varying degrees of sweetness and hop character. They have a definite—but not marked—estery character and a burnt-coffee-like taste of roasted malt. Porters are known as "robust" and "brown."

Robust porter is a medium- to full-bodied balanced beer that has a noticeably coffee-like dryness and a malty, sweet flavor. Chocolate and black malts add a sharp bitterness, but do so without roasted or charcoal notes. Hop bitterness is medium to high, with hop aroma ranging from none too medium. The color is deep with red hues, but not opaque.

Brown porter is a bit lighter than robust, with light to medium body, and is generally lower in alcohol. The malt sweetness is low to medium, with a slight burnt-grain character, and well-balanced with hops. Color is deep, with reddish tones. Hop aroma and flavor can be nonexistent to medium.

Ingredients – Ingredients used for brewing porters include pale ale as the base malt; crystal (the most highly-roasted versions), chocolate, brown, and black malts; and roasted barley. Flaked corn (less than 10%), flaked barley, and some invert sugar (10–20%) are often used. London porters often include black treacle (blackstrap molasses). Crystal malt can make up 10% of the total grist bill and is responsible for much of the mouthfeel. Chocolate malts are more popular in Britain, while American brewers prefer black malts.

The hops of choice for bittering porters include Goldings, Fuggle, Northern Brewer, Galena, Eroica, Talisman, Cluster, Chinook, Target, and Challenger. Cascade, Perle, Chinook, Hallertauer, Styrian Goldings, and Eroica are used as aroma hops.

Moderately carbonated water is preferable, with appreciable amounts of chloride. Chloride ions contribute palate fullness and are desirable in porter, helping to smooth out rough edges.

Some commercial porters are brewed with bottom-fermenting yeasts; however, most modern versions use top-fermenting yeasts. The top-fermenting yeasts produce relatively higher levels of esters than bottom-fermenting yeasts. The yeast should be well-attenuated to minimize residual sweetness, with medium flocculation characteristics.

Commercial Examples (Robust) – Fuller's London Porter, Samuel Smith Taddy Porter, Salopian Entire Butt English Porter, Burton Bridge Burton Porter, Hambleton Nightmare Porter, Harvey's Tom Paine Original Old Porter, St. Peters Old-Style Porter, Shepherd Neame Original Porter, Nethergate Old Growler Porter

Commercial Examples (Brown) – Great Lakes Edmund Fitzgerald Porter, Meantime London Porter, Deschutes Black Butte Porter, Anchor Porter, Smuttynose Robust Porter, Boulevard Bully Porter, Rogue Mocha Porter, Avery New World Porter, Great Divide Saint Bridget's Porter

Vital Statistics (Robust)
Original gravity: 1.048–1.065
Apparent extract/final gravity: 1.012–1.016
IBUs: 25–50
Color SRM: 22–35
Alcohol by volume: 4.8–6.5%

Vital Statistics (Brown)
Original gravity: 1.040–1.052
Apparent extract/final gravity: 1.008–1.014
IBUs: 18–35
Color SRM: 20–30
Alcohol by volume: 4.0–5.4%

Stouts

Stouts were first produced in the early nineteenth century as high-gravity porters called "stout porters." Within time, brewers began to drop the word "porter." Stouts are very dark, almost black in color. Their color is achieved with roasted malt and/or with roasted barley, dark caramel malt, or even some chocolate malt. Stouts, compared to porters, are higher in gravity, lower in attenuation, and somewhat higher in relative bitterness. Stouts do not have a unified style, but rather are a family of sub-styles that have evolved over the years. Their sub-styles include imperial, sweet, and oatmeal stouts.

Imperial Stout

Britain and Continental Europe, including Finland, Prussia, Poland, and Russia brewed this style of stout. It was especially popular in Russia, where the British supplied the Imperial Court. As a result, the style came to be known as "Russian" or "imperial" stout. Imperial stout is a robust, full-bodied, and stronger version of dry stout, with tar-like notes and "burnt" fruitiness. The high gravity leads to notable esters and fruitiness that are balanced with the extremely rich malty flavor. It is highly hopped for bitterness, aroma, and flavor that can be subtle to overwhelmingly hop-floral, -citrus, or -herbal. This style, with its high alcohol and high hopping rate, can be aged for added flavor and complexity.

Ingredients – Stout ingredients include pale ale malt; crystal, amber, chocolate, and black malts; and roasted barley. Invert sugar, licorice, honey, or molasses may be employed, too. Recommended hop varieties for stout include Target, Fuggle, East Kent Goldings, Hersbrucker, Cluster, Willamette, Northern Brewer, and Bullion or Brewer's Gold. Dry hopping is not appropriate. Courage's Imperial stout is hopped with Wye Target. Water with 150 to 200 mg/l (or ppm) carbonate hardness is recommended to help neutralize the extra acidity extracted from the roasted (burnt) grains.

Commercial Examples – Three Floyd's Dark Lord, Bell's Expedition Stout, Samuel Smith Imperial Stout, Stone Imperial Stout, North Coast Old Rasputin Imperial Stout, Scotch Irish Tsarina Katarina Imperial Stout, Southampton Russian Imperial Stout, Rogue Imperial Stout, Thirsty Dog Siberian Night, Bear Republic Big Bear Black Stout, Great Lakes Blackout Stout, Founders Imperial Stout, Brooklyn Black Chocolate Stout

Vital Statistics
Original gravity: 1.075–1.115
Apparent extract/final gravity: 1.018–1.030
IBUs: 50–90
Color SRM: 30–40
Alcohol by volume: 8.0–12.0%

Oatmeal Stout

Oatmeal stout is a variation of sweet stout but do not specifically taste of oats. It has oatmeal added to increase body and enhance flavor. These beers have a hint of nuttiness, with coffee, chocolate, and roast flavors.

Ingredients – Pale, caramel and dark roasted malts and grains. Oatmeal (5–10%+) used to enhance fullness of body and complexity of flavor. Hops are primarily used for bittering. Ale yeast are typically used. Water source should have some carbonate hardness.

Commercial Examples – Samuel Smith Oatmeal Stout, Young's Oatmeal Stout, McAuslan Oatmeal Stout, Tröegs Oatmeal Stout, Maclay's Oat Malt Stout, Anderson Valley Barney Flats Oatmeal Stout, Goose Island Oatmeal Stout, Wolaver's Oatmeal Stout

Vital Statistics
Original gravity: 1.048–1.065
Apparent extract/final gravity: 1.010–1.018
IBUs: 25–40
Color SRM: 22–40
Alcohol by volume: 4.2–5.9%

Sweet Stout

British stouts are very dark in color, are often thought of as sweet, and lack most of the hop bitterness and roasted barley character of their dry counterpart. Sweet stouts are medium- to full-bodied, and are black opaque in color. They are usually much lower in alcohol and original gravity than dry stouts and possess a unique chocolate-caramel malt flavor. They are often called "farm stouts," "milk stouts," and "cream stouts."

Ingredients – Ingredients used to brew sweet stouts include two-row pale ale, crystal, chocolate, and black patent malts. Crystal, chocolate, and black malts are at levels not exceeding 15, 10, and 2%, respectively. Some brewers use up to 10% flaked corn. To achieve the sweet character, brewing sugars such as black treacle (blackstrap molasses) or other sweeteners (e.g., sucrose or lactose sugar) are added during boiling or just before bottling. Hops should be used with the emphasis on bittering, to balance the sweetness, and not for their aromatic character.

Mackeson is brewed from pale ale and chocolate malts and lactose, which represent at least 9% of the total grist. The lactose is added as powder to the kettle. It is hopped with Targets.

Commercial Examples – Mackeson's XXX Stout, Watney's Cream Stout, Farson's Lacto Stout, St. Peter's Cream Stout, Widmer Snowplow Milk Stout, Marston's Oyster Stout, Samuel Adams Cream Stout, Sheaf Stout, Hitachino Nest Sweet Stout (Lacto), Left Hand Milk Stout

Vital Statistics
Original gravity: 1.044–1.060
Apparent extract/final gravity: 1.012–1.024
IBUs: 20–40
Color SRM: 30–40
Alcohol by volume: 4.0–6.0%

Czech Republic Beer Styles

Lagers

Pilsner

Pilsner, one of the world's first golden lagers, was first produced in the region still known as Bohemia, in the town of Pilsen in former Czechoslovakia. Sometimes the designation is spelled "Pilsener," or it may be abbreviated to "Pils." Pilsner is a golden-colored beer that has good malt and hop character, with a strong, clean, assertive flavor. Hop bouquet is impressive, with floweriness of aroma and dryness of finish. Bohemian Pilsner is malty and well-hopped, with a smooth finish. A caramel taste is often observed, and a hint of diacetyl adds the impression of complexity and sweetness. Light- to medium-bodied Bohemian-style Pilsner really makes its impression with the bitterness, flavor, and aromatic character of the spicy Czech Saaz hop.

Ingredients – Ingredients used in brewing pilsners include very pale Czech malt, which is typically responsible for 80 to 100% of the grain bill. The preferred hop of choice is Saaz, which confers a delicate, fresh, soft character. The softness of the water allows these beers to be highly hopped.

Commercial Examples – Pilsner Urquell, Budweiser Budvar (Czechvar in the US), Czech Rebel, Staropramen, Gambrinus Pilsner, Dock Street Bohemian Pilsner

Vital Statistics
Original gravity: 1.044–1.056
Apparent extract/final gravity: 1.013–1.017
IBUs: 35–45
Color SRM: 3–6
Alcohol by volume: 4.2–5.5%

French Beer Styles

Ales

Bière de garde

Although France is not known for its beer, the northeastern district of French Flanders nevertheless possesses strong brewing traditions, which it shares with its Flemish cousins across the border. Traditionally, bière de garde was made from February through March and was consumed in the summer. A malt accent and ale-like fruitiness characterize bière de garde, and it has an earthy taste ranging in color from deep blond to reddish-brown. Bière de garde may have caramel flavors from a long boil. Bière de garde often appears in champagne bottles.

Ingredients – Ingredients used to brew bière de garde include several malts, with pale ale malt making up the bulk of the grist. Flemish Brewers' Gold is the hops most often used for bittering, with Saaz, Hallertau, Spalt, and Hersbruck added for aroma. Some brewers will boil the wort longer than usual to create a malty caramelization in both flavor and color.

Commercial Examples – Castelain Blonde, Jeanne d'Arc Ambre des Flanders, Lost Abbey Avante Garde (blond), Jenlain (amber), Jenlain Bière de Printemps (blond), St. Amand (brown), La Choulette Bière des Sans Culottes (blond), Ch'Ti Brun (brown), La Choulette (all 3 versions), Biere Nouvelle (brown), Southampton Bière de Garde, Saint Sylvestre 3 Monts (blond), Jade (amber), Brasseurs Bière de Garde (amber)

Vital Statistics
Original gravity: 1.060–1.080
Apparent extract/final gravity: 1.008–1.016
IBUs: 18–28
Color SRM: 6–20
Alcohol by volume: 6.0–8.5%

German Beer Styles

Ales

Altbier

Altbiers made in northern Germany are associated with the area around the city of Düsseldorf. Most altbiers are copper to brownish-amber in color, and light to medium in body, with a pronounced malt character that is not overpowering. Altbier lacks hop aroma but has medium to high bitterness, especially in the finish. The hops must balance but not be assertive. Fruitiness from top-fermentation can be a character, but is often minimized by lagering at very cold temperatures, much colder than for typical British ales. Diacetyl character is minimal to nonexistent in altbiers. They have a smoother palate, less yeastiness, and less acidity than classic British ales. Altbier has a dryish finish despite the rather pronounced malt character, but with no roasty overtones.

Ingredients – Ingredients used for brewing altbiers include Pilsner malt, which serves as the base, with smaller portions of Vienna and Munich malts. Crystal, chocolate, and black patent malts are often used too, but in very small amounts. The amount of crystal malt usually does not exceed 10%. Spalt is the preferred hops, followed by Hallertau, Saaz, Hallertau-Hersbrucker, Perle, Tettnanger, and Cascade hops. Mt. Hood is a popular bittering hop among American brewers, while Tettnanger is well-suited as an aroma hop. The alt style uses top-fermenting ale yeast with an apparent attenuative approaching 80%, and with low ester production. Altbiers are brewed with excessively clean, almost antiseptic ale strains. Water hardness generally ranges between 215 to 235 mg/l, expressed as calcium carbonate.

Commercial Examples – Diebels Alt, Zum Uerige, Grolsch Autumn Amber, Zum Schluessel, Im Fuchschen, Schumacher, DAB Traditional, Otter Creek Copper Ale, Hannen Alt, Schwelmer Alt, Alaskan Amber, Long Trail Ale

Vital Statistics
Original gravity: 1.0464–1.054
Apparent extract/final gravity: 1.010–1.015
IBUs: 35–50
Color SRM: 11–17
Alcohol by volume: 4.4–5.2%

Kölschbier

Kölschbier is a local style found in Cologne and is usually ordered simply as "Kölsch." Kölsch is very pale in color and is noted for its delicacy rather than for any robust distinctiveness. It is clean-tasting, light-bodied (very well attenuated), soft, and drinkable. These beers are faintly fruity (less than a

Pilsner) and are slightly acidic, with a medium-hoppy dryness and often with a slightly herbal taste in the finish. Essentially, their character is reminiscent of a hoppier, slightly stronger, slightly fruity cousin to a standard American light lager. The perceived bitterness may be quite similar to that found in an altbier because of the higher level of attenuation and lighter flavor. Bottle-conditioned examples may be called "wiess."

Ingredients – Ingredients used in brewing kölschbier include Pilsner malt or pale ale two-row malt. Sometimes up to 20% wheat malt is used, which heightens the beer's fruitiness and assists in body and color control. Reportedly, if wheat malt is used, then more often than not it is brewed with adjuncts such as corn or rice. Spalt is the preferred hops, followed by Hallertauer, Tettnanger, and Perle. A highly attenuative ale yeast strain is used (attenuation approaching 85%), with low ester production. A faint fruitiness is considered integral for this style. The water in Köln is soft, with low levels of calcium, magnesium, and bicarbonates.

Commercial Examples – PJ Früh, Hellers, Malzmühle, Gaffel, Paeffgen, Peters, Reissdorf, Non-German versions: Eisenbahn Dourada, Goose Island Summertime, Harpoon Summer Beer, Capitol City Capitol Kölsch, Shiner Kölsch

Vital Statistics
Original gravity: 1.044–1.050
Apparent extract/final gravity: 1.007–1.011
IBUs: 20–30
Color SRM: 3–5
Alcohol by volume: 4.4–5.2%

Lagers

Bocks

Traditional German bocks are not too bitter and do not have hop aroma or flavor of any consequence. Hops are used only to offset the sweetness of the malt. Many brewers believe that the most important factor in producing quality bock beers is melanoidins. Melanoidins are at the heart of most of the aromas needed for making this style of beer. Melanoidins are colored compounds that provide many of the malty and bready aromas and flavors that distinguish bocks. These beers have a residual dimethyl sulfide (DMS) character, which adds to the "lager" flavor and enhances the malt character. Most of their fusel oils are below the threshold of perception except for isoamyl and phenol alcohols that are responsible for the banana and the rose flavors, respectively. There are no esters, and there should not be any diacetyl. The bock family of beers also includes many different sub-styles, such as dopplebock, dunkler bock, eisbock, and heller bock.

Bock – Dopplebock

Dopplebock originated from the monastic brewers in Munich. Dopplebocks, literally "double-bock," are a stronger version of bock beers, with an original gravity decreed by German law to not fall below 18° Plato. They are very full-bodied, light to dark brown in color, and often have a roasted-malt flavor. Bitterness levels are similar to or slightly higher than those of dunkler bock. The malty sweetness that is evident in aroma and flavor can be intense. The level of dimethyl sulfide is higher in dopplebock than dunkler bock. In Germany, dopplebocks usually have names ending with an -ator suffix, e.g., Paulaner Salvator.

Ingredients – Pale malt is the main ingredient of bocks, followed by Munich, Vienna, and roasted malt. Small amounts of crystal, dextrin, chocolate, and/or black malt are added for flavor or color. Hallertauer, Perle, Hersbrucker, and Tettnanger are the hops of choice. The water used in brewing dopplebocks is usually high in calcium carbonate.

Commercial Examples – Paulaner Salvator, Ayinger Celebrator, Weihenstephaner Korbinian, Andechser Doppelbock Dunkel, Spaten Optimator, Tucher Bajuvator, Weltenburger Kloster Asam-Bock, Eggenberg Urbock 23°, Bell's Consecrator, Samuel Adams Double Bock

Vital Statistics
Original gravity: 1.072–1.112
Apparent extract/final gravity: 1.016–1.024
IBUs: 16–26
Color SRM: 6–25
Alcohol by volume: 7.0–10.0%

Bock – Dunkler Bock

The origin of dunkler bock is widely recognized to be Einbeck, located in what is today the northern German state of Lower Saxony. Dunkler bock is medium- to full-bodied, dark amber to brown in color, smooth, and strong in alcohol. These beers have a hint of malty sweetness in aroma, and flavor that can include some toasted, chocolate-like undertones, especially in the darker bocks. Dunkler bock is sometimes associated with a goat (the reason: "bock" in German means "goat"). Christmas bocks often are brewed to be consumed under the astrological sign of Capricorn.

Ingredients – Munich malt accounts for about 75 to 90% of the total malt bill for this style, followed by Pilsner malt in amounts not exceeding 25%. Often, crystal malts are added for sweetness and roundness. These usually amount to no more than 10%, with Lovibond ratings ranging from 90 to 120°L. Some brewers use small amounts of chocolate and/or black malts for proper flavor, complexity, and color. Typically, these malts make up at most 1 to 2% of the malt bill. Dextrin and wheat malts are added also. Roasted barley is never used in Germany due to the Reinheitsgebot prohibition against unmalted cereal ingredients, which includes unmalted barley.

In bock beers, a high hop rate is needed to compensate for the malt sweetness. The spicy Hallertauer is typically used in traditional bocks, followed by Hersbrucker, Saaz, and Tettnanger. American brewers use Northern Brewer and Perle hops as well as Mount Hood and Liberty. The strain of yeast chosen should be highly flocculent and moderately attenuative, capable of working in a high-gravity environment to produce a clean, malty flavor. High carbonate waters are necessary to buffer the brewing water from acidification by the addition of large quantities of dark malts.

Commercial Examples – Einbecker Ur-Bock Dunkel, Pennsylvania Brewing St. Nick Bock, Aass Bock, Great Lakes Rockefeller Bock, Stegmaier Brewhouse Bock

Vital Statistics

Original gravity: 1.064–1.072
Apparent extract/final gravity: 1.013–1.019
IBUs: 20–26
Color SRM: 14–22
Alcohol by volume: 6.0–7.2%

Bock – Eisbock

The strongest of the bocks, eisbock is created by freezing the beer and removing some of the ice. This has the result of concentrating the beer, thereby making it heavier, more alcoholic, and sweeter-tasting. The beer is very full-bodied, is amber to black in color, and has low bitterness, with fruity and ale-like flavors.

Ingredients – Pils and/or Vienna malt are used along with Munich malts and occasionally a tiny bit of darker colored malts such as Cara. Water hardness varies from soft to moderately carbonate.

Commercial Examples – Kulmbacher Reichelbräu Eisbock, Eggenberg Urbock Dunkel Eisbock, Southampton Eisbock, Niagara Eisbock, Capital Eisphyre

Vital Statistics

Original gravity: 1.078–1.120
Apparent extract/final gravity: 1.020–1.035
IBUs: 25–35
Color SRM: 18–30
Alcohol by volume: 9.0–14.0%

Bock – Heller Bock/Maibock

Heller bock possesses the same characteristics as traditional bock (maltiness, strength, and palate fullness), except for the toasted chocolate character.

However, they are lighter in color—gold to light amber. They are medium- to full-bodied, emphasizing the malty flavors, whereas dunkler bocks and dopplebocks embody some darker caramel as well as smoky or chocolate flavors. The maltiness comes from the lighter malts such as Pilsner and Munich. Hop bitterness is usually low, but higher than for the dunkler bocks.

Although German breweries produce and sell heller bock year around, Maibock is sold only in May. Maibock, a more assertive version of the heller bock, is hopped more, aged longer, and has a little more color.

Ingredients – Pale malts are used primarily to achieve a more intense and pure malty character without caramel or roasted malts. In addition to pale malt, light-colored Munich and Vienna malts are used. In Heller bocks, higher hopping rates should be used, with waters low in carbonates.

Commercial Examples – Ayinger Maibock, Mahr's Bock, Hacker-Pschorr Hubertus Bock, Hofbräu Maibock, Capital Maibock, Einbecker Mai-Urbock, Gordon Biersch Blonde Bock, Smuttynose Maibock

Vital Statistics
Original gravity: 1.064–1.072
Apparent extract/final gravity: 1.011–1.018
IBUs: 23–35
Color SRM: 6–11
Alcohol by volume: 6.3–7.4%

Dortmunder Export

Beers of this style are produced throughout Germany, but it is considered a local specialty in the city of Dortmund. Generally, Dortmunder is a strong pale lager that is characterized by more bitterness and less maltiness than Munich heller but far less bitterness and more malt body than German Pilsners. Neither hops nor malt is distinctive in this style, but are both medium in flavor and in good balance. The color is very pale, like that of Pilsner, but the beer has a higher gravity than other mainstream pale lagers. It is also slightly lower in carbonation, and less foamy. There are no traces of diacetyl or esters in the beer.

Ingredients – Ingredients, aside from the pale malt, that are used to brew Dortmunder export include pale crystal malts. Hallertauer Mittelfrüh, Hersbrucker, Saaz, Hallertauer, Spalt, and Tettnanger are the preferred hops. In general, the yeast should be highly attenuative. One of the most distinctive features of the Dortmunder style is the brewing water. With considerable levels of carbonate and sulfate present, the water serves to accentuate the beer's bitterness.

Commercial Examples – DAB Export, Dortmunder Union Export, Dortmunder Kronen, Ayinger Jahrhundert, Barrel House Duveneck's

Dortmunder, Great Lakes Dortmunder Gold, Gordon Biersch Golden Export

Vital Statistics

Original gravity: 1.048–1.056
Apparent extract/final gravity: 1.010–1.015
IBUs: 23–30
Color SRM: 4–6
Alcohol by volume: 4.8–6.1%

Märzen/Oktoberfestbier

"Octoberfest" or "märzen" are terms used interchangeably. They originally referred to a brewing process in which the beer was brewed in March and served in October. This style of beer is amber-red in color, slightly above average in gravity and alcohol, and moderately hopped. Märzen has no lingering hop bitterness. It is medium-bodied, and the balance is decidedly towards maltiness, with just enough bitterness to keep the beer from tasting too sweet.

Ingredients – European Pilsner malt is recommended for this style, but U.S. two-row lager malts are just as suitable. Smaller portions of Vienna and Munich malts are also used, as well as crystal malts. Mainly aroma hops such as Saaz, Hallertauer Mittelfrüh, Tettnanger, Spalter, Hersbrucker, Styrian Goldings, and Fuggle are employed. American brewers use Mt. Hood and Willamette hops. Moderately carbonate and low sulfate waters are preferable for märzen.

Commercial Examples – Paulaner Oktoberfest, Ayinger Oktoberfest-Märzen, Hacker-Pschorr Original Oktoberfest, Hofbräu Oktoberfest, Capital Oktoberfest, Victory Festbier, Great Lakes Oktoberfest, Spaten Oktoberfest, Goose Island Oktoberfest, Gordon Biersch Märzen, Samuel Adams Oktoberfest

Vital Statistics

Original gravity: 1.050–1.057
Apparent extract/final gravity: 1.012–1.016
IBUs: 20–30
Color SRM: 7–14
Alcohol by volume: 4.8–5.7%

Munich Dark

This is a traditional style in Munich that is labeled simply "dunkels," meaning "dark." In the United States, they are often labeled as "dark beer." Munich dark is usually dark-amber to dark-brown in color. It is distinctly toasted (not burnt), with a nutty, chocolate-like malt sweetness in aroma and flavor. The mouthfeel is typically dextrinous, and often a mild bitterness and/or slight astringency are present. Buttery notes are also sometimes present, but not as a primary flavor. The best examples have a spicy maltiness that is neither sweet nor roasty dry. Dunkel, the dark counterpart to Munich helles, is usually more bitter because of the contribution of roasted malts used in the formulation.

Ingredients – The grain bill for Munich darks usually consists of more than two-thirds Munich malt, with the balance of the grist composed of pale malt—typically a two-row Pilsner malt—and a small portion of color malt (e.g., black and caramel malts). Hallertauer Mittelfrüh, Hersbrucker, Spalt, and Saaz are commonly preferred hops. Malt-accentuating yeast strains are recommended at low fermentation temperatures to bring out the characteristics of the malt.

Commercial Examples – Ayinger Altbairisch Dunkel, Hacker-Pschorr Alt Munich Dark, Paulaner Alt Münchner Dunkel, Hofbräu Dunkel, Weltenburger Kloster Barock-Dunkel, Ettaler Kloster Dunkel, König Ludwig Dunkel, Löwenbräu Dunkel, Harpoon Munich-type Dark Beer, Gordon Biersch Dunkels, Ettaler Dunkel, Hartmann Dunkel, Kneitinger Dunkel

Vital Statistics
Original gravity: 1.048–1.056
Apparent extract/final gravity: 1.010–1.016
IBUs: 18–28
Color SRM: 14–28
Alcohol by volume: 4.6–5.6%

Munich Helles

Helles, often referred to as "light Munich," is a mildly hopped, malty, well-balanced pale to golden- to straw-colored beer. It is not as dry as Pilsner but is closer to the Dortmunder style, though lower in alcohol and with some sweetness. The malt sweetness, often described as almost a caramel taste, is the mark of this beer. Helles is maltier and less hoppy than dunkel; yet, like dunkel, it is sweet, lightly hopped, and has a nose that is straight malt.

Ingredients – Ingredients used to brew Munich helles include two-row pale malt and Munich and crystal malts at levels not generally exceeding 2 to 5%. The hops commonly used are Hallertau Mittelfrüh, Saaz, Hallertauer, Spalter, Perle, and Tettnanger. Soft water with a low mineral content and low carbonate content is essential to accentuate the malt and the hop flavors.

Commercial Examples – Weihenstephaner Original, Hacker-Pschorr Münchner Gold, Paulaner Premium Lager, Bürgerbräu Wolznacher Hell Naturtrüb, Stoudt's Gold Lager, Mahr's Hell, Spaten Premium Lager

Vital Statistics
Original gravity: 1.045–1.051
Apparent extract/final gravity: 1.008–1.012
IBUs: 16–22
Color SRM: 3–5
Alcohol by volume: 4.7–5.4%

Pilsner

The most well-known type of lager beer in Germany is Pilsner. Pilsner was first brewed in German-speaking Bohemia, a province in the Austrian Empire. Sometimes the designation is spelled "Pilsener" or it may be abbreviated to "Pils." Pilsner is not regarded as a regional style; there are outstanding examples produced throughout Germany. Pilsner is a golden-colored beer that has good malt and a definite hop accent in both its flowery bouquet and its dry finish. German Pilsner is lighter, paler, and drier than the Czech counterpart, with a clean, refreshing bitterness and a flowery, medium hop bouquet. These beers are light in body and have low sweetness. The biggest difference between German and Czech Pilsners is their terminal gravity—the German style is fermented out completely and typically has a much lower gravity.

Ingredients – Ingredients include very pale European Pilsner malt, which is typically responsible for 80 to 100% of the grain bill. Crystal malts are used for improving foam stability and mouthfeel, imparting sweet, mild taste and caramel smoothness. The lower grades (10 and 40°L) are more suitable for Pilsners, with the grades at the lower end of the range recommended for German Pilsners. Generally, no more than 5% of crystal malt is used in Pilsners. Some brewers believe that crystal malt is a better choice than Munich malt for enhancing malt qualities of a full-bodied Pilsner. Adjuncts are used too, usually in quantities up to 20%. Some North American brewers use unmalted barley to enhance foam retention and to impart a smooth finish to the beer. No unmalted grains can be used in Germany because of the Reinheitsgebot.

Traditional Pilsner is brewed using a variety of aroma hops such as Saaz, Hallertauer, Hallertauer Mittelfrüh, Tettnanger, Styrian Goldings, Spalt, Perle, and Hersbrucker. High-alpha varieties such as Northern Brewer can be used for bittering, but they have a high cohumulone content that imparts a coarse, clinging bitterness. The yeast accentuates the sulfury compounds that contribute to the malt aroma of the beer. Some Pilsner yeasts exhibit a slight estery character, imparting a very subtle fruity note to the aroma; however, this is secondary, and some yeast even exhibit a subtle, though noticeable characteristic of diacetyl. Pilsners should be brewed from water low in total alkalinity—ideally under 50 mg/l and certainly not higher than 75 mg/l.

Commercial Examples – Victory Prima Pils, Bitburger, Warsteiner, Trumer Pils, König Pilsener, Old Dominion Tupper's Hop Pocket Pils, Spaten Pils, Jever Pils, Left Hand Polestar Pilsner, Holsten Pils

Vital Statistics
Original gravity: 1.044–1.056
Apparent extract/final gravity: 1.008–1.013
IBUs: 30–45
Color SRM: 3–7
Alcohol by volume: 4.0–5.4%

Rauchbier

Rauchbier literally means "smoked beer." Rauchbier is darkish-amber and opaque in color with a blend of smoke and malt, with a varying balance and intensity. The smokiness from the beech wood imparts a bacony flavor to the beer. The malt character can be low to moderate, and be somewhat sweet, toasty, or malty. Märzen-like qualities should be noticeable, particularly a malty, toasty richness. Hop aroma may be very low to none. There are no fruity esters, diacetyl or DMS.

Ingredients – German Rauchmalz (beech wood-smoked Vienna-type malt) typically makes up 20–100% of the grain bill, with the remainder being German malts typically used in a Märzen. Some breweries adjust the color slightly with a bit of roasted malt.

Commercial Examples – Schlenkerla Rauchbier Märzen, Kaiserdom Rauchbier, Saranac Rauchbier, Eisenbahn Rauchbier, Spezial Rauchbier Märzen

Vital Statistics
Original gravity: 1.050–1.056
Apparent extract/final gravity: 1.012–1.016
IBUs: 20–30
Color SRM: 12–23
Alcohol by volume: 4.8–6.5%

Schwarzbier (Black Beer)

Schwarzbier means "black beer" in German that is medium to very dark brown in color, often with deep ruby to garnet highlights. It is a medium-bodied, malt-accented dark brew with very mild, almost bittersweet, notes of chocolate, coffee, and vanilla. In spite of its dark color, it comes across as a soft and elegant brew that is rich, mild, and surprisingly balanced. It never tastes harsh, toasty or acrid. The beer is often referred to as a Schwarzpils, a "black Pils," but, unlike a Pilsner, which can be assertively bitter, the hop bitterness in Schwarzbier is always gentle and subdued.

Ingredients – German Munich malt and Pilsner malts are used for the base, supplemented by a small amount of roasted malts (such as Cara) for the dark color and subtle roast flavors. Noble-type German hop varieties and clean German lager yeasts are preferred.

Commercial Examples – Köstritzer Schwarzbier, Kulmbacher Mönchshof Premium Schwarzbier, Krušovice Cerne, Einbecker Schwarzbier, Samuel Adams Black Lager, Original Badebier, Gordon Biersch Schwarzbier

Vital Statistics
Original gravity: 1.046–1.052
Apparent extract/final gravity: 1.010–1.016
IBUs: 22–32
Color SRM: 17–30
Alcohol by volume: 4.4–5.4%

Vienna

This style has its origins during the emergence of the Austro-Hungarian Empire, with the development of brewing in Vienna. Vienna style lager is amber-red to copper in color, with a soft maltiness of aroma and palate. It should have a dry finish, with a little sweetness on the palate. This beer is a light- to medium-bodied, with low to medium bitterness and a very mild hop flavor and aroma. Vienna is reminiscent of Octoberfest but somewhat drier, bitter, with less of a sweet malt character. This distinctive style owes much of its character to the method of malting. Vienna malt provides the dominant toasty flavor, and aroma, as well as the unique color.

Ingredients – Ingredients include two-row Pilsner malt (approximately 75% of the total grain bill), augmented with crystal malt and sometimes with Munich and amber malts. The preferred hops are Saaz, Styrian Goldings, Tettnanger, Spalt, Hallertauer, and Hallertauer Mittelfrüh. A clean and neutral yeast strain is preferred. A variety of water supplies can be used to brew Vienna-style beers, but water cannot have high sulfate content. Carbonate waters are preferred since they have a neutralizing effect on the acids in the dark malts, thereby adding to the softness of the finished beer.

Commercial Examples – Great Lakes Eliot Ness, Boulevard Bobs 47 Munich-Style Lager, Negra Modelo, Gordon Biersch Vienna Lager

Vital Statistics
Original gravity: 1.046–1.052
Apparent extract/final gravity: 1.010–1.014
IBUs: 18–30
Color SRM: 10–15
Alcohol by volume: 4.5–5.7%

Wheat Beers

Berliner Weisse

Berliner Weisse, which is commonly produced in northern Germany in the vicinity of Berlin, is known as the champagne of beers. The Berliner Weisse style has a pronounced sour taste, and is very pale, effervescent, and lightly hopped. The mouth-puckering sourness is characterized by an intense vinegary taste, caused by lactic and acetic acids, and is complemented by ester fruitiness. The acidic characteristic is central to Berliner Weisse. Modern versions are less sour, with no esters. Berliners often add to their weisse a dose of sweetened raspberry syrup (which turns it red) or woodruff syrup (which turns it green) to balance the acidity. Weisse is often considered a summer drink.

Ingredients – Berliner Weisse is brewed with about 60 to 75% wheat malt and a smaller amount of pale malt. Common bittering hops are Northern Brewer and Perle. The hopping rates are extremely low. Berliner Weisse is fermented with ale yeast and *Lactobacillus delbruckii* which creates an unforgettable beer. Medium to hard water with high carbonate hardness is typically used.

Commercial Examples – Schultheiss Berliner Weisse, Berliner Kindl Weisse, Nodding Head Berliner Weisse, Southampton Berliner Weisse, Weihenstephan 1809, Bahnhof Berliner Style Weisse, Bethlehem Berliner Weisse

Vital Statistics

Original gravity: 1.028–1.032
Apparent extract/final gravity: 1.003–1.006
IBUs: 3–8
Color SRM: 2–4
Alcohol by volume: 2.8–3.8%

Dunkles Weizen

Dunkles weizen is very popular in lower Bavaria and the Bavarian Forest. Dunkles weizen is the dark version of paler weizenbiers, but with a more pronounced malty aroma and flavor. Also, dunkles weizen usually has a little less of the characteristic phenolic, estery, fruity notes. The combination of wheaty tartness and the richness of dark malts make this style full of flavor and complexity. The alcohol content may be slightly less than other weizens since there is less fermentable extract in dark malt. A very mild sourness is acceptable for this style. These beers are similar in flavor to hefe weizen, but the malt aroma is more pronounced.

Ingredients – By German law, at least 50% of the grist must be malted wheat, although some versions use up to 70%; the remainder is usually Munich and/or Vienna malt. A traditional decoction mash gives the appropriate body without cloying sweetness. Weizen ale yeasts produce the typical spicy and fruity character, although extreme fermentation temperatures can affect the balance and produce off-flavors. A small amount of noble hops is used only for bitterness.

Commercial Examples – Weihenstephaner Hefeweissbier Dunkel, Ayinger Ur-Weisse, Schneider Weisse (Original), Franziskaner Dunkel Hefe-Weisse, Ettaler Weissbier Dunkel, Hacker-Pschorr Weisse Dark, Edelweiss Dunkel Weissbier, Erdinger Weissbier Dunkel

Vital Statistics
Original gravity: 1.044–1.056
Apparent extract/final gravity: 1.008–1.016
IBUs: 10–18
Color SRM: 14–23
Alcohol by volume: 4.3–5.6%

Hefe Weizen

Hefe weizen is the beer of choice in Bavaria. Some weizens are bottle-conditioned and contain some yeast sediment; accordingly, they are labeled as "mit hefe" ("with yeast") or as "hefe weizen." These beers are pale-to-golden colored, light- to medium-bodied, and highly effervescent, with a slight maltiness. Typical of hefe weizen is its phenolic taste, reminiscent of wood smoke or vanilla, and its aroma commonly described as clove-like. The signature pour into tall glassware produces a thick head of foam.

Ingredients – Typically, the malt grist consists of at least 50% wheat malt and a smaller amount of malted barley. Pale malts help achieve a light color, but many brewers use a proportion of dark or caramel malts to increase flavor and body. A very popular strain of top-fermenting yeast is Weihenstephan #68. Generally, hefe weizen beers are traditionally hopped very lightly.

Commercial Examples – Weihenstephaner Hefe-Weissbier, Widmer Hefe-Weizen, Maisel's Hefe Weisse

Vital Statistics
Original gravity: 1.044–1.052
Apparent extract/final gravity: 1.008–1.014
IBUs: 8–14
Color SRM: 2–8
Alcohol by volume: 4.3–5.6%

Kristall Weizen

Kristall weizen is usually identified simply as a weizenbier (though sometimes as a weissbier or a weisse). This style is commonly made in southern Germany (in the states of Bavaria and Baden-Wuerttemberg), but it is also made in other regions throughout Germany. Weizen beer is completely different from Berliner Weisse. Weizen beer lacks the acidity and sourness of Berliner

Weisse but often has a phenolic (clove-like) and estery-fruity (banana) aroma produced by the yeasts. The phenol 4-vinyl guaiacol that is responsible for the clove-like flavor can reach levels between 0.3 and 0.6 mg/l. This flavor is particularly noticeable because weizens are traditionally hopped very lightly, resulting in a smooth, clean, delicate flavor. They are pale- to golden-colored, light- to medium-bodied, and highly effervescent.

Ingredients – Typically, the malt grist consists of 50 to 70% wheat malt and a small portion of pale malt. Pale malts help achieve a light color, but many brewers prefer also using dark or crystal malts (no more than 5%) also to increase flavor and body. Malt is emphasized over hops in weizenbiers, with hopping rates seldom exceeding 15 IBUs. Bittering hops are most commonly used in Weizenbiers, but lower-grade aroma hops can also be used to improve the quality of the hop bitterness. Brewers usually have no preference in the variety of the hops. Yeast should produce the characteristic clove-like taste, and sometimes vanilla notes. Soft to medium-hard water and a low pH are preferable.

Commercial Examples – Weihenstephaner Hefeweissbier, Schneider Weisse Weizenhell, Paulaner Hefe-Weizen, Hacker-Pschorr Weisse, Plank Bavarian Hefeweizen, Franziskaner Hefe-Weisse, Andechser Weissbier Hefetrüb, Barrelhouse Hocking Hills HefeWeizen

Vital Statistics
Original gravity: 1.044–1.052
Apparent extract/final gravity: 1.008–1.014
IBUs: 8–15
Color SRM: 2–8
Alcohol by volume: 4.3–5.6%

Weizenbock

Weizenbock is stronger and more robust than dunkle weizen. The phenolic character is still evident in weizenbock, but more emphasis is placed on the malt character of the beer. Weizenbocks are usually more liberally hopped than the paler weizenbiers. A very mild sourness is acceptable for this style, and it has a dry crispness not usually associated with bock beers. Both blond and dark versions are found.

Ingredients – A high percentage of malted wheat is used (by German law must be at least 50%, although it may contain up to 70%), with the remainder being Munich- and/or Vienna-type barley malts. A traditional decoction mash gives the appropriate body without cloying sweetness. Weizen ale yeasts produce the typical spicy and fruity character. Too warm or too cold fermentation will cause the phenols and esters to be out of balance and may create off-flavors. A small amount of noble hops is used only for bitterness.

Commercial Examples – Schneider Aventinus, Plank Bavarian Dunkler Weizenbock, Plank Bavarian Heller Weizenbock, AleSmith Weizenbock, Erdinger Pikantus, Mahr's Der Weisse Bock, Victory Moonglow Weizenbock

Vital Statistics
Original gravity: 1.064–1.090
Apparent extract/final gravity: 1.015–1.022
IBUs: 15–30
Color SRM: 12–25
Alcohol by volume: 6.5–8.0%

Irish Beer Styles

Ales

Red Ales

Irish ales, a minor category, range in color from light red-amber to light brown. These ales have a pleasant toasted malt character and a candy-like caramel sweetness. These ales are lightly hopped with low levels of fruity-ester and aroma. Diacetyl should be absent. Irish ales are similar to Scottish ales but are a bit lighter and paler.

Ingredients – Ingredients used for brewing Irish ales include pale malt, along with smaller portions of crystal malt and roasted barley. Smithwick's uses 20% corn syrup—which is used quite often among brewers of Irish ale—and 3% roasted barley. The hops used for bitterness are typically Challenger, Northern Brewer, Northdown, and Target; Fuggle and Goldings are added for aroma.

Commercial Examples – Three Floyds Brian Boru Old Irish Ale, Beamish Red Ale, Great Lakes Conway's Irish Ale, Smithwick's Irish Ale, Kilkenny Irish Beer, O'Hara's Irish Red Ale, Caffrey's Irish Ale, Goose Island Kilgubbin Red Ale, Boulevard Irish Ale, Harpoon Hibernian Ale

Vital Statistics
Original gravity: 1.044–1.060
Apparent extract/final gravity: 1.010–1.016
IBUs: 17–28
Color SRM: 6–18
Alcohol by volume: 4.0–6.2%

Stouts

Dry Stout

Ireland is one of the first countries to brew stout; there it is considered a national beverage. Ireland's dry stouts are markedly aromatic, with rich maltiness, and intense hop flavors. Hop bitterness is medium to high. The beer is extra-dark, black opaque ale, with low to medium body and a creamy brown head. The degree of sweetness and dryness will vary in dry stouts, yet they are all have the unique and special character of roasted barley, which produces a slightly roasted (coffee-like) trait. Fruity esters are minimal and overshadowed by malt, high hop bitterness, and roasted barley character.

Ingredients – Ingredients used for brewing stouts include two-row pale ale malt as the base malt, along with crystal (80–120°L), dextrin, chocolate, and black malts. Dark roasted barley is used, too, which is critical to the distinctive character of Irish stouts. Unmalted barley (no more than 10%), cooked or flaked, is also used. Beamish Stout includes malted wheat to improve the creaminess of texture and head. Liberal hopping rates emphasize the dryness of Irish stouts. Hops are used for bittering only and are used primarily to balance the malt character of the beer. Hop aroma is not appropriate to this style and should be kept to a minimum. The hops used are Fuggle, East Kent Goldings, Target, Hallertauer, and Hersbrucker. American brewers often use Clusters and Challenger.

Commercial Examples – Guinness Draught Stout, Murphy's Stout, Beamish Stout, Three Floyd's Black Sun Stout, O'Hara's Celtic Stout, Russian River O.V.L. Stout, Dorothy Goodbody's Wholesome Stout, Orkney Dragonhead Stout, Brooklyn Dry Stout

Vital Statistics
Original gravity: 1.036–1.050
Apparent extract/final gravity: 1.007–1.011
IBUs: 30–45
Color SRM: 25–40
Alcohol by volume: 4.0–5.0%

Scottish Beer Styles

Ales

Scottish Ale

If England is famed for the bitter hops flavor of its "bitters," Scotland is famed for its full-bodied, malty ales. Scotch ales are sweet and very full-bodied, with malt and roast malt flavors predominating. They are deep burnished-copper to brown in color. Scottish ales are invariably rich and mouth filling because they are quite high in unfermentables. They have a maltier flavor and aroma, darker colors, and a more full-bodied and smokier character than British ales. Bitterness and hoppiness are not dominant factors in Scottish ales, and they are less hoppy than their British counterparts. They are similar to British bitters, but are less estery and are generally darker, sweeter, and maltier. Some Scottish heavy ales exhibit a peat or smoke character present at low to medium levels.

Scottish ales are often known by names simply as, e.g., "Scottish Light 60/" ("/" means shillings), "Heavy 70/," and "Export 80/." The strong Scotch ales are designated with higher values, ranging from 90/- to 160/-. The significant differences are reflected in their maltier flavor, relatively darker colors, and occasional faint smoky character. The "shilling" designation is believed to be from the old method of taxing in which the tax rate was based on the gravity of the beer.

Scottish Light Ale 60/ is light-bodied beer with a golden amber to deep brown color. Bitterness as well as aroma is not perceptible while diacetyl (butterscotch) and sulfuriness are acceptable at very low levels.

Scottish Heavy Ale 70/ is a medium-bodied beer with a golden amber to deep brown color. They are moderate in strength and dominated by a smooth, sweet maltiness balanced with low, but perceptible levels of hop bitterness.

Scottish Export Ale 80/ is a medium-bodied beer with a golden amber to deep brown color. They are sweet, caramel-like, and malty with a low to medium bitterness.

Scottish Strong Ale 90/ is a full-bodied beer with a deep-burnished copper to brown color. There can be medium diacetyl present. These beers are much less hopped than English versions, and therefore are maltier, with some kettle caramelization. Slight roasted malt qualities may be provided by the limited use of dark roasted malt or roasted barley.

Ingredients – The base malt preferred is Scotch two-row pale ale malt. Darker malts, notably black, amber, brown, and chocolate, are added. Black appears in more commercial recipes than either chocolate malt or roasted barley. Roasted barley, from 1 to 3%, is used, which is responsible for most of the color and flavor of Scotch ales. Some brewers use crystal malt in amounts no more than 10% of the grist charge, but others believe the use of crystal malt adds excessive fruitiness to the final product. Wheat—malt or torrified—is often found in recipes for this beer. Scottish ales usually include treacle (molasses) or brown sugar.

The Goldings varieties and Fuggle are the hops most often used in brewing Scotch ales. Hops are primarily used to provide a balance to the malt sweetness and are not dominant factors in this style. In order for the malt flavor to express itself, the yeasts should not exhibit overt fruitiness or other estery characteristics typically produced by ale yeasts. Another major trait of these yeasts is that they are relatively flocculent and under-attenuative (65–70%), flocculent yeasts tending to enhance maltiness. Soft to medium-hard water is preferable for brewing Scotch ales.

Commercial Examples Ale 60/ – Belhaven 60/-, McEwan's 60/-, Maclay 60/- Light (all are cask-only products not exported to the US)

Commercial Examples Ale 70/ – Caledonian Amber Ale in the US, Belhaven 70/-, Orkney Raven Ale, Maclay 70/-, Tennents Special, Broughton Greenmantle Ale

Commercial Examples Ale 80/ – Orkney Dark Island, Caledonian 80/- Export Ale, Belhaven 80/- (Belhaven Scottish Ale in the US), Southampton 80 Shilling, Broughton Exciseman's 80/-, Belhaven St. Andrews Ale, McEwan's Export (IPA)

Commercial Examples Ale 90/ – Traquair House Ale, Belhaven Wee Heavy, McEwan's Scotch Ale, Founders Dirty Bastard, MacAndrew's Scotch Ale, AleSmith Wee Heavy, Orkney Skull Splitter

Vital Statistics (Ale 60/–)

Original gravity: 1.030–1.035
Apparent extract/final gravity: 1.010–1.013
IBUs: 10–20
Color SRM: 9–17
Alcohol by volume: 2.5–3.2%

Vital Statistics (Ale 70/–)

Original gravity: 1.035–1.040
Apparent extract/final gravity: 1.010–1.015
IBUs: 10–25
Color SRM: 9–17
Alcohol by volume: 3.2–3.9%

Vital Statistics (Ale 80/–)

Original gravity: 1.040–1.054
Apparent extract/final gravity: 1.010–1.016
IBUs: 15–30
Color SRM: 9–17
Alcohol by volume: 3.9–5.0%

Vital Statistics (Ale 90/–)

Original gravity: 1.070–1.130
Apparent extract/final gravity: 1.018–1.030
IBUs: 17–35
Color SRM: 14–25
Alcohol by volume: 6.5–10.0%

References

1. Allen, Fal and Dick Cantwell. 1988. *Barley Wine: History, Brewing Techniques, Recipes*. Boulder, Colorado: Brewers Publications.
2. Allen, Fal and Dick Cantwell. 1998. "Barleywine – The Revival of the Tradition of Big Beers." *Brewing Techniques* 6 (5).
3. Arnott, Paul. 1997. "Brewing at Chimay – or pick up thy glass and taste." *Brewers' Guardian* 126 (4).
4. Beer Judge Certification Program, Inc. 2008. "Beer Judge Certification Program (BJCP): Style Guidelines for Beer, Mead and Cider." <http://www.bicp.org>.
5. Bergen, Roger. 1993. "American Wheat Beers." *Brewing Techniques* 1 (1).
6. Bergen, Roger. 1993. "Oktoberfest Alternatives." *Brewing Techniques* 1 (2).
7. Bergen, Roger. 1993. "Porters: Then and Now." *Brewing Techniques* 1 (3).
8. Bergen, Roger. 1993. "A Stout Companion." *Brewing Techniques* 1 (4).
9. Bergen, Roger. 1994. "California Steaming." *Brewing Techniques* 2 (1).
10. Burch, Byron and Paddy Giffen. 1991. "Stout." *Zymurgy – Special Issue: Traditional Beer Styles* 14 (4).
11. Carey, Daniel and Grossman, Ken. 2006. "Fermentation and Cellar Operations." *Fermentation, Cellaring, and Packaging Operations*. Volume 2, edited by Karl Ockert. St. Paul, Minnesota: Master Brewers Association of the Americas.
12. Daniels, Ray. 1996. *Designing Great Beers*. Boulder, Colorado: Brewers Publications.
13. Dawaud, Ensaf. 1987. *Characteristics of Enterobacteriaceae Involved in Lambic Brewing*. P.h.D. dissertation, Katholieke Universiteit Leuven.
14. DeBenedetti, Christian T. 1998. "Inside Orval." *Brewing Techniques* 6 (3).
15. Dornbusch, Horst D. 1998. Altbier: History, *Brewing Techniques*, Recipes. Boulder, Colorado: Brewers Publications.
16. Dorsch, Jim. 1996. "Barley Wine: Feast or Famine." *The New Brewer* 13 (6).
17. Dorsch, Jim. 1998. "Märzen – A Question of Balance." *The New Brewer* 15 (1).
18. Dorsch, Jim. 1999. "Scotch Ale: Idiots Need Not Apply." *The New Brewer* 16 (1).
19. Ecimovich III, Victor. 1991. "Bock." *Zymurgy – Special Issue: Traditional Beer Styles* 14 (4).

20. Eckhardt, Fred. 1991. "German Style Ale." *Zymurgy – Special Issue: Traditional Beer Styles* 14 (4).

21. Eckhardt, F. 1993. *The Essentials of Beer Style*. Portland, Oregon: Fred Eckhardt Associates.

22. Ensminger, Peter A. 1997. "The History and Brewing Methods of Pilsner Urquell—Divining the Source of the World's Most Imitated Beer." *Brewing Techniques* 5 (3).

23. Fink, Dan. 1991. "American Light Lager." *Zymurgy – Special Issue: Traditional Beer Styles* 14 (4).

24. Fix, George. 1989. *Principles of Brewing Science*. Boulder, Colorado: Brewers Publications.

25. Fix, George. 1991. "Vienna, Oktoberfest and Märzen." *Zymurgy – Special Issue: Traditional Beer Styles* 14 (4).

26. Fix, George and Laurie Fix. 1991. *Vienna, Märzen, Oktoberfest*. Boulder, Colorado: Brewers Publications.

27. Fix, George. 1993. "Belgian Malts: Some Practical Observations." *Brewing Techniques* 1 (1).

28. Foster, Stephen. 1991. "Pale Ale." *Zymurgy – Special Issue: Traditional Beer Styles* 14 (4).

29. Foster, Stephen. 1991. "Porter." *Zymurgy – Special Issue: Traditional Beer Styles* 14 (4).

30. Foster, Stephen and Greg Noonan. 1991. "English Bitter and Scottish Ale." *Zymurgy – Special Issue: Traditional Beer Styles* 14 (4).

31. Foster, Terry. 1990. *Pale Ale*. Boulder, Colorado: Brewers Publications.

32. Foster, Terry. 1992. *Porter*. Boulder, Colorado: Brewers Publications.

33. Giffen, Paddy and Byron Burch. 1991. "Stout." *Zymurgy – Special Issue: Traditional Beer Styles* 14 (4).

34. Guinard, Jean-Xavier. 1990. *Lambic*. Boulder, Colorado: Brewers Publications.

35. Hardwick, William A., Dirk E. J. van Oevelen, Lawrence Novellie, and Kiyoshi Yoshizawa. 1995. "Kinds of Beer and Beerlike Beverages." *Handbook of Brewing*, edited by William A. Hardwick. New York, New York: Marcel Dekker, Inc.

36. Hardy, Norm. 1995. "Altbier," edited by Martin Lodahl. *Brewing Techniques* 3 (1).

37. Hersh, Jay S. 1996. "Bavaria's Dark Secret – Shedding Light on the Bavarian Dunkel Style," edited by Martin Lodahl. *Brewing Techniques* 4 (1).

38. Hoag, Don. 1991. "Barley Wine." *Zymurgy – Special Issue: Traditional Beer Styles* 14 (4).

39. Kunze, Wolfgang. 1996. *Technology Brewing and Malting*, Translated by Dr. Trevor Wainwright. Berlin, Germany: VLB Berlin.

40. Kuplent, Florian. 1999. "Berliener Weisbier." *Brewing Techniques* 7 (1).

41. Kuplent, Florian. 1999. Personal Communication.

42. Jackson, Michael. 1983. *The New World Guide to Beer*. Philadelphia, Pennsylvania: Courage Books.

43. Jackson, Michael. 1993. *Michael Jackson's Beer Companion*. Philadelphia, Pennsylvania: The Running Press.

44. Jackson, Michael. 1995. *The Great Beers of Belgium*. London, England: Duncan Baird Publishers.

45. Jackson, Michael and Jean-Xavier Guinard. 1991. "Belgium-Style Specialty." *Zymurgy – Special Issue: Traditional Beer Styles* 14 (4).

46. Lewis, Michael J. 1995. *Stout*. Boulder, Colorado: Brewers Publications.

47. Liddil, Jim. 1997. "Brewing in Styles: Practical Strategies for Brewing Lambic at Home, Part I –Wort Preparation," edited by Martin Lodahl. *Brewing Techniques* 5 (3).

48. Liddil, Jim. 1997. "Brewing in Styles: Practical Strategies for Brewing Lambic at Home, Part II – Fermentation and Culturing," edited by Martin Lodahl. *Brewing Techniques* 5 (4).

49. Liddil, Jim. 1997. "Brewing in Styles: Practical Strategies for Brewing Lambic at Home, Part III – The Finishing Touches," edited by Martin Lodahl. *Brewing Techniques* 5 (5).

50. Line, Dave. 1992. *The Big Book of Brewing*. Ann Arbor, Michigan: G. W. Kent, Inc.

51. Lodahl, Martin. 1994. "Witbier: Belgian White,"edited by Roger Bergen. *Brewing Techniques* 2 (4).

52. Lodahl, Martin. 194. "Old, Strong, and Stock Ales," edited by Roger Bergen. *Brewing Techniques* 2 (5).

53. Lodahl, Martin. 1994. "Belgian Trappists and Abbey Beers," edited by Roger Bergen. *Brewing Techniques* 2 (6).

54. Lodahl, Martin. 1995. "Lambic: Belgium's Unique Treasure." *Brewing Techniques* 3 (4).

55. Maytag, Fritz. 1991. "California Common Beer." *Zymurgy – Special Issue: Traditional Beer Styles* 14 (4).

56. Miller, David. 1990. *Continental Pilsener*. Boulder, Colorado: Brewers Publications.

57. Miller, David. 1991. "Classic Pilsener." *Zymurgy – Special Issue: Traditional Beer Styles* 14 (4).

58. Mosher, Randy. 1995. *Brewer's Companion*. Seattle, Washington: Alephenalia Publications.

59. Mussche, Roger A. 1999. "Spontaneous Fermentation – the Production of Belgian Lambic, Gueuze and Fruit Beers." *Brewers' Guardian* 128 (1).

60. Narziss, Ludwig. 1980. *Abiss der Beierbrauerei*. Stuttgart, Germany: Ferdinand Enke Verlag.

61. Noonan, Gregory J. 1987. "Bock Beer." *The New Brewer* 4 (1).

62. Noonan, Gregory J. 1987. "Scotch Ales." *The New Brewer* 4 (2).

63. Noonan, Gregory J. 1987. "Weiss/Weizen Beer." *The New Brewer* 4 (3).

64. Noonan, Gregory J. 1988. "Pilsener." *The New Brewer* 5 (1).

65. Noonan, Greg and Stephen Foster. 1991. "English Bitter and Scottish Ale." *Zymurgy – Special Issue: Traditional Beer Styles* 14 (4).

66. Noonan, Gregory J. 1993. *Scotch Ale*. Boulder, Colorado: Brewers Publications.

67. Noonan, Gregory J. 1996. *New Brewing Lager Beer*. Boulder, Colorado: Brewers Publications.

68. Norton, Dave. 1991. "American Dark." *Zymurgy – Special Issue: Traditional Beer Styles* 14 (4).

69. Norton, Dave. 1991. "Dortmund, Export." *Zymurgy – Special Issue: Traditional Beer Styles* 14 (4).

70. Nummer, Brian A. 1996. "Brewing with Lactic Acid Bacteria." *Brewing Techniques* 4 (3).

71. Oliver, Garrett. 1996. "Barleywines – The Power and the Glory." *The Malt Advocate* 5 (1).

72. Papazian, Charlie. 1991. "Munich Helles." *Zymurgy – Special Issue: Traditional Beer Styles* 14 (4).

73. Papazian, Charlie. 1994. *The Home Brewer's Companion*. New York, New York: Avon Books.

74. Papazian, Charles. 2006. "Beer Styles: Their Origins and Classification." *Handbook of Brewing*, edited by Fergus G. Priest and Graham G. Stewart. Boca Raton, Florida: CRC Press, Taylor & Francis Group.

75. Protz, R. 1993. *The Real Ale Drinker's Almanac*. Glasgow, Britain: Neil Wilson Publishing, Ltd.

76. Rajotte, Pierre. 1992. *Belgian Ale*. Boulder, Colorado: Brewers Publications.

77. Rhodes, Christine P. and Pamela B. Lappies, eds. 1995. *The Encyclopedia of Beer*. New York, New York: Henry Holt and Company.

78. Richman, Darryl. 1991. "Brown Ale." *Zymurgy – Special Issue: Traditional Beer Styles* 14 (4).

79. Richman, Darryl. 1994. *Bock*. Boulder, Colorado: Brewers Publications.

80. Richman, Darryl. 1995. "Bock and Doppelbock." *Brewing Techniques* 3 (2).

81. Satula, David. 1997. "Mild Ale – Back from the Brink of Extinction," edited by Martin Lodahl. *Brewing Techniques* 5 (6).

82. Scheer, Fred. 1991. "Barvarian Dark." *Zymurgy – Special Issue: Traditional Beer Styles* 14 (4).

83. Schönfeld, Franz. 1938. *Obergärige Biere und ihre Herstellung*. Berlin, Germany: Verlag Paul Parey.

84. Slosberg, Pete. 1995. "The Road to an American Brown Ale." *Brewing Techniques* 3 (3).

85. Smith, Quentin B. 1991. "Specialty Beer." *Zymurgy – Special Issue: Traditional Beer Styles* 14 (4).

86. Snyder, Stephen. 1996. *The Beer Companion*. New York, New York: Simon & Schuster, 1996.

87. Tomlinson, Thom. 1994. "India Pale Ale, Part II: The Sun Never Sets," edited by Roger Bergen. *Brewing Techniques* 2 (3).

88. Warner, Eric. 1991. "German Wheat Beer." *Zymurgy – Special Issue: Traditional Beer Styles* 14 (4).

89. Warner, Eric. 1992. *German Wheat Beer*. Boulder, Colorado: Brewers Publications.

90. Williams, Forrest. 1998. "The Queen of Köln: A Visit to the Court of Germany's Kölschbier," edited by Martin Lodahl. *Brewing Techniques* 6 (1).

Chapter Twenty Two

Government Regulations

The brewing industry is subject to extensive government regulations at both the federal and state levels, as well as to regulation by a variety of local governments. Some of the regulations imposed at the federal and state level involve production, distribution, labeling, advertising, trade and pricing practices, credit, container characteristics, and alcoholic content. Federal, state and local governmental entities also levy various taxes, license fees and other similar charges and may require bonds to ensure compliance with applicable laws and regulations. Specific alcohol taxation (as opposed to more general sales taxes) is primarily a federal and state right although some states permit some additional local taxation. The brewing industry must also comply with numerous federal, state, and local environmental protection laws.

Federal Regulations

Until recently, nearly all federal regulations involving alcohol were issued by the Treasury Department Bureau of Alcohol, Tobacco, and Firearms (BATF), established by the Federal Alcohol Administration Act of 1935 and the 1968 Gun Control Act. However, in 2002, under the Homeland Security Act, the Bureau was divided. The part remaining in the Department of the Treasury was renamed the Alcohol and Tobacco Tax and Trade Bureau (TTB). A new Bureau of Alcohol, Tobacco, Firearms and Explosives (ATF) was formed in the Department of Justice. The TTB is responsible for administering and regulating the operations of distilleries, wineries, and breweries, as well as importers and wholesalers in the industry. Some of the specific functions TTB is responsible for as related to beer brewing include:

1. Brewery application approval
2. Excise tax collection
3. Labeling and advertising approval
4. Homebrewing

Brewery Application Approval

To qualify as a brewer you must complete and submit to TTB the appropriate forms along with any other required documentation. TTB will usually complete our screening and processing within sixty days of receipt of a completed Brewers Notice packet.

The following forms are required prior to beginning operations

- Brewer's Notice, Form 5130.10
- Brewer's Bond, Form 5130.22 or Brewer's Collateral Bond, Form 5130.25
- Surety company representative Power of Attorney
- Diagram (plat/plan) of the brewery premises
- Legal description of the brewery premises
- Statement describing the security at the brewery
- Personnel Questionnaire, Form 5000.9
- Environmental Information, Form 5000.29
- Supplemental Information on Water Quality Considerations, Form 5000.30
- Signing Authority:
  - Power of Attorney, Form 5000.8
  - Signing Authority for Corporate Officials, Form 5100.1 (if not granted on the Brewer's Notice)
  - Corporate resolution or specific notification in organizational documents
- Organizational Documents:
  - Certificate and Articles of Incorporation (for Corporations)
  - Certificate and Articles of Organization (for LLCs)
  - Partnership Agreement (for Partnerships)
  - By-Laws (for Corporations)
  - Operating Agreement (for LLCs)
  - Trade Name Registration (if required by state or local government)
  - Certificate to Operate in a Foreign State (if organized in a different state)
  - List of Stockholders of 10% or more of stock, Directors, Officers, members/managers, and partners, including addresses, with titles and amount of shares/interest held
- Statement advising that the premises is covered under the National Historic Preservation Act (if applicable)
- Controlled group listing, if applicable. Include a breakdown of the 60,000 barrel allocation for the reduced rate.
- Source of Funds Documentation

- Lease Agreement (if applicable)
- Legible photocopy of the Driver's License or official State ID card of the primary contact person who will be interviewed by phone by TTB regarding the application

Note: All TTB applications are now processed at the National Revenue Center in Cincinnati and TTB is systemizing and streamlining the process.

Excise Tax Collection

U.S. Government involvement in the beer industry also includes taxation. The current federal excise tax on beer, in effect since January 1, 1991, is $18 per barrel for 31 gallons. However, a reduced tax rate applies, at a rate of $7 per barrel, to the first 60,000 barrels of beer removed for consumption or sale by brewing companies that do not produce more than 2,000,000 barrels of beer per calendar year. The federal excise tax regulations also include other rules, including for removals without tax payment and inter-brewery purchases.

Labeling and Advertising Approval

The TTB implements and enforces a broad range of statutory and compliance provisions to ensure that alcohol products are created, labeled, and advertised in accordance with Federal laws and regulations. Brewers must follow the labeling and advertising requirements found at 27 CFR Part 7, *Labeling and Advertising of Malt Beverages* and 27 CFR Part 16, *Alcoholic Beverage Health Warning Statement*.

Label Approval

The TTB has exacting rules relating to label statements and design that are strictly enforced in the beer industry. Every commercial beer label and label change must be submitted for approval upon ATF form 5100.31, *Application for and Certification/Exemption of Label/Bottle Approval (COLA)*. Depending on the type of packaging and whether the beer is domestic or imported, the following information may be required on the label:

- Brand Name – name under which the malt beverage is marketed
- Class and Type – the specific identity of the malt beverage, e.g., beer, ale, porter
- Name and Address – city and State of the bottler/packer
- Net Contents – must be expressed in English units of measure, e.g., fluid ounces, pints
- Alcohol Content – optional unless required or prohibited by State law. Mandatory when any alcohol comes from added flavors.
- Government Warning Statement

COLAs Online System – TTB's COLAs Online System is an Internet based system that allows registered Industry Members (IM) the option to apply for Certification/Exception of Label/Bottle Approval online. COLAs Online also gives registered IM the ability to track the status of their electronic label submissions (COLAs). Currently, COLAs Online does not allow for the tracking of paper submissions of COLAs. In addition, COLAs Online serves as the sole internal database for TTB's Advertising, Labeling and Formulation Division to track all work related documentation, including all COLA submissions received for approval either in hard copy or soft copy. COLAs Online also stores all approved COLAs and provides the general public with detail information on COLAs, along with a printable version of the approved COLAs, if available. COLAs Online can be accessed at: https://www.ttbonline.gov/colasonline/.

Some of the advantages of filing COLAs electronically are as follows:

- Label applications are submitted instantaneously, reducing the time and cost associated with filing paper label applications.
- Users receive an immediate electronic confirmation of receipt of label application submission and a unique TTB ID number for tracking purposes.
- Users have fewer rejections for omissions of data because the system has business rules that will not allow users to submit applications that have certain incorrect or incomplete portions.
- The system is generally accessible 24 hours per day, 7 days a week unless undergoing routine system maintenance.
- If reasons for rejection are disclosed during the initial review of an e-Application, users are given the opportunity to make corrections to the application within fifteen calendar days after which the system automatically rejects the application.
- Corrected e-Applications receive priority consideration over all other e-filed applications received that same day.

Advertising Approval

While TTB must approve alcohol beverage labels prior to bottling or removal from customs custody through the certificate of label approval (COLA) procedures, TTB has taken a voluntary approach relative to the approval of the alcohol industry's advertisements. In the past, TTB invited industry members to voluntarily submit new or questionable advertising for clearance prior to using such material.

However, in view of the high volume of new products entering the marketplace, increased competition, and expanded alcohol advertising through the Internet, cable and satellite television, and the print media, TTB is taking a more proactive approach toward monitoring and reviewing alcohol beverage advertising.

Under TTB's new *Alcohol Beverage Advertising Program*, TTB plans to contact alcohol producers and importers and request that they provide the following information as to specified brands: current, selected advertising materials such as posters, flyers, point of sale materials, press releases, Web sites, Web page materials, and magazine and periodical ads, as well as television and radio advertisements. TTB will examine this advertising to determine compliance with mandatory and prohibited statement requirements. Examples of advertising areas that TTB will review include, but are not limited to, the following:

TTB Industry Circular 2004–6

- The presence of mandatory information, such as the name and address of the responsible advertiser and product class/type information;
- Statements or depictions that are inconsistent with approved product labels;
- Statements that are false, misleading, or deceptive;
- Statements, designs, or the use of subliminal representations that are obscene or indecent;
- False or misleading statements that are disparaging of a competitor's product;
- Prohibited uses of the word "pure" for distilled spirits products;
- Misleading or false curative or therapeutic claims;
- The form and use of mandated and optional alcohol content statements;
- Misleading references to carbohydrates, calories, fat, protein, and other macronutrients or "components"; and
- Specific health claims and health related statements.

Reportedly TTB will select industry members for advertising examination based on various indicators such as prior compliance history, advertising methods, and market impact, as well as random sampling factors. Of course, TTB will also continue to respond to specific complaints and referrals.

TTB does offer a voluntary pre-clearance review program as a way to help avoid improper advertising.

Homebrewing

Any adult may produce beer, without payment of tax, for personal or family use and not for sale. An adult is any individual who is 18 years of age or older. If the locality in which the household is located requires a greater minimum age for the sale of beer to individuals, the adult shall be that age before commencing the production of beer. This exemption does not authorize the production of beer for use contrary to state or local law. The production of beer per household, without payment of tax, for personal or family use may not exceed 200 gallons per calendar year if there are two or more adults residing in the household, or 100 gallons per calendar year if there is only one adult residing in the household. Beer made for personal or family use may not be sold or offered for sale.

State Regulations

In addition to meeting Federal regulations, individuals and businesses must comply with state regulations too. These state regulations, which vary widely from state to state, may be more restrictive than Federal regulations and must be met in addition to Federal requirements unless the Federal law pre-empts the State law wherein they desire to do business. For example regarding pre-emption, the Government Warning label on alcohol beverages pre-empts the states from imposing a similar requirement. (See 27 U.S.C. Section 216.) Likewise, the state may not authorize a bottle size for beer that is not also authorized at the Federal level. Most States have commissions or agencies, which oversee persons and businesses involved in the production, distribution, and selling of alcohol beverage products.

Alcoholic Beverage Control Boards

Following national Prohibition, the 21st Amendment to the Constitution provides states with broad powers and authority to regulate the production, importation, distribution, retail sale, and consumption of alcohol beverages inside their borders. (This is in addition to Federal requirements.) Each state created its own unique system of alcohol beverage control. There are two general classifications—open and "control" states.

Open States

The larger group (now referred to as "license" or "open" states) license all aspects of private production, distribution, and sales. License states have set up a hierarchical business licensing system known as the three-tier system. Under such a system, all alcohol must move from supplier to wholesaler to retailer, rather than directly to the consumer. The state maintains control through the approval and sale of licenses as well as the oversight of licensees' business practices and collection of taxes.

Control States

The smaller group (now referred to as "control" or "monopoly" states) opted to become wholesalers and retailers themselves for wine and spirits. The "control" states have a direct interest in the levels of revenue—profit produced by alcohol sales—but they also have public health and safety concerns and obligations. Thus, they may exert more extensive controls on the conditions of sale and promotion. Some extend this to how alcohol is advertised. In addition, some states program extensive public alcohol education activities and focus on prevention of sales to minors.

There are currently 18 control and monopoly states in the United States (see list below). Out of the 18 states that regulate alcohol wholesaling, only 9 (Alabama, Idaho, New Hampshire, Oregon, North Carolina, Pennsylvania, Virginia, Washington, and Utah) run liquor establishments. The others either permit ABC licensed private stores to sell liquor or contract the management and operations of the store to private firms, usually for a commission.

Control or Monopoly States

1. Alabama
2. Idaho
3. Iowa
4. Maine
5. Michigan
6. Mississippi
7. Montana
8. New Hampshire
9. North Carolina
10. Ohio
11. Oregon
12. Pennsylvania
13. Utah
14. Vermont
15. Virginia
16. Washington
17. West Virginia
18. Wyoming

Note: Montgomery County, Maryland, is the only non-state government to participate in this control system.

Licenses

On the state and local level, the license process varies widely. In some areas, the state is the lead agency for all licenses (manufacturer's and retailer's licenses), and local approval is not necessary except to confirm proper zoning. The state license process may involve investigation and inspections, but often the state will not issue the manufacturer's license until the approved Brewer's Notice is given by TTB. In other states, the local licensing authority acts as the lead agency and issues the retailer's license, while the state issues the manufacturer's license.

Retailer's Licenses

Some states may control the number and type of beer retailers by issuing retail licenses and by determining to which retailer's credit can be extended. Some also determine permissible locations for the sale of beer: on-premise, in restaurants and bars; and off-premise, in grocery stores, gas stations, liquor stores, and drug stores. If beer is sold on-premise, the state can regulate whether it is sold in packages, by the drink, or "to go." The state may also determine the maximum alcohol content and pricing method of beer sold at retail. In addition, the state can determine permissible days and/or hours for the sale of beer. States can also regulate persons selling or serving spirits, wine or beer to consumers. This includes the minimum legal age that a person may sell or serve. With the exception of labeling, advertising, and containers, the TTB does not enforce laws about selling or serving spirits, wine or beer to consumers.

Taxation

Every state imposes an excise tax on beer that is levied as a dollar amount on a specified volume (in liquid measure—e.g., gallons). On January 1, 2008, state excise taxes ranged from $0.02 per gallon in Wyoming to $1.07 per gallon in Alaska. In addition to the state excise taxes the brewer may be faced with other taxes such as taxes based on the alcohol content of the beer, a local tax, wholesaler taxes, and on-premise sales taxes. States frequently require a bond for the payment of the state's excise taxes on the beer you produce.

For more information regarding state excise taxes, and other taxes assessed by each state go to Appendix C: State Beer Excise Tax Rates.

Container Deposit Laws

Certain states, including California, Connecticut, Delaware, Iowa, Maine, Massachusetts, Michigan, New York, Oregon, and Vermont, and a small number of local jurisdictions, have adopted beverage packaging laws and regulations that require deposits on beverage containers.

Local Regulations

Many states permit local jurisdictions to regulate and separately tax beer sales, and even to prohibit the sale of beer within their jurisdiction. Jurisdictions in which the sale of alcoholic beverages is prohibited are called "dry" states. Over half of all states have dry cities or counties, and about 4.3% of the U.S. population lives in dry counties. Many cities and counties that are not dry regulate operations and/or impose taxes on the sale of beer. In addition to the federal and state excise taxes, some states have local taxes, on-premise taxes, wholesale taxes, and private club taxes. Georgia, Illinois, Louisiana,

Maryland, New York, and Ohio have cities or counties that impose local beer taxes. As might be expected, taxes can potentially represent the largest single-cost item in a glass of beer.

References

1. Barsby, Steve L. and Associates. 1999. *Beer Wholesalers: Their Role and Economic Performance*, 3rd ed. Alexandria, Virginia: National Beer Wholesalers Association.

2. Elzinga, K.G. 1990. *The Beer Industry, The Structure of American Industry*. ed. W. Adams MacMillan: New York, New York.

3. Finnegan, T., ed. 1997. *Modern Brewery Age Blue Book*. Modern Brewery Age Publishing: Stamford, Connecticut.

4. Harney, Amy K. 1995. *Malt Beverages*. Washington, D.C.: Office of Industries, United States International Trade Commission, USITC Publication 2865.

5. Katz, P.C. 1991. "Brewing Industry in the United States." *Brewers Almanac*, The Beer Institute, Washington, D.C.

Appendixes

A. Hop Varieties, page 423

B. Carbon Dioxide Volume Table, page 429

C. State Beer Excise Tax Rates, page 431

D. Conversion Factors, page 433

Appendix A: Hop Varieties

BITTERING HOPS

VARIETY	ALPHA ACIDS (w/w) %	BETA ACIDS (w/w) %	TOTAL OIL (w/w) %	CO-HUMULONE (w/w) %	CHARACTERISTICS/USAGE
Admiral	13–16.2	4.8–6.1	1–1.7	37–45	Admiral is a high-alpha variety considered to have a more pleasing, less harsh bitterness than Target.
Brewer's Gold (UK)	7–10	3–4.5	1.8–2.5	40–45	Pungent and very bitter; used in British ales and can be used in heavier style German lagers
Bullion (UK)	6–9	3.2–4.7	3.2	36	Pungent; used in British ales as well as heavier style German lagers
Centennial (US)	8.5–12	2.5–5	1.5–2.3	29–30	Floral and citrus notes; used in American pale ales, American wheat, stouts, and porters
Chinook (US)	12–14	3–4	1.5–2.5	29–34	Spicy; used in everything from pale ales to lagers; best for porters and stouts
Cluster (US)	5.5–8.5	4.5–5.5	0.4–0.8	36–42	Pungent, quite spicy; general purpose bittering hop for any style with medium and well-balanced bittering potential and no undesirable aroma properties
Eroica (US)	9.5–14	2.5–5	0.8–1.3	36–42	Unbalanced aroma characteristics; pungent and very bitter; used only as base bittering hop, especially dark British ales; can be used in wheat beers; pronounced eh-roy'-ih-cah
Galena (US)	12–14	7–9	0.9–1.2	32–42	Pleasant hopiness; an excellent high-alpha acids hop with balanced bittering properties combined with an acceptable aroma profile; suitable for any style, particularly British ales

VARIETY	ALPHA ACIDS (w/w) %	BETA ACIDS (w/w) %	TOTAL OIL (w/w) %	CO-HUMULONE (w/w) %	CHARACTERISTICS/USAGE
Herald	9–13	4.8–5.5	1.0–1.9	35–37	It combines strong bitterness with a citrusy, grapefruit character. It has been quickly adopted by many English craft breweries.
Northern Brewer (GM)	6–10	3–5	1.6–2.1	28–33	Clean bitter character, fragrant; considered superior to US variety; used in Belgian ales, German lagers and California common beers
Northern Brewer (US)	6.5–10	1.5–5	1.5–2	20–30	Clean bitter character, fragrant; used in Belgian ales, German lagers and California common beers
Nugget (US)	12–14	4–6	1.7–2.3	24–30	General purpose bittering hop with good aroma profile; bittering for all types of beers except for light lagers; herbal undertones
Olympic (US)	11.5–13.5	3.5–5.5	1.6–1.8	27–34	Suitable for general bittering; can leave a lingering aftertaste
Perle (US)	7–9.5	4–5	0.7–0.9	27–32	Minty bitterness; a hop with German type aroma properties; used successfully in almost any beer; pronounced perh'-luh
Pride of Ringwood (AS)	6–11	5.3–6.5	1–2	33–39	Sharp, spicy; used in duplicating Australian beers; similar to Cluster
Super Styrians (Slovenia)	8–10	3.6–4.5	0.8–1	25–30	General purpose bittering
Wye Target (UK)	9.5–13	4.3–5.9	1.2–1.4	35–40	Some aromatic qualities; used in all styles of English ales and lagers

Appendix A: Hop Varieties   425

VARIETY	ALPHA ACIDS (w/w) %	BETA ACIDS (w/w) %	TOTAL OIL (w/w) %	CO-HUMULONE (w/w) %	CHARACTERISTICS/USAGE
AROMA HOPS					
Cascade (US)	4.5–7	4.5–7	0.8–1.5	33–40	Pleasant, floral, spicy with strong citrus notes; used for dry hopping; well balanced bittering potential; used in American pale ales. California common, stouts, porter, American wheat; lager brewers generally find floral aroma and taste to be undesirable
Crystal (US)	2–4.5	4.5–6.5	1.0–1.5	20–26	Mild and pleasant; used in lagers but primarily German styled-lagers
Fuggle (US)	4–5.5	1.5–2	0.7–1.2	25–32	Pleasant and spicy; used in British-style ales, milds, bitters, India pale ales, porters, stouts; pronounced fuh'-gull
Fuggle (UK)	4–5.5	2–3	0.7–1.1	25–30	Pleasant and spicy; used in British-style ales, milds, bitters, India pale ales, porters, stouts; considered superior to US Fuggle; pronounced fuh'-gull
Goldings (UK)	4–6.5	2.5–3.5	0.8–1	20–32	Slightly flowery; sweet and spicy; used in all British-style ales, milds, bitters, India pale ales, porters, stouts, especially paler beers; perfect for dry hopping British-style ales
Goldings (Slovenia)	4–6	2–3	0.5–1.3	26–30	Fairly aromatic; used in British ales
Hallertauer (US)	3.5–4.5	3.5–4.5	0.6–1	18–24	Slightly flowery; used in German lagers and ales; pronounced holler tower

VARIETY	ALPHA ACIDS (w/w) %	BETA ACIDS (w/w) %	TOTAL OIL (w/w) %	CO-HUMULONE (w/w) %	CHARACTERISTICS/USAGE
Hallertauer Magnum	11–16	5–7	1.6–2.6	21–29	Sometimes just called Magnum, is a bittering hop cultivar with a clean German flavor and aroma profile, excellent for bittering any kind of German beer. It is one of the most popular bittering hops in Germany, and is also grown in the United States.
Hallertauer Mittelfrüh (GM)	3.5–5.5	3.5–5.5	1	21	Used in lagers (perhaps with the exception of pilsners), especially German lagers; pronounced mih'-tehl-frue
Hallertauer Tradition (GM)	5–7	4–5	1–1.4	26–29	Truly noble in aroma and character; used in lagers (perhaps with the exception of pilsners), especially German lagers
Hersbrucker (GM)	2–5.5	3–5.5	0.7–1.3	19–25	Fresh, delicate, spicy; somewhat flowery; used in German lagers and ales; wheat beer; pronounced holler tow -- hehrs'-bruh-kehr
Hersbrucker (US)	3.5–5.5	5.5–7	0.6–1.2	20–30	Somewhat spicy; used in German lagers and ales
Hüller Bitterer (GM)	4.5–7	4–6	1.2	30	Used in German lagers and ales; pronounced hue'-ler
Liberty (US)	3.5–6.5	3–4	0.7–1.2	24–28	Spicy; aroma pleasant and clean with marked similarities to the German Hallertauer Mittelfrüh; used in German-style lagers
Lubelski or Lublin (PL)	3–5	2.3–3.8	1	27	Pleasant hoppy notes; used in Czech-style pilsners and Belgian strong ales; pronounced lew-behl'-skee or lewb'-lihn

VARIETY	ALPHA ACIDS (w/w) %	BETA ACIDS (w/w) %	TOTAL OIL (w/w) %	CO-HUMULONE (w/w) %	CHARACTERISTICS/USAGE
Mt. Hood (US)	5–8	5–7.5	1.0–1.3	22–23	Spalt Spalter has a mild, spicy aroma that is often compared to Czech Saaz or Tettnang Tettnanger.; used in lagers
Spalt Spalter	4–5.5	4–5.5	0.5–1.1	22–28	Slightly spicy; used in German lagers
Saaz (CZ)	2.5–4.5	2.5–4	0.4–0.7	22–28	Very mild and pleasant hoppy notes; a gentle fruity flavor, taste and smell that can be described as elegant; this character is mainly due to the unique oil content of this hop; used in making Pilsners as well as for Belgian pale ales; pronounced sahz or sahts
Spalter (GM)	4–5.5	4–5.5	0.5–1.1	22–28	Slightly spicy; used in German lagers, altbier; pronounced shpawl'-tehr
Spalter (US)	3–6	3–5	0.5–1	20–25	Mild, delicate, clean; used in German lagers and premium American lagers; pronounced shpawl'tehr
Spalter Select (GM)	4–6	3.5–4.5	0.5–1	21–25	Used in German lagers; pronounced shpawl'tehr
Sterling	4.5–9	4–6	0.6–1.9	21–24	It is considered to have a very good noble hop style of spicy, floral, slightly citrusy character and is often recommended as a substitute for Saaz..
Tettnanger (GM)	3.5–5.5	3.5–5	0.6–1.1	23–29	Spicy, herbal, delicate, flowery; used in German lagers and ales; pronounced teht'-nang-ehr

VARIETY	ALPHA ACIDS (w/w) %	BETA ACIDS (w/w) %	TOTAL OIL (w/w) %	CO-HUMULONE (w/w) %	CHARACTERISTICS/USAGE
Tettnanger (US)	4–5	3.5–4.5	0.4–0.8	20–25	Spicy, herbal, delicate, flowery; used in German lagers and ales; pronounced teht'-nang-ehr
Willamette (US)	4–7	3–4	1–1.5	30–35	Pleasant, aromatic and slightly spicy; used in British-style ales, milds, bitters, India pale ales, porters, stouts

NOTES: The values are approximations and do not necessarily represent representative actual values. Information provided has been compiled from a number of sources including U.S. Hops Resource Guide by Hop Growers of America, The Practical Brewer by Herbert L. Grant, Using Hops-The Complete Guide to Hops for the Craft Brewer by Mark Garetz, The Brewer's Companion by Randy Mosher, The New Brewing Lager Beer by Gregory Noonan and numerous other technical articles.

Appendix B: Carbon Dioxide Volume Table   429

°C	\multicolumn{30}{c|}{POUNDS PER SQUARE INCH (PSI) – SATURATION PRESSURE}
	1	2	3	4	5	6	7	8	9	10	11	12	13	14	15	16	17	18	19	20	21	22	23	24	25	26	27	28	29	30
-1.1	1.82	1.92	2.03	2.14	2.23	2.36	2.48	2.60	2.70	2.82	2.93	3.02	3.13	3.27	3.35															
-0.6	1.78	1.88	2.00	2.10	2.20	2.31	2.42	2.54	2.65	2.76	2.86	2.96	3.07	3.19	3.30															
.00	1.75	1.85	1.95	2.05	2.15	2.27	2.38	2.48	2.59	2.70	2.80	2.90	3.01	3.10	3.20	3.32														
.06	1.71	1.81	1.91	2.01	2.10	2.23	2.33	2.43	2.53	2.63	2.74	2.84	2.96	3.05	3.15	3.25														
1.1	1.67	1.78	1.86	1.97	2.06	2.18	2.28	2.38	2.48	2.58	2.69	2.79	2.89	3.00	3.09	3.19	3.30													
1.7		1.73	1.83	1.93	2.02	2.14	2.24	2.34	2.43	2.52	2.63	2.73	2.83	2.93	3.02	3.12	3.22													
2.2		1.68	1.79	1.88	1.98	2.09	2.19	2.29	2.38	2.47	2.57	2.67	2.77	2.86	2.96	3.05	3.15	3.24												
2.8			1.74	1.84	1.94	2.04	2.14	2.24	2.33	2.42	2.52	2.62	2.71	2.80	2.90	3.00	3.09	3.18	3.27											
3.3			1.70	1.80	1.90	2.00	2.10	2.20	2.29	2.38	2.48	2.57	2.66	2.75	2.85	2.94	3.03	3.12	3.21											
3.9				1.76	1.86	1.96	2.06	2.15	2.25	2.34	2.43	2.52	2.61	2.70	2.80	2.89	2.98	3.07	3.16	3.25										
4.4				1.72	1.83	1.92	2.01	2.10	2.20	2.30	2.39	2.47	2.56	2.65	2.75	2.84	2.93	3.01	3.10	3.19	3.28									
5.0					1.79	1.88	1.97	2.06	2.16	2.25	2.34	2.43	2.52	2.60	2.70	2.79	2.88	2.96	3.05	3.14	3.23									
5.5					1.75	1.85	1.94	2.02	2.12	2.21	2.30	2.39	2.48	2.56	2.65	2.74	2.83	2.91	3.00	3.09	3.18	3.26								
6.1					1.72	1.81	1.90	1.99	2.08	2.17	2.26	2.34	2.43	2.52	2.61	2.69	2.78	2.86	2.95	3.04	3.13	3.21								
6.7					1.69	1.78	1.87	1.95	2.04	2.13	2.22	2.30	2.39	2.47	2.56	2.64	2.73	2.81	2.90	2.99	3.07	3.16	3.24							
7.2					1.66	1.75	1.84	1.91	2.00	2.08	2.17	2.26	2.34	2.42	2.51	2.60	2.69	2.77	2.86	2.94	3.02	3.11	3.19							
7.8					1.62	1.71	1.80	1.88	1.96	2.04	2.13	2.22	2.30	2.38	2.47	2.55	2.64	2.72	2.81	2.89	2.98	3.06	3.15	3.23						
8.3					1.59	1.68	1.76	1.84	1.92	2.00	2.09	2.18	2.26	2.34	2.42	2.50	2.59	2.67	2.76	2.84	2.93	3.02	3.09	3.18						
8.9					1.56	1.65	1.73	1.81	1.89	1.96	2.05	2.14	2.22	2.30	2.38	2.46	2.54	2.62	2.71	2.79	2.88	2.96	3.04	3.13						
9.4					1.53	1.62	1.70	1.79	1.86	1.93	2.01	2.10	2.18	2.25	2.34	2.42	2.50	2.58	2.67	2.75	2.83	2.91	3.00	3.07	3.15					
10.0					1.50	1.59	1.66	1.74	1.82	1.90	1.98	2.06	2.14	2.21	2.30	2.38	2.46	2.54	2.62	2.70	2.78	2.86	2.94	3.02	3.10	3.17				

POUNDS PER SQUARE INCH (PSI) – SATURATION PRESSURE

°C	1	2	3	4	5	6	7	8	9	10	11	12	13	14	15	16	17	18	19	20	21	22	23	24	25	26	27	28	29	30
10.6					1.47	1.57	1.64	1.71	1.79	1.87	1.95	2.02	2.10	2.18	2.26	2.34	2.42	2.49	2.57	2.65	2.74	2.82	2.90	2.97	3.05	3.13	3.19			
11.1					1.43	1.54	1.61	1.68	1.76	1.84	1.92	1.99	2.06	2.14	2.22	2.30	2.38	2.45	2.53	2.61	2.68	2.76	2.84	2.92	3.00	3.06	3.13	3.22		
11.7					1.41	1.51	1.59	1.66	1.74	1.81	1.89	1.96	2.03	2.10	2.18	2.26	2.34	2.41	2.49	2.57	2.64	2.71	2.79	2.86	2.94	3.01	3.09	3.16		
12.2					1.40	1.47	1.56	1.63	1.71	1.78	1.86	1.93	2.00	2.07	2.15	2.22	2.30	2.37	2.45	2.52	2.59	2.66	2.74	2.81	2.89	2.96	3.04	3.10	3.17	
12.8						1.45	1.53	1.60	1.68	1.75	1.82	1.89	1.97	2.04	2.12	2.19	2.26	2.33	2.40	2.47	2.54	2.62	2.69	2.76	2.83	2.89	2.97	3.04	3.11	3.18
13.3						1.43	1.50	1.57	1.65	1.72	1.79	1.86	1.93	2.00	2.08	2.15	2.22	2.29	2.36	2.43	2.50	2.57	2.64	2.71	2.78	2.85	2.92	2.99	3.06	3.13
13.9							1.47	1.54	1.62	1.70	1.77	1.83	1.90	1.97	2.04	2.11	2.18	2.25	2.32	2.39	2.46	2.53	2.60	2.66	2.73	2.80	2.87	2.94	3.00	3.08
14.4							1.45	1.51	1.59	1.67	1.74	1.80	1.87	1.94	2.01	2.08	2.15	2.21	2.28	2.35	2.42	2.48	2.55	2.62	2.69	2.75	2.82	2.88	2.95	3.02
15.0								1.50	1.56	1.64	1.71	1.77	1.84	1.91	1.98	2.04	2.11	2.17	2.24	2.31	2.38	2.43	2.50	2.57	2.64	2.70	2.77	2.84	2.91	2.97
15.6									1.54	1.62	1.69	1.75	1.82	1.88	1.95	2.01	2.08	2.14	2.21	2.27	2.34	2.40	2.47	2.53	2.60	2.66	2.73	2.79	2.86	2.92

Appendix C: State Beer Excise Tax Rates

State	Excide Tax Rates ($ per gallon)	Other Taxes
Alabama	$0.53	$0.52/gallon local tax
Alaska	1.07	
Arkansas	0.16	under 3.2% - $0.16/gallon; $0.008/gallon and 3% off - 10% on-premise tax
California	0.20	−1.0%
Colorado	0.08	−1.2%
Connecticut	0.19	
Delaware	0.16	
Florida	0.48	2.67¢/12 ounces on-premise retail tax
Georgia	0.32	$0.53/gallon local tax
Hawaii	0.93	$0.54/gallon draft beer
Idaho	0.15	over 4% - $0.45/gallon
Illinois	0.185	$0.16/gallon in Chicago and $0.06/gallon in Cook County
Indiana	0.115	
Iowa	0.19	
Kansas	0.18	over 3.2% - {8% off- and 10% on-premise}, under 3.2% - 4.25% sales tax
Kentucky	0.08	11% wholesale tax
Lousiana	0.32	$0.048/gallon local tax
Maine	0.35	additional 5% on-premise tax
Maryland	0.09	$0.2333/gallon in Garrett County
Massachusetts	0.11	0.57% on private club sales
Michigan	0.20	under 3.2% - $0.077/gallon. 9% sales tax
Minnesota	0.15	
Mississippi	0.4268	
Missouri	0.06	
Montana	0.14	
Nebraska	0.31	
Nevada	0.16	
New Hampshire	0.30	
New Jersey	0.12	
New Mexico	0.41	
New York	0.11	$0.12/gallon in New York City
North Carolina	0.53	
North Dakota	0.16	7% state sales tax, bulk beer $0.08/gal
Ohio	0.18	

State	Excide Tax Rates ($ per gallon)	Other Taxes
Oklahoma	0.40	under 3.2% - $0.36/gallon; 13.5% on-premise
Oregon	0.08	
Pennsylvania	0.08	
Rhode Island	0.10	$0.04/case wholesale tax
South Carolina	0.77	
South Dakota	0.27	
Tennessee	0.14	17% wholesale tax
Texas	0.19	over 4% - $0.198/gallon, 14% on-premise and $0.05/drink on airline sales
Utah	0.41	over 3.2% - sold through state store
Vermont	0.265	6% to 8% alcohol - $0.55; 10% on-premise sales tax
Virginia	0.26	
Washington	0.261	
West Virginia	0.18	
Wisconsin	0.06	
Wyoming	0.02	
District of Columbia	0.09	8% off- and 10% on-premise sales tax

Source: Compiled by Federation of Tax Administrators, Washington, D.C. (as of January 1, 2008).

Appendix D: Conversion Factors

To Convert:	To:	Multiply by:
Air, grams	liters of air	0.7735
Air, liters	grams of air	1.2292
Air (ml)/gallon (U.S.)	parts per million of air	0.3427
Air (ml)/liter	parts per million of air	1.2971
Air (ml)/ounce (U.S.)	parts per million air	46.86
	parts per million oxygen	9.6730
Air (ml)/ounce (Imp.)	parts per million oxygen	10.0694
Air (ppm)	liters air/gallon (U.S.)	2.917
	milliters air/liter	0.771
	milliters air/ounce (U.S.)	0.0228
Atmospheres	pounds per square inch	14.696
	bar	1.0132
	kilopascals	101.325
	mm/Hg	760.002
	kg/cm^2	1.03323
Bar	atmosphers	0.9869
	pounds per square inch	14.5
	mm/Hg	750.062
	kg/cm^2	1.01972
Beer barrels (Imp.)	barrels (U.S.)	1.373
	cubic meters	0.1636591
	gallons (Imp.)	36.0
	gallons (U.S.)	43.23
	hectoliters	1.673
Beer barrels (U.S.)	barrels (Imp.)	0.728
	hectoliters	1.1734
	cubic feet	4.144
	gallons (Imp.)	26.23
	gallons (U.S.)	31.0
British Thermal Units	kilocalories	0.2520
	Wh	0.2931
	KJ	1.055
	ft. lb.	778.2
Bushels (U.S.) of barley	pounds	34.0
Bushels (U.S.) of malt	pounds	48.0
Carbon dioxide (cu ft)	ounces of carbon dioxide	957.5
	pounds of carbon dioxide	0.1227
Carbon dioxide, grams	milliters of carbon dioxide	506.0
Carbon dioxide, liters	grams of carbon dioxide	1.976
	cubic feet of carbon dioxide	0.0352
Carbon dioxide, pounds	cubic feet of carbon dioxide	8.1499
C. dioxide (%) by vol.	percent carbon dioxide by weight	0.1943

To Convert:	To:	Multiply by:
C. dioxide (%) by wt.	percent carbon dioxide by volume	5.1470
Centigrade (°C)	Fahrenheit (°F)	(°C x 9/5) + 32
Centiliters	cubic centimeters	10.0
	cubic inches	0.6102545
	ounces (U.S.)	0.33815
Centimeters	feet	0.0328
	inches	0.3937
Centimeters, cubic	cubic feet	0.000035314
	cubic inches	0.0610
	drams (US)	0.027051
	gallons (Imp)	0.000219
	gallons (US)	0.0002642
	milliliters	1.0
Centimeters/minute	inches/minute	0.3937
Centimeters/second	feet/minute	1.969
	feet/second	0.03281
	meters/minute	0.6
	meters/second	0.01
Clark degree	parts per million	14.25
	grains/gallon (U.S.)	0.833
	°French	1.43
	°German	0.8
	millival	0.7
Cups	liters	0.2366
	milliliters	236.0
	ounces (U.S.)	8.0
Drams (apothecary)	drams (avoirdupois)	0.21942
Drams (avoirdupois)	grams	1.7718
Fahreheit (°F)	Centigrade (°C)	(°F x 5/9) - 32
Feet	meters	0.3048
Feet, cubic	beer barrels (U.S.)	0.2413
	cubic centimeters	28,316.85
	cubic meters	0.028317
	gallons (Imp.)	6.25
	gallons (U.S.)	7.481
	liters	28.32

Appendix D: Conversion Factors

To Convert:	To:	Multiply by:
Feet/minute	centimeters/second	0.508
	feet/second	0.01667
	meters/minute	0.305
	meters/second	0.00508
	miles per hour	0.01136
Feet/second	centimeters/second	30.48
	meters/minute	18.29
French degree	parts per million	10.0
	grains/gallon (U.S.)	0.583
	°Clark	0.70
	°German	0.56
Gallons (Imp.)	cubic feet	0.16
	cubic inches	277.3
	liters	4.546
	gallons (U.S.)	1.2009
	cubic meters	0.0046
	ounces (Imp.)	160.0
	ounces (U.S.)	153.72
Gallons (U.S.)	beer barrels	0.0323
	cubic feet	0.1337
	cubic inches	231.0
	cups	16.0
	liters	3.7853
	gallons (Imp.)	0.8327
	cubic meters	0.0038
	ounces (U.S.)	128.0
	pounds of water at 20° C	8.3216
German degree	parts per million	17.9
	grains/gallon (U.S.)	1.044
	°Clark	1.24
	°French	1.78
Grains avoirdupois	grains troy	1.0
Grains per gallon (U.S.)	parts per million	17.1497
	°Clark	1.2
	°French	1.71
Grains per gallon (U.S.)	°German	0.958
Grains troy	grams	0.064799
Grams	grains	15.43
	kilograms	0.001
	ounces avoirdupois	0.0353

To Convert:	To:	Multiply by:
	pounds avoirdupois	0.0022
Grams/liter	grains/gallon (Imp.)	70.114
	grains/gallon (U.S.)	58.4
	grains/ounce (U.S.)	0.46
	grains/ounce (Imp.)	4.546
	grams/gallon U.S.)	3.785
	ounces (U.S.)/gallon	0.134
	pounds/gallon (U.S.)	0.0083
	teaspoons per gallon	0.8
Hectoliters	beer barrels (U.S.)	0.8522
	bushels	2.8375
	cubic feet	3.5316
	liters	100.0
	gallons (U.S.)	26.4178
Hundredweight (long)	kilograms	50.8024
Hundredweight (short)	kilograms	45.3592
Inches	centimeters	2.54
Inches, cubic	cubic centimeters	16.387
	cubic feet	0.00058
	gallons (Imp.)	0.0036
	gallons (U.S.)	0.0043
	liters	0.0164
	ounces (U.S.)	0.554
	ounces (Imp.)	0.6653
Kilocalaries	KJ	4.186
	British thermal units	3.968
	Wh	0.2931
	ft lb	778.2
Kilograms	centimeters (cu)	1000.0
	cubic inches	61.023
	gallons (U.S.)	0.26417
	grams	1000.0
	hundredweigth (long)	0.01968
Kilograms	hundredweight (short)	0.02204
	ounces avoirdupois	0.009842
	pounds	2.205
	tons (long)	35.2739
	tons (metric)	0.001
	tons (short)	0.0011023
Kilograms/hectoliter	pounds/barrel (U.S.)	2.59
	pounds/gal	0.0834

Appendix D: Conversion Factors

To Convert:	To:	Multiply by:
Kilograms/square atm.	atmospheres	0.96
	pounds/square inch	14.22
Kilometers	yards	1,093.61
Liters	cubic feet	0.035
	cubic inches	61.03
	fluid ounces (Imp.)	35.196
	fluid ounces (U.S.)	33.814
	gallons (Imp.)	0.2199
	gallons (U.S.)	0.2642
	hectoliters	0.01
	pints (Imp.)	1.7598
	pints (U.S.)	2.1134
	quarts (Imp.)	0.8799
	quarts (U.S.)	1.0567
Liters/minute	cubic feet/minute	0.0353
Liters/second	gallons (U.S.)/second	0.2642
Meters	feet	3.28
	inches	39.57
Meters, cubic	barrels (U.S.)	8.39
	barrels (Imp.)	6.11
	cubic feet	35.313
	gallons (Imp.)	220.0
	gallons (U.S.)	264.17
	liters (U.S.)	1000.0
Meters/second	feet/second	3.281
Microns (micrometer)	centimeters	0.0001
Milliliters	ounces (Imp.)	0.0352
	ounces (U.S.)	0.0338
	inches	0.0394
Milliliters	cubic centimeters	1.0
	cubic inches	0.06102
	grams of water at 20° C	0.997151
Ounces, fluid	liters (Imp.)	0.0208
	liters (U.S.)	0.0296
Ounces (Imp.)	gallons (Imp.)	0.0063
	milliliters	28.41
	ounces (U.S.)	0.9608
Ounces (U.S.)	gallons (U.S.)	0.0078

To Convert:	To:	Multiply by:
	cubic inches	1.805
	kilograms	0.0284
	liters	0.0295
	milliliters	29.57
	ounces (Imp)	1.041
Oxygen (ml)/oz. (U.S.)	parts per million oxygen	48.408
Oxygen (ml)/oz. (Imp.)	parts per million oxygen	50.352
Ounces (avoirdupois)	grains	437.5
	grams	28.35
Oxygen (ppm)	milliters air/ounce (Imp.)	0.0993
	milliters air/ounce (U.S.)	0.1034
	milliliters oxygen/ounce (Imp.)	0.0199
	milliliters oxygen/ounce (U.S.)	0.0207
Parts per million	grains/gallon (U.S.)	0.0584
	grams/liter	0.001
ounces/barrel	0.0042	
	milligrams/liter	1.0
	°Clark	0.07
	°French	0.10
	°German	0.056
Pints (Imp.)	liters	0.568
	pints (U.S.)	1.2009
Pints (U.S.)	liters	0.4732
	pints (Imp.)	0.8327
	cubic inches	28.875
	cups	2.0
Pounds per square inch	atmosphers	0.068
	mm/Hg	70.308
	kg/cm	0.0703086
Pounds	grams	453.6
	grains	7000.0
	kilograms	0.45
	ounces	16.0
Pounds/square inch	kilograms/square centimeter	0.0703
Quarts (Imp.)	liters	1.1365
Quarts (U.S.)	cups	4.0
	liters	0.946
Tablespoons	milliliters	14.787
	ounces (U.S.)	0.5
Tons (long)	kilograms	1,016.01

To Convert:	To:	Multiply by:
	tons (metric)	1.01605
	tons (short)	1.10231
Tons (metric)	short tons	1.1023
	kilograms	1000
	pounds	2204.62
Yards	feet	3
	inches	36
	meters	0.9144
	miles	0.000568

Glossary

ABSOLUTE ALCOHOL – The total amount of alcohol contained in a beverage.

ACETALDEHYDE – A volatile compound derived from the degradation of sugars during fermentation through decarboxylation of pyruvic acid. It decreases as the fermentation progresses and as the beer ages.

ACETIC ACID – Acetic acid is a by-product of yeast metabolism that forms through oxidation of alcohols during the fermentation process giving the beer a vinegar taste and smell.

ACETOBACTER – Acetobacter is a microorganism that turns ethyl alcohol to acetic acid during fermentation.

ACID – Any compound that yields hydrogen ions (H$^+$) in solution having a pH lower than 7.0. It is also used to refer to a beer exhibiting a sour acidic smell or flavor.

ACIDITY – Refers to the pH of the water when the pH measures less than 7.0. An increasing hydrogen-ion concentration leads to increasing acidity as the pH decreases from 7.0

ACIDIC – Descriptive term used of a beer having a biting, sour taste reminiscent of vinegar or acetic acid.

ADJUNCT – Any substitute unmalted cereal grain or fermentable ingredient added to the mash for the purpose of providing extract at a lower cost (cheaper form of carbohydrate) than that available from malt. Corn, oats, wheat, rice, and barley are used for this purpose and are either in their native form (i.e., meal, grits, flour or dry starch) or pre-gelatinized form (i.e., flakes, refined starches).

AFTERTASTE – The taste, odor and tactile sensations that linger after the beer has been swallowed.

AGAR – Agar is a gelatinous substance chiefly used as a solid substrate to contain culture medium for microbiological work. It is nonnitrogenous and more heat-stable than gelatin.

ALBUMIN – A name for a certain group of water-soluble proteins, which coagulate when, heated. Albumins are hydrolyzed to peptides and amino acids by proteolytic enzymes.

ALCOHOL – A synonym for ethyl alcohol or ethanol.

ALCOHOL AND TOBACCO TAX AND TRADE BUREAU – The Alcohol and Tobacco Tax and Trade Bureau, shortened to Tax and Trade Bureau or TTB, is a bureau of the United States Department of the Treasury responsible the revenue and regulatory functions related to wine, beer, and distilled spirits. On January 24, 2003, the Homeland Security Act of 2002 (the Act) split functions of the Bureau of Alcohol, Tobacco and Firearms (ATF), into two new organizations with separate functions. First, the Act established The Alcohol and Tobacco Tax and Trade Bureau (TTB) under the Department of the Treasury. Second, the Act transferred certain law enforcement functions from Treasury to the Department of Justice. The ATF was transferred to the Justice Department and was renamed the Bureau of Alcohol, Tobacco, Firearms and Explosives.

ALCOHOLIC – Flavor effect of ethanol and higher alcohols.

ALCOHOL BY VOLUME (v/v) – A measurement of the alcohol content of a solution in terms of the percentage volume of alcohol per volume of beer. To calculate percent alcohol by weight multiply percent alcohol by volume times 0.796 since alcohol only weighs 79.6% as much as water. For example, a beer that is 4% alcohol by volume is only 3.2% alcohol by weight.

ALCOHOL BY WEIGHT (w/v) – A measurement of the alcohol content of a solution in terms of the percentage weight of alcohol per volume of beer.

ALDEHYDE – An organic compound that is a precursor to ethanol in normal fermentation. Ethanol, in the presence of air, can revert back to unpleasant-tasting aldehydes, typically papery/cardboardy/sherry notes. Some aldehydes contribute to the bouquet of beer.

ALE – A generic name for beers produced by top fermentation, usually by infusion mashing, as opposed to lagers produced by bottom fermentation, usually by decoction mashing. Ales tend to have higher alcoholic contents, more robust flavor and deeper hues than lagers. It is the style predominant in the British Isles.

ALE YEAST – It is known as "top-fermenting" type yeast (*Saccharomyces cerevisiae*) because of its ability to form a layer of yeast on the surface during primary fermentation.

ALGAECIDE – It is an agent that kills algae.

ALKALINITY – Refers to the pH of water when the pH measures greater than 7.0. A decreasing hydrogen-ion concentration leads to increasing alkalinity as the pH increases from 7.0.

ALPHA ACID – One of the two resins in hops that is responsible for the bitterness.

ALPHA ACID CONTENT – The percentage of alpha acid in the hop cone varies widely among hop varieties from levels of 3–4% in aromatic type varieties to 8–14% in the bitter hops.

ALPHA ACID UNIT (AAU) – A measurement of the potential bitterness of hops expressed in terms of their percentage alpha acid content.

ALPHA-AMYLASE – A diastatic enzyme that breaks down long glucose chains (starch) into smaller molecules of glucose, maltose, maltotriose, maltotetraose, and complex carbohydrates called dextrins during mashing.

AMINO ACIDS – The basic building blocks of any protein are amino acids, which join by forming a peptide bond. One way to classify proteins is according to the number of amino acid molecules they contain. The simplest are the di- and tripeptides, which contain two and three amino acids, respectively. Proteins with 10 amino acids are given the general name "polypeptides," and proteins that have over 100 amino acids are called "peptones." Larger proteins (over 1,000 amino acids) are referred to either as "albumins, water soluble," or "globulins, water insoluble."

AMYL ACETATE – An ester derived from acetic acid and responsible for the "fruity" or "banana" odor in beer.

AMYLASE – Enzymes that convert starch to malt sugar.

ANAEROBIC – Conditions in which there is not enough oxygen for metabolic function. Anaerobic microorganisms are those which can function without the presence of oxygen.

ANION – An ion in solution having a negative charge. When applied to the composition of the water, it designates ions such as Cl^-, HPO_4^-, and SO_4^{2-}, which are common forms. In chemical notation, the minus sign indicates the number of electrons the compound will give up.

ANTIOXIDANT – A reducing agent added to bottled beer to delay or prevent oxidation. Ascorbic acid (vitamin C) and the sulfites (sulfur dioxide, potassium metabisulfite, and bisulfite) are used for this purpose.

APPARENT ATTENUATION – The gravity of finished beer and sometimes called terminal or final gravity. It is apparent because it does not represent the extract lost during fermentation, since the drop in gravity caused by the transformation of sugars is added to that of alcohol, which is lighter than water (0.79 at 15°C). Brewers commonly use the term attenuation without qualification, which invariably means apparent attenuation since it, is easier to measure apparent attenuation rather than real attenuation. Real attenuation is obtained by multiplying the apparent attenuation by 0.81. Formula: A = (B - b/B x 100

    A = apparent attenuation in percent

    B = original gravity in degrees B (or Plato)

    b = gravity of beer devoid of CO_2

APPARENT EXTRACT – See APPARENT ATTENUATION.

AROMA – The fragrance of beer that originates from the natural odors of its ingredients – barley, malt and hops.

AROMA HOPS – Hop varieties known for their fine aroma and flavoring properties; also called noble hops.

ASTRINGENT – A flavor effect demonstrating a drying, puckering feeling often associated with sourness caused by tannins, oxidized tannins (phenols), and various aldehydes (in stale beer). Boiling to long, long mashes, oversparging or sparging with hard water can result in high tannin levels.

ATTENUATE – Reduction of the extract/density by yeast metabolism.

ATTENUATION – The percentage reduction in the wort's density caused by the conversion of sugars into alcohol and carbon dioxide gas through fermentation.

AUTOLYSIS – The process of self-digestion of the body content of a yeast cell by its own enzymes.

BACTERIA – Microscopic single-cell plants which reproduce by fission or by spores, identified by their shapes: coccus, spherical, bacillus, rod-shaped, and spirillum, curved.

BACTERICIDE – Capable of inhibiting the growth or reproduction of bacteria.

BACTERIOSTATIC – Bacteria inhibiting.

BALANCE – Refers to the overall harmony of flavors in a beer. More specifically, it usually refers to the levels of hops and malts. For example, if a beer's taste is predominately malt oriented, it is said to be balanced toward malts.

BALLING – A standard for the measurement of the density of solutions expressed as a percentage of the weight of the solute on (grams of sucrose per 100 grams of solution). Degrees Balling (°B) is sometimes used as a substitute for degrees Plato (°P). However, the two are not equivalent since the Balling scale was calibrated at 17.5°C and not at 20°C as is the Plato scale. The difference between the two scales is about 0.05% wt/wt units. Both degrees P and B are temperature dependent. Traditionally used in Europe, it has largely been replaced by the Plato scale.

BARM – Liquid yeast appearing as froth on fermenting beer.

BARREL – A unit of measure used by brewers. In the United Kingdom, a barrel holds 36 imperial gallons (1 imperial gallon = 4.5 liters). In the United States, a barrel holds 31.5 gallons (1 US gallon = 3.8 liters).

BATCH FERMENTATION – The traditional method of fermentation where each batch is fermented separately as opposed to continuous fermentation which is a continuous process.

BEER – A generic name for alcoholic beverages produced by fermenting malt with or without other cereals and flavored with hops.

BEERSTONE – Brownish-gray deposits left on fermentation equipment composed of calcium oxalate and organic residues.

BETA ACID – One of the two resins in hops that are responsible for bitterness; however, it contributes very little to the bitterness of beer. They are present in larger quantities than alpha acids but because of their low solubility (soluble in water only if oxidized) their content level is usually insignificant.

BETA-AMYLASE – A diastatic enzyme that breaks down long glucose chains into fermentable sugars, a process called saccharification. When mash temperatures are optimized for beta-amylase the wort is highly fermentable.

BETA-GLUCANS – A carbohydrate known to increase wort viscosity when not properly degraded causing filtration and haze problems. Beta-glucans have been known to show a color reaction when using the iodine test even though satisfactory degradation of malt sugars has been achieved. The grain kernels may need to be strained out to minimize this effect.

BICARBONATE – A salt that contributes to the alkalinity of water.

BIOCIDE – A chemical agent that kills living organisms.

BIOCHEMICAL OXYGEN DEMAND – Biochemical Oxygen Demand, or BOD, is a measure of the quantity of oxygen consumed by microorganisms during the decomposition of organic matter. The BOD of brewery waste varies widely among breweries and depends to a large extent on water usage and how much of the high-BOD load is treated or separated prior to discharge. The main components of brewery BOD are wort and beer residues from the brewhouse, cellars, and packaging operations. Cleaning chemicals, such as nitric and phosphoric acids, are major sources of nutrients. BOD results are expressed as mg O_2/liter water.

BITTER – A flavor effect associated with hops experienced on back of tongue.

BITTERING HOPS – Hops with high alpha acid content used to add bitterness, but not aroma, to beer.

BITTERING UNITS – A formula developed by the American Homebrewers Association to measure the total amount of bitterness in a given volume of beer multiplying the alpha acid content (in percent) by the number of ounces. Example: 3 ounces of hops at 9% alpha acid for five gallons: 3 x 9 = 27 BU per 5 gallons.

BITTERNESS – Bitterness is a taste sensation experienced between the tongue and palette. In beer it primarily comes from the contents of the hops, but can also result from salts naturally occurring in brewing water and/or from vegetative fermenting elements within the malt.

BLACK MALT – Malted barley roasted at high temperatures. Used in stouts and dark beers to contribute dark color and a burnt flavor.

BODY – A mouth-filling property of beer that gives it substance and consistency.

BOILING – Following extraction of the carbohydrates, proteins, and yeast nutrients from the mash the clear wort must be conditioned by boiling.

BOTTLE-CONDITIONED – Secondary fermentation and maturation in the bottle induced by adding yeast and sugar to the beer in the bottle.

BOTTOM FERMENTATION – One of the two basic methods of fermentation for beer, as opposed to top fermentation, where the dead yeast cells rise to the surface of the wort during fermentation. Beers brewed in this fashion are commonly called lagers or bottom-fermented beers.

BOUQUET – The overall smell of beer caused by odors that originate during fermentation and maturation.

BREWER'S GRAVITY – It is the specific gravity in brewer's pounds. Formula: (S.G. – 1000) x 0.36.

BREW KETTLE – A large vessel (or copper in the UK), similar in shape to a mash tun, made of copper or stainless steel in which the wort is boiled.

BREW PUB – An establishment that brews it own beers for consumption on its premises.

BRIGHT BEER TANKS – Tanks for storing beer after it has been filtered.

BRIX – A specific gravity scale based on the Balling scale but designed for use at 15°C.

BUNG – A wooden plug for a beer barrel.

BUFFER – A substance capable of resisting changes in the pH of a solution.

BUFFER CAPACITY – Buffer capacity is the ability of water to resist a change in pH.

BUREAU OF ALCOHOL, TOBACCO AND FIREARMS – The Bureau of Alcohol, Tobacco, Firearms and Explosives (abbreviated ATF, sometimes BATF) is a specialized law enforcement and regulatory organization within the United States Department of Justice. Its responsibilities include the investigation and prevention of federal offenses involving the unlawful use, manufacture, and possession of firearms and explosives, acts of arson and bombings, and illegal trafficking of alcohol and tobacco products.

CALCIUM DISODIUM EDTA – A foam-stabilizing additive used in commercial beers.

CANDI SUGAR – Candi sugar is made by superheating and then cooling a highly concentrated sugar solution. Pale candi syrup is much darker than sucrose or invert sugar syrup. Belgian brewers prefer to use candi sugar, in either solid or syrup form, because it contributes to good head retention in a high-gravity, lightly hopped beer.

CABOHYDATES – Organic compounds including insoluble cellulose, soluble hemicellulose, starch and sugars.

CAMRA – CAMRA stands for the "Campaign for Real Ale," a British consumers' group that is concerned with changes, primarily in the quality of British beers.

CARBONATES – Alkaline salts whose anions are derived from carbonic acid.

CARBONATION – The process of injecting or dissolving carbon dioxide gas in beer.

CARBON DIOXIDE – A gas produced during fermentation. Carbon dioxide produces the characteristic tingle in the beer and is also extremely important because of the way its level can affect the perception of other flavors.

CASKS – Traditionally casks were made of wood though most modern casks are made with aluminum or stainless steel and are barrel-shaped. In the U.S., drought vessels are commonly called "kegs" made from metal and are cylindrical in shape. Unlike a cask, a keg has only one hole, which is in the center of the head. Casks have two holes, one hole is situated in one of the heads for drawing beer and the other hole is in the middle of the arched side for drawing air.

CASK-CONDITIONED – Beer conditioned in a cask that is neither filtered nor pasteurized.

CENTRIFUGATION – A clarification method using centrifugal force to clarify the wort during the cooling stage and the finished beer during conditioning.

CHELATING AGENT – Any of several chemical compounds that will inactivate metals ions e.g., calcium and magnesium by forming a water-soluble complex with the ions.

CHEMICAL OXYGEN DEMAND – Chemical oxygen demand (COD) does not differentiate between biologically available and inert organic matter and it is a measure of the total quantity of oxygen required to oxidize all organic material into carbon dioxide and water. COD values are always greater than BOD values, but COD measurements can be made in a few hours while BOD measurements take much longer. The results are expressed as mg O_2/liter water.

CHILLING – Boiled wort is chilled rapidly with a heat exchanger to the temperature of primary fermentation.

CHILL HAZE – Haziness caused by a combination and precipitation of protein matter and tannin molecules during when beer is chilled.

CHILLPROOFING – A treatment applied to finished beer during conditioning to remove proteins or polyphenols and improve physical stability.

CHLOROPHENOLIC – A flavor effect associated with strong and unpleasant tasting chemical compounds formed by the combination of chlorine with a phenolic compound. Some are carcinogenic.

CLOYING – A beer that is overly sweet to the point of being unpleasant.

COLD BREAK – Cold break is the precipitation of protein and tannin material to a fine coagulum during wort cooling.

COLIFORM – Water-borne bacteria often associated with pollution.

COLLOIDAL STABILITY – It is the ability of a beer to resist turbidity or haziness when exposed to cold temperatures.

COLOR – Three-color scales are commonly used in brewing SRM, Lovibond, and EBC. SRM, or Standard Reference Method, and Lovibond are the same scale, but brewers tend to describe their malts in degrees Lovibond and their finished beer in SRM. The EBC (European Brewery Convention) is an entirely different scale.

CONDITIONING – A process that follows primary fermentation when the beer undergoes maturation, clarification, and stabilization before being filtered. Cold conditioning imparts a clean, round taste whereas warm conditioning develops complex flavors.

CONDITIONING TANKS – They are stainless steel tanks used for conditioning, carbonating, and storing beer. The tanks can either be vertical or horizontal and are normally pressure rated. They are often referred to as bright beer tanks.

CONGELATION – A method for producing stronger beers by freezing the beer, and removing some of the ice. This has the result of concentrating the beer, making it heavier, more alcoholic, and sweeter-tasting.

CONVERSION TIME – It is the time to complete conversion of starch to sugar during mashing.

COPPER – An old term that refers to the brewkettle which used to be made from copper; some brewers still use coppers.

CONTINUOUS FERMENTATION – Continuous fermentation involves recycling part of the fermented beer back to the wort at the start of the fermentation process and requires a continuous supply of wort into the system. The result is a continuous flow of beer out the other end of the process.

CROP – The yeasts that are skimmed off the surface of top fermenting ale during primary fermentation.

DECARBONATE – To remove carbonate and bicarbonate ions from water, either by boiling or by adding chemicals.

DECOCTION MASHING – Mashing is carried out in a series of controlled temperature stages by removing a portion of the grist from the mash and boiled in the mash kettle and latter returned to the mash tun thus gradually raising the temperature of the entire mash. The process usually is repeated two or three times with each successive step, or decoction, used to raise the temperature of the main mash. Decoction mashing typically achieves a high rate of extraction and increased amount of malt character.

DEFLOCULATION – The action of breaking up solid chunks into smaller particles by chemical action and mechanical action.

DEIONIZED WATER – Water that has had ions removed by an ion-exchanger.

DENSITY – The measurement of the weight of a solution, as compared with the weight of an equal volume of water. Density can either be expressed as degrees Plato or as specific gravity. Degrees Plato is most commonly used by professional brewers in the United States and Continental Europe whereas specific gravity is more than likely to be used by homebrewers.

DETERGENT – A cleaning agent, compounded from chemicals that acts as a wetting agent and emulsifier.

DEXTRINS – A complex, unfermentable and tasteless carbohydrates that contribute to the final gravity and body of beer; some dextrins remain undissolved in the finished beer giving it a malty sweetness. In recent years have been referred to alpha-glucans.

DEXTRINIZATION – The enzymatic process by which alpha amylase degrades soluble starch molecules into dextrin molecules.

DEXTROSE – A common monosaccharide, the building block for dextrins, starches, and other sugars.

DIACETYL – A volatile compound produced in beer by the oxidative decarboxylation of acetohydroxyl acids produced by yeasts. Diacetyl is an important to beer flavor and aroma producing a butterscotch flavor to beer. It is arguably a desired flavor note in English ales but a major flaw if detected in lagers and lighter beers.

DIASTASE – An enzyme capable of changing starches into maltose and later into dextrose.

DIASTATIC POWER – Diastatic power is a measure of the total amylase content of a given sample of malt; usually expressed in degrees Lintner.

DIKETONE – Aromatic, volatile, compound perceivable in minute concentration, from yeast or *Pediococcus* metabolism.

DIMETHYL SULFIDE (DMS) – A major sulfur compound of lagers not found in British ales because their malts are highly modified at very high temperatures. DMS is released during boiling as a gas that dissipates into the atmosphere. The precursor of DMS, S-methylmethionine, remains present in the wort and converts to DMS if the wort is not cooled rapidly enough or if it is allowed to sit after cooling.

DIMPLE JACKETS – A type of heat transfer surface custom designed to produce maximum efficiency and uniformity through single- or multi-zone controlled flow of heating or cooling media. Available in ammonia-Freon (AF) or water-glycol (WG) styles.

DISACCHARIDES – Sugar group; two monosaccharide molecules joined by the removal of a water molecule. Common disaccharides are sucrose, maltose, and lactose.

DISINFECTANT – A chemical agent that is capable of inhibiting growth of microorganisms or otherwise eliminating them from an area or surface.

DISPERSION – The breaking up soil clumps into smaller particles, which are less likely to redeposit on the surface.

DOSAGE – The addition of yeast and/or sugar to the cask or bottle to assist in secondary fermentation.

DOUGHING-IN – The gradual addition of water to crushed malt to prevent the formation of dry spots in the mash. Also is referred to as mashing-in.

DRAFT BEER – Beer drawn from casks or kegs rather than canned or bottled. It is often referred to as a draught beer.

DRY HOPPING – The addition of loose dry hops during wort boiling or conditioning to increase the aroma and hop character of the finished beer without affecting its bitterness. In Britain, dry hopping more specifically refers to the addition of fresh hops to a cask of draft beer when it racked from the primary fermenter.

DRY MALT – Dry malt is malt extract in dry, powdered form as opposed to liquid or syrup malt.

EMULSIFICATION – The chemical action of breaking up fats and oils into small particles to facilitate removal from the surface.

ENDOSPERM – The nonliving part of the barley grain, which contains starch and protein to feed the growing acrospire.

ENZYME – An organic protein substance that acts as catalyst for specific chemical reactions, converting a specific set of reactants (called substrates) into specific products.

ESSENTIAL OIL – The aromatic volatile liquid from the hop.

ESTERS – Volatile flavor compounds which form through this interaction of organic acids with alcohols during fermentation. They are often fruity, flowery or spicy. Some of the more common esters are ethyl acetate, isoamyl acetate, and ethyl hexanoate. Ethyl acetate imparts a fruity solvent overtone and is typically found in British ales. Isoamyl acetate imparts a banana flavor and ethyl hexanoate has a very strong flavor reminiscent of apples. They associated with wild-yeast fermentation, although esters also arise in high-gravity brewing even with pure cultures. Some top fermenting yeast strains are prized for their ability to create esters.

ETHANOL – Ethanol is a flavor enhancer in addition to its more direct roles of producing a warming effect and contributing to perceived sweetness and body.

EUROPEAN BREWING CONVENTION METHOD (EBC) – In Europe, the standard method for determining beer color is the European Brewing Convention method (EBC). Color measured as degrees EBC. To convert to Standard Research Method (SRM) color values use the following equation for a reasonable approximation: SRM = 0.375 x EBC + 0.46.

EXTRACT – The total amount of soluble constituents from the malt is usually reported as a percent on the basis of weight. Extract can be expressed on a weight to volume basis using the following formula: kg extract/hl = (°P) x (SG). For example, 1.2 hectoliters of wort at 13°P with a SG of 1.050 has 13.65 kg of extract/hl or 16.38 of extract for 1.2 hectoliters. If 22 kilograms of grain is used the yield is (16.38 x 100)/22 = 74.5%.

EXTRACT YIELD – Extract is a measure of the amount of sugar recovered from the malt after mashing. The extract value is based on a laboratory mash.

FATTY ACIDS – A group of carboxylic acids, all of which impart a foul, soapy flavor to beer, contribute to its staling and affect its head retention.

FERMENTATION – The chemical conversion of fermentable sugars into ethyl alcohol and carbon dioxide gas, through the action of yeast resulting in a drop in the density.

FINAL DEGREE OF ATTENUATION – The maximum apparent attenuation attainable by a particular wort.

Formula: $A = (B - b/B \times 100)$

A = final degree of attenuation

B = original gravity in degrees B (or deg. P)

b = final gravity in degrees B (or deg. P).

FINES – The finely crushed, flour-like portion of the draft.

FINING – A clarifying process that adds organic or mineral settling agents during boiling of wort and conditioning to precipitate colloidal matter through coagulation or adsorption.

FINISH – The long-lingering aftertaste of a beer, usually dominated by a balance between malty sweetness and hops aromas. The less malty the finish, the "drier" is the beer.

FINISHING HOPS – Hops added to the wort near the end of the boiling to impart hop aroma and character to the beer as opposed to bittering hops that contribute to its bitterness and are added during the boil.

FLASH PASTEURIZATION – Flash pasteurization is the heat treatment of beer in bulk for subsequent packaging.

FLAVOR STABILITY – The ability of a beer to retain the quality it had when it was bottled until it is consumed.

FLOCCULATION – The phenomenon by which yeast cells aggregate into masses toward the end of the fermentation process.

FOAM STABILITY – Measured as rate of foam collapse in freshly poured beer under controlled conditions.

FRUCTOSE – A highly fermentable monosaccharide that occurs naturally in malt typically making up from 1 to 2% of wort carbohydrates.

FUNGICIDE – An agent that kills fungi (molds).

FUSEL ALCOHOLS – Alcohols of high boiling point then ethanol, which are derived from keto-acids during the yeast protein synthesis. The formation of higher alcohols varies with yeast strain and yeast growth, fermentation temperature (an increase in temperature is followed by an increase in the formation of alcohols), and fermentation method. Fusel alcohols impart a harsh, clinging bitterness and are produced through the metabolism of amino acids. They are also referred to as higher alcohols.

GELATINIZATION – Gelatinization is the physical breakup and dispersal of starch molecules in water, making them more accessible to enzymatic attack.

GERMINATION – Sprouting of the barley kernel, to initiate enzyme development and conversion of the malt.

GLUCOSE – An important building block of carbohydrates typically making up 8 to 10% of wort sugars. Pure, commercial glucose, sometimes called dextrose, always contains a certain amount of dextrins which, being unfermentable, remain in the beer and give it a sweet, mellow flavor. Syn: dextrose; corn sugar.

GRAIN BILL – The list of grains and their amounts used for a recipe.

GRANT – A grant is a small vessel to which wort is directed from the mash tun for visual and hydrometer inspection.

GREEN BEER – After primary fermentation, beer is called green beer.

GRIST – The crushed malts and adjuncts that are mixed with hot water to form the mash.

GYLE – That portion of unfermented beer wort that is reserved for or added to finished beer for conditioning (carbonation).

HEFE – German word for yeast and the first part of the term Hefeweizen, the name for the German wheat beer.

HETEROFERMENTATIVE – Organisms whose metabolic activities lead to a wide range of products including acids, alcohols, and carbon dioxide.

HOMEBREW BITTERNESS UNITS (HBU) – Homebrew Bitterness Units are a measure of the total amount of bitterness potential in a given volume of beer. Homebrew Bitterness Units are calculated by multiplying the percent of alpha acid in the hops by the number of ounces. For example, if 2 ounces of Northern Brewer hops (9% alpha acid) and 3 ounces of Cascade hops (5% alpha acid) were used in a 10-gallon batch, the total amount of bitterness units would be 33: (2 x 9) + (3 x 5) =18 + 15. Bitterness units per gallon would be 3.3 in a 10-gallon batch or 6.6 in a 5-gallon batch, so it is important to note volumes whenever expressing Homebrew Bitterness Units. HBUs are not related to IBUs (International Bitterness Units). HBU is a measure of alpha acids added to the boil, whereas IBU represents a measurement of the intensity of the bitterness of the beer.

HOMOFERMENTATIVE – Organisms whose metabolic activities lead to only a single product, usually an organic acid.

HOPBACK – Sieve-like vessel used to strain out the hops. It is known as a hop jack in the United States.

HOPPY – Characteristic odor of the essential oil of hops.

HOP EXTRACT – Bitter resins and hop oils extracted from hops by organic solvents, usually methylene chloride or hexane.

HOP OILS – Hop oils are responsible to a large extent for the characteristic aroma of hops, and thus are an important aspect in brewing value. Like all essential oils, they are very volatile and are largely lost by steam evaporation during the boil.

HOP PELLETS – Hop pellets are nothing more than whole hops mechanically processed by compressing and extruding the hops into pellets.

HOP RATE – The quantity of hops to be added to a given volume of wort during boiling.

HOT BREAK – The coagulation and precipitation of protein matter during the boiling stage.

HUMULONE – One of the three major components of alpha acids found in hops that contributes to the bittering power.

HYDROMETER – A very simple device to measure the specific gravity of liquids. Because the hydrometer was first calibrated using a sucrose solution, the hydrometer is also called a saccharometer.

INFUSION MASHING – A traditional method for top-fermenting beer involving a single-step, single-temperature method employed to mash highly modified malt.

INTERNATIONAL BITTERNESS UNITS (IBU) – An international measurement system that represents the intensity of the beer's bitterness. This is a system of measuring bitterness devised by brewing scientists and is an accepted standard throughout the world. Homebrewers usually do not have the sophisticated equipment to measure actual IBUs and often use a system of Homebrew Bitterness Units to closely approximate the desired bitterness in their beer.

INVERT SUGAR – Processed common sugar (sucrose) obtained industrially by the inversion of sucrose with dilute acid, usually sulfuric acid, into equal parts of glucose and fructose. It does not contain dextrins and can be used as an adjunct or for priming.

JUMP MASHING – A mash process by which the brewer starts the mash at a relatively low temperature, often around 38°C, which is called an acid rest, and than increases the temperature to what is called the mash-out temperature, usually around 77°C.

KILNING – The process of drying germinated barley.

KRÄEUSEN – The large head of foam which forms on the surface of the wort during the early stages of fermentation. Pronunciation: kroysen.

KRÄEUSENING – The addition of actively fermenting wort to a larger volume of wort or beer to induce fermentation. It imparts a crisp, spritzy character.

LACTOSE – An unfermentable disaccharide.

LAGER – A generic term for any beer produced by bottom fermentation, traditionally by decoction mashing, as opposed to top-fermented beer produced by infusion mashing, called ales.

LAGERING – After primary fermentation the lager beer is stored in cold cellars at near-zero temperatures for periods of time ranging from a few weeks to several months for maturing and clarification.

LAGER YEAST – It is also known as "bottom-fermenting yeast (*Saccharomyces pastorianus*) because of its tendency not to flocculate, or form a head of yeast on the surface of the brew.

LAUTER TUN – A vessel fitted with a false slotted bottom equipped with a sparging (hot water delivery) system to wash the mash to recover the extract. Unlike mash tuns they are wider and shallower and are equipped with rakes.

LAUTERING – Process of removing spent grains or hops from the wort. This is simply done by the utilization of a strainer and a subsequent quick, hot water rinse (sparging) of the caught spent grains and hops.

LIGHT STRUCK – An unpleasant flavor in beer caused by exposure to light causing undesirable chemical reactions of hydrogen sulfide and other sulfur compounds, which impart a skunk-like flavor.

LIPIDS – Lipids are fatty acids derived from yeast metabolism, oxidized hops and malt. They affect ability of beer to form a foam head and play an important role in beer staling. Lipids cause soapy flavors, and when oxidized contribute to stale flavors. On the positive side, they contribute to yeast viability.

LIQUEFACTION – The process by which alpha-amylase enzymes degrade soluble starch into dextrin.

LIQUOR – The name given, in the brewing industry, to water used for mashing and brewing.

LOVIBOND METHOD – A standard method for determining beer color used both in the United States and in Europe. It has since been replaced by the more official Standard Research Method (SRM) and the European Brewing Convention (EBC) method by the United States and the European Union, respectively. Color measured as degrees Lovibond.

LUPULIN GLANDS – The tiny yellow sacs found at the base of the petals of the hop cone. They contain the alpha acids, beta acids, and hop oils.

LUPULONE – See Beta acid.

MAILLARD REACTION – A browning reaction, where situations of highly concentrated amino acids and reducing sugars exists producing compounds, collectively called melanoidins that provide the characteristic malty aroma and flavor to beers. Melanoidin production is most active in the malting process and continues to some extent in decoction mashing and to a somewhat lesser extent in boiling. For dark beers, it is important to use malt with a great deal of melanoidins already in it.

MALT – Barley or other grain that has been malted.

MALT EXTRACT – Wort that has been concentrated into a thick syrup or dry powder.

MALTING – The process of converting barley into malt by soaking, sprouting, and then germinating the barley (or other grain) to develop its enzyme content.

MALTOSE – A fermentable disaccharide of two glucose molecules, and the primary sugar produced by the breakdown of barley starch (diastatic hydrolysis of starch). It generally comprises 46 to 50 of the sugars in grain wort.

MANWAYS – Opening in tank that provides access to interior of tank.

MASH – A mixture of ground barley malt, cooked adjuncts (if used), and water that has undergone the conversion of grain starch to sugar.

MASH-IN – The initial stage of mashing; the process of mixing grist and water. It is also referred to as doughing-in.

MASHING – The process of mixing ground malt and cooked adjuncts (if used) with water in the mash tun to degrade haze-forming proteins and to further convert grain starches to fermentable sugars and non-fermentable carbohydrates (dextrins). The two main methods of mashing are infusion and decoction.

MASH-OFF – The final stage of mashing where the temperature is raised to about 76°C to terminate enzymatic activity and improve flow of sugar solution.

MASH TUN – A vessel used for mashing as well as for wort separation.

MATURATION – Refers to maturing of beer flavor. A secondary fermentation may or may not be included.

MEALINESS – Hardness of kernels at completion of malting – soft mealy kernels indicate good modification.

MELANOIDINS – Dark colored organic compounds which form during kilning and wort boiling through a complex series of chemical reactions (called Maillard reactions) involving amino acids and sugars.

MODIFICATION – The physical and chemical changes occurring in barley during malting.

NEPHELOMETER – An instrument used for measuring the turbidity (haze) in liquids.

NITROGEN CONTENT – The percentage of the weight of barley or malt which is nitrogen. Protein content of the grain is about 6.25 times the nitrogen content.

NOBLE HOPS – See Aroma hops.

NOSE – The total sensation of aroma and bouquet.

OLIGOSACCHARIDES – Sugars of more than three molecules, less complex than dextrins.

ORIGINAL GRAVITY – It is a measure of the total amount of dissolved solids present in the beer wort before fermentation. This is expressed in degrees Plato or specific gravity, which is equivalent to the total percentage of fermentable and unfermentable sugars. Abbr.: O.G.

OXIDATION – Any chemical reaction in which oxygen combines with another substance, often times causing off-flavors; wet cardboard, papery, winery, sherry.

PALATE FULLNESS – See Body.

PASTEURIZATION – The application of heat to bottled, canned, kegged or bulk beer for a specific period of time for the purpose of stabilizing it biologically by killing microorganisms to prolong the shelf life of packaged beer.

pH – It is a numerical designation given to the acidity and alkalinity of an aqueous solution, on a scale of 1 to 14. Technically, pH is common logarithm of the reciprocal of the hydrogen-ion concentration of a solution. A pH value of 7 (pure water at 25°C) indicates neutrality, while values below 7 (7 to 1) indicate acidity and above 7 (7 to 14) indicate alkalinity.

PEPTIDASE – A proteolytic enzyme that breaks down large and medium-sized protein molecules into amino acids during mashing.

PEPTIDES – A class of proteins consisting of between two and thirty amino acid molecules bound by peptide bonds. Peptides enhance the viscosity, or fullness, of beer, may also oxidize to high-molecular weight polypeptides.

PEPTIZATION – The spontaneous dispersion of protein soils throughout the cleaning solution without mechanical agitation.

PEPTONIZING – The action of proteolytic enzymes upon protein, yielding albumin/proteoses, peptides and amino acids.

PHENOL – This structure is the building bloc of a large group of phenolic compounds (often referred to as polyphenols or simply "phenols"). Phenols are derived mostly from malt and to a lesser degree hops. Yeast produce phenols but the amounts are quite small. Generally speaking, *S. cerevisiae* produce more phenols than *S. pastorianus* and, wild strains of yeast or bacteria are often the cause of phenolic flavors. Spices used in specialty beers are known to contain phenolic compounds.

PHYSICAL STABILITY – A measure of the susceptibility to precipitation, suspension, or haze formation in bottled beer.

PHYTASE – An enzyme that reduces the mash pH be creating phytic acid from the phytin of the malt. This aids in saccharification, wort clarification, and fermentation.

PINT – Standard measure in which beer is served in the UK. Equivalent to 0.568 of a liter.

PITCHING – Adding yeast to the sterile wort to initiate fermentation.

PLATO, DEGREES – Commercial brewers' standard for the measurement of the density of solutions expressed as an equivalent weight of the solution (grams of sucrose per 100 grams of solution, measured at 20°C). It expresses the weight of any substance in solution as a percentage of the weight of that solution, based upon sucrose dissolved in pure water. A density that measures one °P means that 1% of the weight of the measured liquid is dissolved sugar. In other words, a density of 15°P indicates that there would be 15 grams of dissolved sugars in enough water to make 100 grams of solution.

POLYPHENOLS – Complex molecules that are responsible for astringent and bitter flavors as well as the haze formation in beer. In small concentrations phenolic compounds add character to beer whereas high concentrations impart off-flavors described as phenolic, medicinal or pharmaceutical. They are a major influence on color in beer and are found in greater concentrations in alkaline waters and dark beers. Tannins and lignins are well-known examples of polyphenol compounds.

PHOSPHATES – These molecules are the source of phytic acid created during malting and during decoction mashes. They contribute to the acidulation of the mash.

POLYSACCHARIDES – Carbohydrate complexes formed by the union of two or more monosaccharides.

PPM – Parts per million (mg/ml). It is a measurement of particles of matter in solution.

PRIMARY ATTENUATION – The attenuation measured at the end of primary fermentation.

PRIMARY FERMENTATION – The initial fermentation, during which most of the carbohydrates in the wort are assimilated and converted to ethyl alcohol and carbon dioxide gas. If no secondary fermentation is done, than all the carbohydrates are assimilated during primary fermentation.

PRIMING – The addition of small amounts of fermentable sugars (preferably corn sugar or syrup) to fully fermented beer before bottling or kegging to induce secondary fermentation and thus carbonate the beer.

PRIMING SOLUTION – A solution of sugar in water added with yeast to the aged beer at bottling to induce fermentation (bottle-conditioning).

PROTEASE – A malt enzyme that develops in barley during germination and is capable of degrading complex proteins into polypeptides and amino acids.

PROTEIN – A complex chain of amino acids held together by peptide links. Proteins are responsible for the head retention and body of beer and partially for its haziness.

PROTEIN REST – A stage of the mashing process during which complex proteins are decomposed by proteolytic enzymes to progressively less complex fractions.

PROTEINASE – This proteolytic enzyme works to break large protein molecules down into medium-sized proteins during mashing.

PROTEOLYSIS – The reduction of protein by proteolytic enzymes to fractions.

RACKING – The process of transferring beer from one vessel to another.

REAL ATTENUATION – The attenuation of beer after alcohol and carbon dioxide is removed. Real attenuation will always be a lower number than the apparent attenuation.

In the brewing industry only the apparent attenuation is used to measure the progress of fermentation. Formula: A = (B-b)/B x 100 where A equals real attenuation and B equals original gravity in degrees B or degrees P and b equals specific gravity of beer devoid of alcohol and carbon dioxide. Real attenuation (extract) may also be calculated from the apparent attenuation:

RA = AA + (alcohol by weight x 0.46)

Thus, a beer with an apparent extract of 2.6, and 3.7% alcohol would have a real extract:

RE = 2.6 (3.7 x 4.6) = 2.6 + 1.7 = 4.3

REINHEITSGEBOT – "Purity Law" applied to all German brewers making beer for consumption in their own country. It only allows water, hops, malt (barley and wheat), and yeast for brewing beer. Chemical additives, sugar, rice, corn, and unmalted barley are prohibited.

RESIN – Noncrystalline (amorphous) plant excretions.

ROPY FERMENTATION – Viscous, gelatinous blobs, or "rope," from bacterial contamination.

RUH BEER – A nearly fermented beer, ready for lagering with some yeast present following the completion of the primary fermentation.

RUN-OFF – The complete lautering operation (straining of wort from grains); the time of run-off is an important factor.

SACCHARIFICATION – A stage of the mashing process during which complex glucose chains are broken down into fermentable sugars, mainly maltose.

SACCHAROMETER – A type of hydrometer for measuring the sugar concentration of a solution by determining the specific gravity.

SALT – Any numerous compounds made up of two or more ions.

SANITARY TANKS – Vessels cleanable to sanitary standards (e.g., FDA, 3A USDA standards).

SAPONIFICATION – A chemical reaction between strong alkalines and fats resulting in formation of soluble soaps and glycerol.

SCALE – A precipitate that forms or surfaces in contact with water as the result of a physical or chemical change.

SECONDARY FERMENTATION – Fermentation following the transfer of the green beer from the primary fermenters with some yeast and fermentable carbohydrates present, during which residual carbohydrates are assimilated.

SEQUESTRANT – A chemical agent that forms a water-soluble metallic complex and prevents the precipitation of metal ions from solution. A sequestering agent, unlike a chelating agent, losses its effectiveness at high temperatures or after extended periods of time.

SPARGING – It is an operation consisting in spraying hot water over the spent grains in the mash to retrieve the fermentable sugars remaining in the spent grains.

SPECIFIC GRAVITY (sp. gr.) – Specific gravity is the density (weight per unit volume) of substance (fermentable and unfermentable sugars) divided by the density of water, which has a specific gravity of 1.000 at 4°C. The specific gravity of liquids is often measured with a hydrometer, whose weight (a constant) displaces different volumes of liquid as the liquid's density varies. The densities of water and wort vary with temperature. Consequently, the reference temperatures for the wort sample and for the water must be specified along with the value of specific gravity. Brewers often refer to gravity as "gravity," ignoring the decimal, e.g., 1050. Specific gravity can also be expressed in gravity points; thus wort with SG of 1.046 has 46 gravity points. Dividing specific gravity @ 15.6°C by 4 gives the approximate degrees Plato and conversely multiplying °Plato by 4 roughly gives specific gravity. A much more accurate, yet still easily calculated conversion, also for specific gravity @15.6°C, is given by: °P = 259 − (259/SG) or SG = 259/(259 − °P).

STAMPED VESSELS – Regulatory certified vessels (such as ASME-code stamped).

STANDARD REFERENCE METHOD (SRM) – In the United States, the standard method for determining beer color is the Standard Research Method (SRM). This technique was originally set up to approximate the Lovibond scale and is now used as the basis for assigning Lovibond ratings to grains as well as to determine the actual color of finished beer. Color is measured as degrees SRM. It is assumed to be equivalent to the older degrees Lovibond system. To convert to European Brewing Convention (EBC) color values use the following equation for a reasonable approximation: EBC = 2.65 x SRM − 1.2.

STARCH – Any of a group of carbohydrates or polysaccharides which hydrolyzes to yield dextrins and maltose through the action of amylases.

STARTER – A small volume of wort to which yeast is added, in order to activate it before it is pitched into the main batch.

STEEL MASHER – Used for mixing crushed malt with hot brewing water.

STEEPING – A process in malting where the grain seeds are soaked in water to provide moisture for germination.

STEP INFUSION – A method of mashing in which the various temperature rests are employed.

STERILANT – An agent capable of sterilizing an object or surface, i.e., killing or removing all living organisms.

STRIKE HEAT (TEMPERATURE) – The temperature of the mash water before its addition to the grist during the mashing process. The temperature needs to be higher than the desired rest temperature because the temperature will be reduced when the grist and water are combined.

STYLE – A term used to differentiate and categorize beers by various factors such as color, flavour, strength, ingredients, production method, recipe, history, or origin. Beers of the same style have the same general flavor profile.

STUCK FERMENTATION – Fermentation that has stopped prematurely, i.e., before the final gravity has been reached.

SUCROSE – A fermentable sugar (disaccharide) that is produced from sugar cane and occurs naturally in malt consisting of a fructose unit joined with a glucose unit. It generally makes up 8 to 10% of wort sugars. It is not directly fermentable by yeast and must first be hydrolyzed to glucose and fructose by the enzyme invertase secreted by the yeast. Occasionally, it is used in some forms of brewing, such as the candy sugar used in certain Belgian ales.

SURFACTANT – A surface-active agent, usually an organic compound that dissolves in water to solubilize oil materials in the cleaning process.

TANK – Fermentation vessel or conditioning vessel, now usually made of stainless steel. Modern tanks are closed and can be pressurized. In the old days, tanks tended to be open fermenters. In closed tanks, contamination from airborn bacteria during fermentation can be minimized.

TANNINS – Tannins are the most widely known class of phenols and if in high enough concentrations are responsible for bitterness, astringency, and haze. Tannins can be divided into hydrolyzable and condensed tannins. Hydrolyzable tannins come primarily from hops, and most are removed with the break material. Condensed tannins, on the other hand, are quite resistant to degradation.

TEMPORARY HARDNESS – It is a form of hardness in water caused by the presence of soluble bicarbonates of calcium and magnesium.

TERMINAL EXTRACT – It is the density of the fully fermented beer.

TERMINAL GRAVITY – The specific gravity of a beer after fermentation is completed.

THERMOPHILIC – "Heat loving" bacteria operating at unusually high temperatures.

TRISACCHARIDES – Sugars formed from the combination of three monosaccharides.

TOTAL DISSOLVED SOLIDS (TDS) – The mineral ion content in solution.

TRISACCARIDE – A sugar composed of three monosaccharides joined by the removal of water molecules.

TRUB – The sediment (trubaceous matter) formed from boiling and cooling the wort.

TUNNEL PASTEURIZATION – It is a method of pasteurization for bottled and canned beer.

TURBIDITY – Sediment in suspension; hazy, murky water.

ULLAGE – The empty space at the top of a bottle, cask or barrel between the liquid and the top of the container. It is also referred to as headspace or airspace.

UNDERLETTING – Filling the space beneath the false bottom of the lauter tun with hot water before running off the sweet wort.

UNITANK – A stainless steel vessel in which the beer can be both fermented and lagered. Unitanlks have cone-shaped bottoms for allowing excess yeast to be purged out of the brew.

VDK (vicinal diketones) – Diacetyl (2,3-butanedione) and 2,3-pentanedione are important contributors to beer flavor and aroma. Sometimes these two ketones are grouped and reported as the vicinal diketone (VDK) content of beer.

VENTING – Once cask conditioned ale is delivered to a pub it must be set up in its serving position and then left undisturbed until the cask is empty. The publican must also "vent" the cask — allowing the cask to breathe and secondary fermentation to take place. Secondary fermentation of the beer in a closed cask ensures that the beer becomes completely saturated with CO_2.

VISCOSITY – Viscosity is a measure of the resistance of a fluid to being deformed by either shear stress or extensional stress. It is commonly perceived as "thickness", or resistance to flow.

VITREOSITY – Vitreosity, an inverse of mealiness, is a measure of malt's glassiness.

VOLATILE – Readily vaporized especially esters, essential oils, and higher alcohols.

WATER HARDNESS – The degree of hardness of water caused by the presence of mineral elements dissolved into it. It is expressed in parts of calcium carbonate per million (ppm) of water.

WETTING AGENT – It is a substance that reduces the surface tension of water to increase penetration of cleaning solutions to all surfaces.

WHIRLPOOL – A whirlpool is used to physically eliminate cloudy protein elements.

WHOLE (LEAF) HOPS – Hops that have been dried but retain their natural shape and bulk.

WILD YEAST – Any yeast strain that is not deliberately selected and introduced into the beer by the brewer.

WORT – The solution of malt sugars, proteins, and other substances that is produced by mashing.

WORT CLARITY – Clarity of wort obtained from lautering.

WORT GELATIN – Culture medium made up from wort as a nutrient source and gelatin to solidify it, for surface culturing yeast.

YEAST – Yeasts are single-celled microorganisms that grow and reproduce by budding. They are biologically classified as fungi, which are responsible for converting the fermentable sugars into the alcoholic beverage called beer. There are literally hundreds of varieties and strains of yeasts. Generally, there are two types of beer yeast: Ale yeast ("top-fermenting" type, *Saccharomyces cerevisiae*) or lager yeast ("bottom-fermenting" type, *Saccharomyces pastorianus*).

YEAST HEAD – The yeast-containing froth at the surface of top fermenting ale at the end of primary fermentation. It is sometimes recovered by skimming or suction to be used for pitching further worts.

YIELD EXTRACT – The percentage of extractable dry matter in the grist; i.e., the total amount of dry matter that passes into solution in the wort during mashing.

ZYMURGY – The science/art of yeast fermentation.

Index

Symbols
α-amylase, 155
ß-amlyase, 155
ß-glucan, 155

Numbers
2,3-Pentanedione, 68, 240

A
Aeration, 227
Aerobic wastewater treatment, 353
Acetaldehydes, 68
Acetic acid bacteria, 337
Acetobacter spp., 337
Acid-based detergents, 123
Acid-based sanitizers, 131
Acid malt, 158
Acid rest, 153
Acidification of mash, 158
Acidification of water
 boiling, 100
 calcium hydroxide, 101
 carbon dioxide, 101
 food grade acids, 101
Acrospire, 24
Activated sludge process, 353
Adjunct cooker, 171
Adjuncts
 classification of adjuncts
  flakes, 112
  grits, 112
  flours, 111
  refined starches, 112
  torrified cereals, 112
 gelatinization temperatures, 111
 employment in brewing
  cooker mash adjuncts, 110
  kettle adjuncts, 111
  mash tun adjuncts, 111
 types of cereal adjuncts
  barley, 114
  corn, 113
  oats, 116
  rice, 114
  wheat, 115
  sugars, 116
  syrups, 117
Adolf Coors Company, 6
Aeration of wort, 227
 measuring oxygen levels, 230
 methods of oxygenation, 229
 oxygen requirements, 228
 point of injection, 228
Aerobic wastewater treatment, 353
 activated sludge process, 353
 advantages and disadvantages, 356
 attached growth process, 354
 lagoons, 355
 sludge treatment and disposal, 356
Agar slants, 75

Alcohol and Tobacco Tax and Trade
  Bureau (TTB), 411
Alcohol Beverage Control Boards, 416
Aldehydes, 198, 212, 227
Alexis barley variety, 22
Alkaline-based detergents, 122
Alkaline-based disinfectants, 128
Alkalinity in water, 95
Alpha acids, 41
Alpha-amylase, 155
Altbier (German), 388
Aluminum
  cleaning and sanitation, 139
Amber ale (American), 362
Amber malt, 35
American Society of Brewing Chemists,
  (ASBC), 149
Amino acids
  malting, 23
  mashing 154
  yeast byproducts, 71
  yeast nutritional requirements, 65
Anheuser-Busch, Inc., 5
Anaerobic wastewater treatment, 356
  advantages and disadvantages, 358
  fluidized bed reactor, 357
  upflow anaerobic sludge blanket, 357
Anionic acids, 131
Aroma hops, 43
ATP bioluminescence method, 342
Attached Growth (Biofilm) Process, 354
Augers, 144
Auxiliary finings, 262

B

Bacteria
  acetic acid, 337
    Acetobacter spp., 337
    Gluconobacter spp., 338
  Enterobacteriaceae, 339
    Obesumbacterium proteus, 339
  Gram-negative, 337
  Gram-positive, 334
  Lactobacillus spp., 334
    L. brevis, 336
    L. delbrueckii, 336
    L. pastorianus, 336

Pectinatus spp., *338*
Pediococcus spp., *336*
  P. damnosus, 337
Zymomonas spp., *338*
  Z. anaerobia, 338
  Z. mobilis, 338
Barley adjuncts, 114
Barley malts
  barley types
    identification, 23
    six-row barley, 21
    two-row barley, 22
  base malts
    mild ale malt, 32
    pale ale malt, 32
    Pilsner malt, 31
    Munich malt, 33
    Vienna malt, 32
  malt analysis
    alpha-amylase (DU), 29
    beta-glucan, 30
    coarse-grind extract, 27
    cold-water extract, 28
    color, 26
    diastatic power, 28
    dimethyl sulfide, 30
    fine grind-coarse grind difference, 27
    fine-grind extract, 27
    free amino nitrogen, 29
    friability, 30
    hot-water extract, 28
    mealiness, 30
    moisture content, 26
    sieving tests, 30
    total protein, 29
    total soluble protein, 29
    viscosity, 31
  malt constituents, 25
  malt extracts, 37
  malting
    germination, 23
    kilning, 24
    steeping, 23
  malt modification, 24
  other malted grains
    rye malt, 37
    wheat malt, 36

specialty malts
  caramel (crystal) malts, 34
  dry-roasted malts, 35
  light-colored malts, 33
  unmalted barley, 36
Barley steeping, 23
Barley wine (British), 376
Base malts, 31
Beer carbonation, 295
Beer clarification, 260
Beer filtration, 275
Beer foam
  carbonation, 314
  fatty acids, 71
  hops, 46
  proteolytic enzymes, 264
  total soluble nitrogen, 29
Beer industry
  advertising, 15
  beer sales, 7
  beer segments, 12
  classification of brewers, 4
  demographics, 18
  distribution, 15
  history, 1
  industry concentration, 2
  off- on-premise sales, 8
  per capita consumption, 11
  top beer brands, 8
Beer stabilization, 263
Beer styles, 361
  American, 362
  Belgium, 368
  British, 376
  Czech Republic, 386
  French, 387
  German, 388
  Irish, 401
  Scottish, 402
Beerstone
  removing, 127
Belgian beer styles, 368
Bentonite, 266
Berliner Weisse (German), 398
Beta acids, 42
Beta-amylase, 155
Beta-glucans
  cold break, 222

malt analysis, 30
malt constituents, 25
mashing 156
unmalted barley, 36
viscosity, 31
Bière de garde (French), 387
Biofiltration towers, 355
Biological acidification of mash, 158
Biscuit malt, 35
Bitter beer (British), 376
Bittering hops, 43
Bitterness units, 54
Black barley, 36
Black beer, 396
Black malt, 35
Blood alcohol content, 11
Bock beer (German), 389
BOD, 346
Boiling systems, 206
Boiling the wort, 195
Bottle-conditioned beers, 259
Bottling
  bottle feeding
    bulk-pack, 306
    pre-pack, 305
  bottle filling
    bottle drying, 315
    counter-pressure, 312
    crowning, 314
    evacuation, 312
    filler bowl operations, 308
    filling valve technology, 311
    post filler operations, 313
    post rinse, 314
    types of bottling machines, 309
  bottle labeling, 316
    types of labelers,
  bottle rinsing, 307
    cleaning treatment, 307
    types of bottle rinsers, 307
  case packing, 318
    case sealing, 319
    types of case packers, 318
    warehousing, 319
  sterilization of beer
    flash pasteurization, 304
    filtration, 304
    tunnel pasteurization, 315

Bottom-fermenting yeast
  lager yeast, 61
  open square fermenters, 248
Brettanomyces spp., *341*
Brewer's window, 156
Brewhouse yield, 37
Brewing industry, 1
Brewing water ions, 96
Brewpubs, 6
Bright beer tanks, 268
  temperature control, 269
    coolants, 269
    cooling jackets, 270
    vessel size, 269
British beer styles, 376
Brown ale (British), 378
Brown malt, 35
Bureau of Alcohol, Tobacco, Firearms and Explosives (ATF), 411
Burton-on-Trent waters, 94

C

Calandria, 207
Calcium in water, 96
Calcium chloride
  wort boiling, 205
Calcium hydroxide treatment of water, 101
Calcium sulfate
  acidification of mash, 158
  kettle additives, 205
  reduction in wort pH, 199
  water treatment, 102
Calculations
  assessing the wort, 191
  brewhouse yield, 37
  hop additions, 216
  raw material requirements, 174
  strike temperature, 176
  wort fermentability, 250
California Common (American), 366
Candida spp., *341*
Candi sugar, 117
Candle filter, 279
Caramel malts, 34
Caramelization
  specialty malts, 33

wort boiling, 196, 197
Carbohydrates
  yeast nutritional requirements, 64
Carbon dioxide
  in water treatment, 101
  levels in beer, 296
  volume table, 429
Carbon filtration of water, 103
Carbonate in water, 98
Carbonation
  levels in beer, 296
  mechanical
    in-line, 298
    in-tank, 299
  methods
    bottle conditioning, 297
    kraeusening, 297
    mechanical, 298
    secondary fermentation, 297
  principles, 295
  safety procedures, 301
Carlsberg flask, 78
Cartridge membrane filters, 286
  advantages and disadvantages, 288
  filtration mechanisms, 286
  operation, 287
  sterile filtration, 288
Cask-conditioned beers, 258
Centrifugation
  clarification, 260
  cold break separation, 227
  hot break separation, 212
Cereal cooker, 171
Chelating agents, 125
Chillproofing agents, 264
  bentonite, 266
  hydrolysable tannins, 265
  polyvinylpolypyrrolidone, 267
  proteolytic enzymes, 264
  silica gels, 265
  tannic acid, 264
Chlorides in water, 98
Chlorine
  acid-based sanitizers, 128
  treatment of brewing water, 105
Chlorine dioxide
  brewery sanitation, 129
  water treatment, 105

yeast washing, 84
Chocolate malt, 35
Clarification
  centrifugation, 262
  fining agents
    isinglass, 260
Clean-in-place systems (CIP), 133
  recovery-type, 137
  single-use, 136
Cleaning detergents
  acid-based, 123
    nitric acid, 124
    phosphoric acid, 124
  additives, 125
    chelating agents, 125
    emulsifiers, 126
    surfactants, 125
  alkaline-based, 122
    sodium hydroxide, 122
    sodium hydroxide/hypochlorite solutions, 123
Closed-keg systems, 321
Coarse-grind extract, 27
COD, 346
COLAs Online System, 414
Cold break
  formation of, 222
    factors affecting quantity of, 223
  removal of cold break, 225
    centrifugation, 227
    DE filtration, 226
    flotation, 226
    starter tank, 225
    whirlpool, 227
Cold glue labeler, 316
Cold-water extract, 28
Colloidal hazes, 263
Colloidal stabilization
  conditioning, 263
  filtration, 282
Complex phosphates, 127
Conditioning, 255
  clarification, 260
  maturation, 255
  stabilization, 263
  tanks, 268
Conditioning tanks, 268
  temperature control, 270

  collants, 270
  cooling jackets, 270
  vessel size, 269
Container deposit laws, 418
Control states, 416
Coolship, 211
Copper
  cleaning and sanitation, 139
Copper finings, 203
Copper in water, 99
Corn adjuncts, 113
Corrosion resistance, 137
Craft brewing
  brewing equipment, 173
Crystal malts, 34
Cylindroconical fermenters, 244
  coolants, 246
  cooling jackets, 247
  foam production, 246
  temperature control, 246
  yeast strains, 245
Czech Republic beer styles, 386

D

Darauflassen, 237
Dark lager (American), 363
Dechlorination in water treatment
  carbon filtration, 103
  heating the water, 103
  potassium metabisulfite, 104
Decoction mashing, 164
  single-step mash, 168
  three-step mash, 165
  two-step mash, 168
Degree of attenuation
  lager yeast, 62
  yeast strain selection, 72, 73
Deep-bed filtration, 289
Degeneration of yeast, 74
Demographics of U.S. beer market, 18
Depth filtration, 275
Detergents, 121
  acid-based, 123
    nitric, 124
    phosphoric, 124
  additives
    chelating agents, 125

emulsifiers, 126
surfactants, 125
alkaline-based, 122
sodium hydroxide, 122
sodium hydroxide/hypochlorite solutions, 123
Dextrin malt, 33
Dextrose
bottle-conditioned beers, 259
sugars, 117
syrups, 116
Diacetyl
fermentation, 240
yeast byproducts, 68
Diacetyl rest, 240
Diastatic power
base malts, 32
malt analysis, 28
mash tun adjuncts, 111
two-row barley, 22
Diatomaceous earth
beer filtration, 280
safety and disposal, 280
solid waste, 348
Diet/light (American), 364
Dimethyl sulfide, 69
Dissolved air flotation (DAF), 351
Domestic specialties, 14
Dopplebock (American), 390
Dortmunder Export (German), 392
Dosing, 236
Double-mash infusion system, 168
Double-pass filtration, 276
Dry hopping, 57
Dry milling, 144
Dry-roasted malts, 35
Dry stout (Irish), 402
Dunkler bock (German), 390
Dunkles Weizen (German), 398

E

EDTA, 125
Eisbock (German), 391
Emulsifiers
complex phosphates, 127
orthophosphates, 126
Enterobacteriaceae, 339

Enzymes
activity during mashing
alpha-amylase, 155
beta-amylase, 155
beta-glucanases, 156
peptidase, 154
proteinase, 154
limit dextrinase, 155
Esters, 67
European Brewing Convention, 149
Evaporation rate
kettle designs, 206
kettle times, 196
wort boiling 198
Excise taxes, 413
Extract yield, 191

F

Faro, 372
Fatty acids, 71
Federal agencies
BATF, 411
OSHA, 137
TTB, 100, 411
Federal Alcohol Administration (FAA) Act, 2
Fermentation
ale fermentations
modern, 240
starter tanks, 239
traditional, 239
diacetyl rest, 240
high-gravity fermentations, 242
lager fermentations, 236
modern, 238
starter tanks, 236
topping up, 237
traditional, 237
systems
ale dropping, 247
ale top skimming, 247
burton union, 249
cylindroconical, 244
dual purpose vessel, 249
open squares, 249
yorkshire squares, 248
yeast collection, 242

fermentation systems, 242
   storage, 243
Filter cake
  mash filters, 189
  types of powder filters, 277
Filtration
  cartridge membrane filters, 286
    advantages and disadvantages, 288
    filtration mechanisms, 286
    operation, 287
    sterile filtration, 288
  colloidal stabilization, 282
  deep-bed filtration, 289
  depth filtration, 275
  powder filter aids, 280
    filtration process, 281
    safety and disposal, 280
    types, 280
  powder filters, 276
    candle, 279
    leaf, 278
    plate and frame, 277
  sheet pad filters, 284
    advantages and disadvantages, 286
    filtration mechanisms, 285
    operation, 285
    sterile filtration, 289
  surface filtration, 275
Fine grind-coarse grind difference, 27
Fine-grind extract, 27
Fining agents, 260
Finish hopping, 202
Flanders brown ale (Belgium), 368
Flakes, 112
Flash pasteurization
  bottling, 304
  kegging, 324
Flocculation
  yeast strain selection, 72
Flours, 111
Flow equalization, 350
Fluidized bed reactor, 357
Foam
  fermentation, 246
  reduced by
    cold break, 224
    degeneration of yeast, 74
    fatty acids, 71

hop stems, 46
Food Chemicals Codex, 100
Free amino nitrogen (FAN)
  malt analysis, 29
  yeast nutritional requirements, 65
French beer styles, 387
Friability, 30
Fructose
  rate of yeast attenuation, 73
  yeast nutritional requirements, 64
Fruit lambics (Belgium), 374
Fusarium, 22
Fusel alcohols, 70

G

Galena barley variety, 42
Gehaltemeter, 300
Gelatinization, 155
German beer styles, 388
Germination, 23
Geuze, 373
Gluconobacter spp., 337
Glucose
  bottle-conditioning, 297
  malt constituents, 25
  rate of attenuation, 73
  yeast nutritional requirements, 64
Glycogen
  yeast life cycle, 63
Glycol, 221, 247
Government regulations
  federal regulations, 411
    brewery application approval, 412
    excise tax collection, 413
    labeling and advertising approval, 413
    homebrewing, 415
  local regulations, 418
  state regulations, 416
    alcohol beverage control boards, 416
    container deposit laws, 418
    licenses, 417
    taxation, 418
Grain silos, 143
Gram-negative bacteria, 335, 337
Gram-positive bacteria, 334, 335

Gravity sedimentation, 350
Green beer
  conditioning, 255
  diacetyl and 2,3-pentanedione, 68
Grist case, 148
Grist profile, 149
Grit removal, 350
Grits, 112
Gueuze (Belgium), 373
Gushing
  causes of,
    molds, 99
    nitrates, 342

H

Hammer mills, 147
Harrington barley variety, 22
Heating jackets, 206
Hefe weizen (German), 399
Heller bock (German), 391
Hemacytometer
  determining pitching rates, 235
  yeast viability, 85
High gravity fermentations, 242
Higher alcohols, 70
Holding hopper, 170
Homebrewing, 415
Homebrew bitterness units (HBUs), 54
Honey malt, 34
Hop additions, 201
Hop back, 210
Hop bitter substances, 41
Hop extracts, 49, 51
Hop plugs, 46
Hop separator (strainer), 210
Hop varieties, 423
Hopjack, 202
Hops
  bitterness levels
    controlling acid levels, 55
    HBUs, 54
    IBUs, 54
  constituents
    oils, 42
    resins, 41
  dry hopping, 57
  isomerization, 42

products
  hop oils, 52
  isomerized, 50
  non-isomerized, 46
stability, 53
utilization, 55
varieties
  aroma hops, 43
  bittering hops, 43
  cultivars, 44
  dual-purpose hops, 44
  growing regions, 44
  whole hops, 44
Horizontal leaf pressure filter, 278
Hot break, 199
  quality, 200
  quantity, 200
Hot trub separation, 211
  brew kettle/whirlpool, 215
  centrifuge, 212
  coolship, 211
  settling tanks, 212
  whirlpool, 213
Hot-water extract, 28
Hot wort separation
  whole hops, 210
    hop back, 210
    hop separator (Strainer), 210
  hot trub separation, 211
    brew kettle/whirlpool
    centrifuge, 212
    coolship, 211
    settling tanks, 212
    whirlpool, 213
Humulus lupulus, 43
Hydrochlorous acid, 129
Hydrogen peroxide, 131
Hydrogen sulfide
  lager yeast, 62
  Pectinatus spp., 338
  vitamin deficiency, 66
  yeast byproducts, 69
Hydrolysable tannins, 265

I

Imperial stout (British), 384
Imports, 7, 13

In-line carbonation, 298
In-tank carbonation, 299
India pale ale
  American, 362
  British, 379
Infusion mashing, 160
International Bitterness Units (IBUs), 54
Iodophores, 132
Irish ale, 401
Irish beer styles, 401
Irish moss, 203
Iron in water, 99
Isinglass, 260
iso-alpha acids, 42
Isomerization, 42, 197
Isomerized hop extracts, 51

K

Kegging
  keg racking machines, 322
    in-line, 323
    rotary, 324
  keg styles
    closed-keg systems, 321
    open-keg systems, 321
    sterilization of beer, 324
  kegging operations
    cold storage, 330
    external washing, 325
    inspection, 325
    internal cleaning, 326
    keg depalletizing, 325
    keg filling, 328
    volumetric/fill by weight, 330
Kettle designs
  direct-fired, 206
  external heating jackets, 206
  external wort boilers, 207
  internal heating systems, 206
Kilning, 24
Keiselguhr, 280
Kolbach index, 29
Kölschbier (German), 388
Kraeusening, 257
Kriek, 377
Kristall weizen (German), 399

L

Lactobacillus amylolyticus, 158
Lactobacillus delbrueckii, 158
Lactobacillus spp., 334
Lagering, 256
Lagoons, 355
Lambic (Belgian), 372
Laminar flow hood, 79
Late kettle hopping,
Lauter tun, 183
Lautering
  collection of first wort, 187
  establishing grain filter bed, 186
  grains-out, 189
  mash rest and underletting, 186
  mash transfer, 184
  recirculation of wort, 186
  sparging and second wort collection, 187
Leaf filters, 278
License states, 416
Light beer, 13
Light-colored malts, 33
Limit dextrinase
  malt germination, 23
  starch degradation, 155
Lipids
  malt constituents, 26
  sparging, 183
Liquefaction, 156
Lyophilization, 76

M

Magnesium in water, 97
Magnesium sulfate in water treatment, 102
Maibock, 391
Maillard reaction
  kilning, 24
  wort boiling, 196
Malt conditioning, 148
Malt extracts, 37
Malt liquor
  beer segments, 12
Malt milling
  dry milling, 145

hammer mills, 147
roller mills, 145
sampling, 148
handling and storage, 143
malt conditioning, 147
safety, 150
sieve analysis, 149
  mashing systems, 149
wet milling, 148
Malt storage, 143
Malting, 22
Malto-dextrin, 117
Maltose
  diastatic power, 28
  malt constituents, 25
  influence of mash thickness, 159
  rate of attenuation, 73
  starch degradation, 155
  yeast nutritional requirements, 64
Maltose/dextrin ratio, 156
Maltotriose
  malt constituents, 25
  mashing temperature, 157
  influence of mash thickness, 159
  mutation of yeast, 73
  rate of attenuation, 73
  starch degradation, 155
  yeast nutritional requirements, 64
Manganese in water, 99
Märzen (German), 393
Mash filters, 189
Mash kettle, 171
Mash mixer, 171
Mash tun
  mashing, 171
  wort separation, 181
Mashing
  acidification, 153
  equipment
    adjunct cooker, 171
    craft brew systems, 173
    holding hopper, 170
    mash kettle, 171
    mash mixer, 171
    mash tun, 172
    premasher, 170
    underback, 173
  factors affecting

malt modification, 158
mash thickness, 159
mash water, 159
pH, 158
temperature, 157
time, 157
protein degradation, 154
starch degradation, 154
  alpha-amylase, 155
  beta-amylase, 155
  beta-glucans, 156
  stages in starch breakdown, 155
systems, 159
  decoction, 164
  double-mash infusion, 168
  infusion, 160
  temperature-controlled, 163
Mash-in, 161
Material corrosion resistance
  aluminum, 139
  copper, 139
  stainless steel
    chemical agents, 138
    passivation, 138
Material safety data sheets, 137
Maturation
  cold storage, 259
  secondary fermentation, 255
Mealiness, 29
Methylene blue staining, 85
Meura 2001 mash filter, 190
Microbiological testing, 342
Microbreweries, 6
Mild ale (British), 380
Mild ale malt, 32
Miller Brewing Co., 5
Mineral salt adjustment of water, 102
Minerals in water, 96
Molds, 342
Molson Coors Brewing Co., 6
Monopoly states, 416
Multi-roll mills, 145
Munich dark (German), 393
Munich helles (German), 394
Munich malt, 33
Mutation of yeast, 73
Myrcene, 43

N

National brewers, 4
National Pollution Discharge
    Elimination System, 348
Nitrates in water, 99
Nitric acid
    acid-based detergents, 124
Nitrites in water, 99
Nitrogen
    yeast byproducts, 71
    yeast nutritional requirements, 65
Noble hops, 44

O

Oatmeal stout (British), 385
Oats
    as in adjuncts, 116
Obesumbacterium proteus, 339
Oktoberfestbier (German), 393
Old ale (British), 380
Open states, 416
Open-keg systems, 321
Optic barley variety,
Organic acids, 71
Orthophosphates, 126
Oxygen requirements for yeast, 227
Ozone treatment of water, 106

P

Pad filters, 284
Pale ale
    American, 363
    Belgium, 368
    British, 381
Pale ale malt, 32
Palletizing, 318
Papain, 264
Passivation, 138
Pasteurization
    flash, 304
    tunnel, 315
Pasteurization unit, 304, 315
Pectinatus spp., 338
Pediococcus spp., 336
Peptidase, 154

Peracetic acid, 131
Per-capita beer consumption, 11
Perlite, 280
Permanent hardness, 96
Peroxyacetic acid, 131
pH
    beer quality, 258
    mash, 158
    water, 94
pH reduction of water, 100
Phosphate in water, 98
Phosphoric acid
    as in acid-based detergents, 124
Phytase
    malt constituents, 26
    mashing (acidification), 153
Pichia spp., 341
Pilsner
    Czech Republic, 386
    German, 395
Pilsner malt, 31
Pitching yeast
    determination of, 235
        centrifugation, 235
        electron cell counting, 235
        hemacytometer, 235
        in-line biomass sensors, 236
        weight-based, 235
    dosing, 236
    microbiological contamination, 236
    rates, 23
    strain, 233
    viability, 234
Plate and frame filter, 277
Plate heat exchangers, 221
Polycar AT, 283
Polyphenols
    cold break, 222
    malt constituents, 26
    stabilization, 263
    wort boiling
        color development, 196
        oxidation, 197
Polyvinylpolypyrrolidone (PVPP)
    conditioning, 267
    filtration, 283
Porter (British), 382
Potassium in water, 98

Powder filters, 276
   candle, 279
   leaf, 278
   plate and frame, 277
Powder filter aids
   diatomaceous earth, 280
   filtration process, 281
   perlite, 280
   safety and disposal, 280
Premasher, 170
Premium beer (American), 365
Pressure-sensitive labeler, 316
Prohibition, 2
Propagation of yeast
   laboratory phase, 77
   plant phase, 78
Propagation plants, 80
Protein rest, 154
Proteinase, 154
Proteins
   cold break formation, 223
   hot break formation, 199
   malt constituents, 26
   malt modification, 24
   mashing – protein degradation, 154
   conditioning – stabilization, 263
   wort boiling – protein precipitation, 196
Proteolytic enzymes, 264
Pub breweries, 6

Q

Quaternary ammonium, 130

R

Rauchbier (German), 396
Raw material requirements, 174
Red ales
   Belgium, 369
   Irish, 401
Regional breweries, 6
Reinheitsgebot
   adjunct use, 110
   conditioning, 266
   mashing, 158

Refined starches, 112
Regional brewers, 6
Residual alkalinity, 95
Resins,
Retailers, 17
Rice, 114
   as in adjuncts, 114
Roasted barley, 36
Robust barley variety, 21
Roller mills, 145
Rotary labeler, 316
Rotating biological contactor process, 355
Rye malt, 37

S

S/T ratio, 6
S-methyl methionine
   malt germination, 24
   wort boiling, 198
Saccharification, 156
Saccharomyces cerevisiae, 61
Saccharomyces pastorianus, 61
Saccharomyces uvarum, 61
Saison (Belgium), 369
Sanitizing agents
   acid-based, 131
      anionic acids, 131
      hydrogen peroxide, 131
      iodophores, 132
      peroxyacetic acid, 131
   alkaline-based, 128
      chlorine, 128
      quaternary ammonium, 130
Sankey keg, 322
Schwarzbier (German), 396
Scottish ale, 402
Scottish beer styles, 402
Screen analysis, 149
Screening, 350
Secondary fermentation
   bottle-conditioned, 259
   cask-conditioned, 257
   kraeusening, 257
   lagering, 256
Settling tanks, 212
Sheet (pad) filters, 284

advantages and disadvantages, 286
filtration mechanisms, 285
operation, 285
Sieving tests, 149
Silica gels
  beer filtration, 283
  conditioning, 265
Silos, 143
Single-pass filtration, 276
Single-step decoction system, 168
Six-roll mills, 145
Six-row barley, 21
Sludge treatment and removal, 356
Smoked beer, 396
Sodium hydroxide
  as in alkaline-based detergents, 122
Sodium hypochlorite, 129
Sodium in water, 97
Solid waste disposal, 347
Solid wastes
  diatomaceous earth slurry, 348
  packaging materials, 348
  spent grains, 347
  spent yeast, 348
  trub, 347
Sorghum, 111
South African Breweries (SAB-Miller), 5
Sparging, 182, 187
Specialty malts, 33
Specialty craft brewers, 6
Spent grains, 347
Spent yeast, 347
Spray balls
  dynamic, 135
  static, 135
Stabilization of green beer, 263
  brewhouse procedures, 268
  chillproofing agents, 263
Stabilized hop pellets, 49
Stainless steel
  cleaning and sanitation, 137
Standard beer (American), 365
Stander barley variety, 21
Starch
  action of alpha-amylase on, 155
  action of beta-amylase on, 155
  action of limit dextrinase on, 155

barley malt, 25
gelatinization, 155
Starch degradation, 154
Starter tanks, 225
Steam
  malt conditioning, 147
  sanitizing agents, 128
  wort boiling
    methods of heating, 209
    systems, 206, 208
Steeping, 23
Sterile filtration
  bottling, 303
  filtration, 288, 289
  kegging, 304
Strainmaster, 190
Strike temperature, 157, 162, 176
Strong golden ale (Belgium), 370
Stouts
  American, 366
  British, 384
  Irish, 402
Subculturing of yeast, 75
Sucrose
  cask-conditioned beers, 258
  influence of mash thickness, 159
  rate of attenuation, 73
  sucrose-based sugars, 117
  sucrose-based syrups, 116
  yeast nutritional requirements, 54
Sugars
  caramel, 117
  dextrose, 117
  invert, 117
  malto-dextrin, 117
  sucrose, 117
Sulfates in water, 98
Sulfur volatiles, 69
Surface filtration, 275
Surfactants, 125
SVK keg, 322
Sweet stout (British), 385
Syrups, 116

T

Tannic acid
  conditioning, 264

wort boiling, 205
Taxes
  federal, 413
  local, 418
  state, 418
  state excise tax rates, 431
Temperature-programmed mash, 163
Temporary hardness, 96
Three-step mash decoction, 165
Three-tier system, 15
Top-fermenting yeast
  ale yeast, 61
  traditional ale top-skimming systems, 247
Torrified cereals, 112
Torulopsis spp., 341
Total dissolved solids (TDS), 346
Total protein, 29
Total soluble nitrogen, 29
Total suspended solids (TSS), 346
Trappist ales (Belgium), 370
Trickling filter process, 355
Trisodium phosphate (TSP), 126
Trub
  characteristics of brewery wastewater, 347
  formation of cold break, 222
  formation of hot break, 199
  removal of cold break, 224
  removal of hot break, 211
TSP, 126
TSS (total suspended solids), 346
Tunnel pasteurization, 315
Two-roll mills, 145
Two-row barley, 22
Two-stage cooling, 221
Two-step mash decoction system, 167
Type 45 hop pellets, 48
Type 90 hop pellets, 48

U

Ultraviolet radiation treatment of water, 104
Underback, 173
Underletting, 182, 186
United States beer styles, 362
Unmalted barley, 36, 115

Upflow anaerobic sludge blanket, 357
U.S. beer industry (see beer industry)

V

Venturi tube
  methods of oxygenation, 229
Vertical leaf pressure filter, 279
Vicinal diketones (VDKs)
  diacetyl rest, 240
  Lactobacillus spp., 335
  yeast byproducts, 68
  yeast lag phase, 63
Vienna beer (German), 397
Vienna malt, 32
Viscosity
  barley malts, 31
Vitamins
  yeast nutritional requirements, 66
Vorlauf, 186

W

Wastewater
  characteristics, 346
  quality, 346
  quantity, 346
Wastewater treatment, 348
  biological treatment, 352
    aerobic, 353
    anaerobic, 356
  chemical, 351
    flocculation, 352
    pH adjustment, 351
  physical, 349
    flow equalization, 350
    gravity sedimentation, 350
    grit removal, 350
    screening, 350
Water
  alkalinity, 95
  evaluation, 93
  hardness, 95
    permanent, 96
    temporary, 96
  minerals
    calcium, 96

carbonate, 98
  chloride, 98
  copper, 99
  iron, 99
  magnesium, 97
  manganese, 99
  minor ions,
  nitrate, 99
  nitrite, 99
  phosphate, 98
  potassium, 98
  principal ions,
  sodium, 97
  sulfate, 98
  zinc, 99
 pH, 94
 residual alkalinity, 95
 treatment
  dechlorination, 103
  microbiological control, 104
  mineral salt adjustment, 102
  reduction in alkalinity, 100
  removal of particulate matter, 100
Water demand for brewing, 93
Weizenbock (German), 400
Wet milling, 148
Wheat
 as in adjuncts, 115
 malt, 36
Wheat beers
 American, 367
 Belgium, 375
 German, 398
Whirlpool
 cold wort clarification, 227
 hot wort clarification, 213
White beer (Belgium), 375
Wholesalers, 16
Wild yeast, 340
 Brettanomyces spp., 341
 Candida spp., 341
 Pichia spp., 341
 Saccharomyces spp., 340
 Torulopsis spp., 340
Wort aeration, 227
 methods of oxygenation, 229
 oxygen requirements, 228
 point of injection, 228

Wort assessment, 191
Wort boiling
 biochemical changes, 195
  color development, 195
  concentration of wort, 198
  dissipation of volatiles, 198
  enzyme inactivation, 195
  isomerization, 197
  protein precipitation, 195
  reduction in wort pH, 199
  sterilization, 195
 formation of hot break, 199
  quality, 200
  quantity, 200
 kettle additives
  acids, 205
  calcium chloride, 205
  calcium sulfate, 205
  copper finings, 203
  flavorings, 206
  hops, 201
  syrups and sugars, 205
 kettle designs
  direct-fired, 206
  external heating jackets, 206
  external wort boilers, 207
  internal heating systems, 206
 methods of heating, 208
Wort clarification (hot), 210
 hot trub separation, 211
  brew kettle/whirlpool, 215
  centrifuge, 212
  coolship, 211
  settling tanks, 212
  whirlpool, 213
 whole hop separation, 209
  hop back, 210
  hop separator, 210
Wort cooling
 cooling systems, 221
 formation of cold break, 222
 microbiological stability, 223
Wort fermentability, 250
Wort lipids, 187
Wort recovery, 215
Wort separation equipment
 lauter tun, 183
 mash filter, 189

mash tun, 182
strainmaster, 190

Y

Yeast
　ale, 61
　attenuation rate, 73
　byproducts
　　acetaldehydes, 68
　　diacetyl, 68
　　dimethyl sulfide, 69
　　esters, 67
　　fatty acids, 71
　　fusel alcohols, 70
　　nitrogen compounds, 71
　　organic acids, 71
　　sulfur volatiles, 69
　culture contamination, 81
　culturing, 77
　degeneration, 74
　flocculation, 72
　lager, 62
　life cycle
　　fermentation phase, 63
　　growth phase, 63
　　lag phase, 63
　　sedimentation phase, 64
　mutation, 73
　nutritional requirments
　　carbohydrates, 64
　　minerals, 66
　　nitrogen, 65
　　vitamins, 66
　pure culture maintenance
　　desiccation, 76
　　freezing in liquid nitrogen, 77
　　lyophilization, 76
　　sub-culturing, 75
　propagation and scale-up
　　laboratory phase, 77
　　plant phase, 78
　replacement, 86
　storage, 86
　strain selection, 71
　viability
　　fermentation tests, 85
　　selective staining, 84
　　slide viability method, 85
　　standard-slide culture method, 85
　washing
　　acid wash, 82
　　acid wash with ammonium persulfate, 83
　　chlorine dioxide, 84
　　distilled or sterile water wash, 82
Yeast collection, 242
　fermentation systems, 242
　　cylindroconical fermenters, 243
　　traditional ale system, 242
　　traditional lager system, 242
　storage, 243
Yeast pitching
　dosing, 236
　methods for determining rates, 235
　　centrifugation, 235
　　electron cell counting, 235
　　hemacytometer, 235
　　in-line biomass sensors
　　weight-based, 235
　microbiological contamination, 236
　pitching rates, 233
　strain, 233
　viability, 234

Z

Zahm meter, 300
Zinc
　in water, 99
　yeast nutritional requirements, 66
Zymomonas spp., 338